Conservation of Biodiversity and Natural Resources

Conservation of Biodiversity and Natural Resources

Dr. M.P. Singh, FMA
Head of the Department Forest Sciences
Birsa Agricultural University
Ranchi-834006 (Jharkhand)

Dr. Bijay S. Singh, Ph.D.
Department of Forest Sciences,
Birsa Agricultural University
Ranchi-834006 (Jharkhand)

Miss Soma Dey
Department of Forest Sciences
Birsa Agricultural University
Ranchi-834006 (Jharkhand)

2013
Daya Publishing House®
A Division of
Astral International Pvt. Ltd.
New Delhi – 110 002

First Impressions, 2004
ISBN 9788170359883

Published by : **Daya Publishing House®**
A Division of
Astral International Pvt. Ltd.
– ISO 9001:2008 Certified Company –
4760-61/23, Ansari Road, Darya Ganj
New Delhi-110 002
Ph. 011-43549197, 23278134
E-mail: info@astralint.com
Website: www.astralint.com

Laser Typesetting : **Classic Computer Services**, Delhi - 110 035

Printed at : **Thomson Press India Limited**

PRINTED IN INDIA

GANDHIJI'S TALISMAN

"I will give you a talisman. Whenever you are in doubt or when the self becomes too much with you, apply the following test :

Recall the face of the poorest and the weakest man whom you may have seen and ask yourself if the step you contemplate is going to be of any use to him. Will he gain anything by it? Will it restore him to a control over his own life and destiny? In other words, will it lead to Swaraj for the hungry and spiritually starving millions?

Then you will find your doubts and your self melting away."

Foreword

The constant shifting of the Earth's outer shell also causes a cycle of volcanic activities, and creates small mountains within the sea of a sudden. In the Pacific ocean, north West word movements of the earth's crust over an abnormally not zone cased the eruption of a series of volcanic island known as the Hawalian chain.

The physical characteristics occupied with their relative inaccessibility and remoteness from the rest of the world and extreme climatic regions contributed to the origin, establishment and diversification of unique biodiversity and cultural diversity of the mountain people. In India, the two mega biodiversity hot spots – the Western Ghats and the Eastern Himalayas – are treasure houses of endemic species and repositories of genetic diversity, millions of people all over the world depend on forests for their physical, cultural and economic survival.

Today, it is an accepted fact that the impact of global ecological phenomena such as global warming is the felt mountains. United Nations, Environment Programme (UNEP) show that nearly 50 glacial lakes in Nepal and Bhutan are fast felling with water due to increase in temperature and subsequent melting of ice in the Himalayas. The breakage of their banks may trigger off a host of catastrophic events in the region.

Destruction of vegetation for various developmental activities too is a problem of great concern. The colossal loss of vegetation due to massive deforestation has lead to catastrophic situations including soil erosion, less of soil productivity, avalanches and tremors. This may ultimately lead to the destruction of the whole ecosystem. The loss of biodiversity of mountains ecosystems might have ecological, ethical, economic and health-related side efforts.

I am pleased to place on record that senior author of this book Mr. M.P. Singh is an eminent Taxonomist of present day.

This book is divided into two parts. It is organised to provide an even and logical, how of concepts and to treat this complex field in a clear and thorough manner. The publication of this book is a rare event.

This one is worth the wait.

New York **Bill Clinton**
Datecd 16.01.2003 **Former U.S. President**

Preface

India has an age old tradition of Natural Resources and the same is reflected not only in the old literature and cultural ethics but also in the constitution, politics, and organisation.

Biodiversity and Natural Resources forms the root of all living system. India is fortunate enough to be ranked sixth among the twelve mega biodiversity countries. Its biological resources include 50,000 species of plants and 81,000 species of animals including ones belonging to lower phylas. However, due to habitate loss and over exploitation owing to burgeoning population. The biodiversity of our country is severely threatened and some species, which one abundantly found, have now become rare and some have become extinct. The Indian Cheetah (*Acinonyx jabatus*) is one such stark example. The concern of the Government towards the conservation of biodiversity and Natural Resources has grown considerably. This is amply reflected in the Wildlife (Protection) act 1972, Forest Conservation, 1980 and the forest policy of 1988. The Indian Government is also a party to a number of International treaties and conservations which include the CITES, Ramper Convention, Biodiversity and Natural Resources Convention, etc. of late, the modalities of patenting of biodiversity under Intellactual Property Rights (IPA) are being worked out to check the piracy of Genetic material. A number of institutions in the world are working towards both *exsitu* and *insitu* conservation of biodiversity and natural resources. The conservation of biodiversity and natural resources is linked with the maintenance of ecological stability and productivity. These attributes are important in sustained development and stable national economy (McNeely, 1988). The concept of biodiversity is very vast and seated at microlevel. In holistic sense, it can be linked with most of the biological processes in nature. Its conservation involves a number of parameters such as number of species, their population dynamics, distribution, habitate, structure, micro habitates, physical environments, climate, present management and past history. Therefore, the efforts for conservation of biodiversity and natural resources should be in tune with the processes and its occurrence in space and time from microlevel to megalevel. The problem of biodiversity and natural resources conservation has become a global issue. It is being realised that the forest existing in a country is not a resource just for that country, but for the whole of the world. The Amazonian Rain Forests have been called the "Lungs of the World" as they serve to puity the global atmosphere by release of oxygen and absorption of pollutants. The present book is based on numerous materials, reports, personal communication with scientists, naturalist, and authors own extensive surveys and researches on the flora of various parts of the nation. The author is also fully concious of the several limitation in this work and study of some phytogeographically significant groups. It is hoped that the publication will be welcomed by all taxonomists, foresters, enviornmentalists, naturalists, politicians and other decision makers of the nation.

A number of people have helped us in different ways for going through the publication. We are particularly thankful to Professor P.S. Ramakrishnan, Department of Life Sciences, Jawahar Lal Nehru University, Delhi for encouraging us to takeup this task, Dr. H.Y. Mohan (Retd.) Professor of

Botany, Delhi University, Delhi for correction of manuscripts of this work. I am thankful to a number of teachers, scientists, for their assistance Dr. K.K. Nag, Dr. N. Dayal, Dr. S.N. Dixit, Dr. S.N. Sachan, Dr. P.K. Khosla, Dr. D.N. Tiwary have kindly spared valuable discussion on some issues. I (first author) would be failing in our duties if we do not record our special thanks to our children (Ankita & Saurave) who have been a constant source of encouragement during the entire course of this hard task for the nation.

Ranchi

M.P. Singh
Bijay S. Singh
Soma Dey

Contents

1
Introduction

Definition, Perspective and Overview

Biodiversity is a popular way of describing the diversity of life on earth : it includes all life-forms and the ecosystems of which they are a part. It forms the foundation for sustainable development, constitutes the basis for the environmental health of our planet, and is the source of economic and ecological security for future generations. In the developing world, biodiversity provides the assurance of food, many raw materials such as fibre for clothing, materials for shelter, fertilizer, fuel and medicines, as well as a source of work energy in the form of animal traction. In the continuing drive to develop efficient and sustainable agriculture for many different conditions, these resources provide raw material for plant and animal breeding as well as the new biotechnologies. In addition, biodiversity maintains the ecological balance necessary for planetary and human survival.

At the United Nations Conference on Environment and Development held in Rio de Janeiro in 1992, biological diversity was defined as : The variability among living organisms from all sources including, *inter alia,* terrestrial, marine and other aquatic ecosystems and the ecological complexes of which they are a part; this includes diversity within species and of ecosystems.

Estimates of the total number of plant, animal and other species found on the planet Earth vary widely, from 3 million to 30 million or more. But in the absence of any central archives, no one even knowns how many species have already been named and recorded. Whereas birds and mammals have been well recorded, the total diversity of organisms such as insects, spiders, fungi, nematodes and bacteria is poorly known.

In many temperate and alpine areas where richness of terrestrial species is comparatively low, there are acceptable floristic and faunistic lists of the vascular plants and vertebrate animals. There are also reasonable estimates of invertebrate animals and non-vascular plants, including fungi. Less well known are soil organisms, bacteria and viruses. In many tropical regions having much higher species diversity, there are no reliable lists of vascular plants and vertebrate animals, and there are only very rough estimates of the kinds of invertebrate animals, crytogams and fungi. Most insects, soil organisms, bacteria and fungi in these regions have yet to be collected and described (Table 1.1).

All cells, organisms, populations, species and possibly even communities originate, grow, reproduce and die. Consequently, the number and abundance of entities is in constant flux. A distinction is usually made between processes that generate diversity, those that maintain diversity and those that reduce diversity. All heritable diversity originates ultimately at the molecular level in the phenomenon on mutation. Genetic recombination also affects individual and population diversity. Diversity is reduced by certain processes (called selection) that eliminate variation (however, balanced polymorphisms and disruptive selection can increase genetic diversity). Though the existence of diversity does not necessarily imply a process of selection, most theories regarding the existence of biological diversity relate it to some process of selection or optimization. Darwinian selection at the organismic level is the oldest and best known of these theories. Less well understood is what maintains diversity in a system (Solbrig, 1991).

New diversity is constantly injected into biological systems through mutation and recombination (Fig 1.1). On the other hand, it is being eliminated by selection. The fate of every variation, from gene mutations to communities, is to disappear eventually. This can be a very fast process or the variant may survive for a very long time. The rapidity with which new mutations arise in relation to the speed with which they are eliminated determines the actual diversity of a system.

Table 1.1 Approximate numbers of known species of micro-organisms and probable world species totals (After Solbrig, 1991).

Group	Known species	World species
Algae	40,000	60,000
Bacteria (Incl. Cyanobacteria)	3,000	30,000
Fungi (incl. lichen-forming and yeasts)	64,200	800,000
Viruses (incl. plasmids and phages)	5,000	130,000
Protoctists (incl. protozoa, excl. algae and fungal protoctists)	30,000	100,000
Totals	**142,000**	**1,120,000**

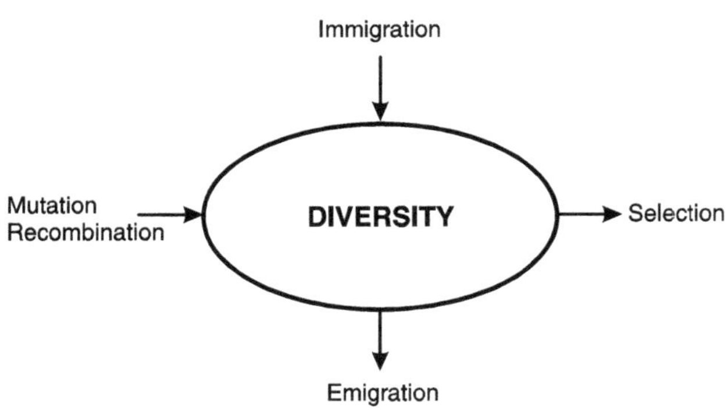

Fig. 1.1 : A generalized diversity model.

It seems that a certain degree of diversity is needed for biological systems to function properly. There may be system feedback that affect the rates of mutation and selection. Also there may exist diversity thresholds below which the system collapses. Diversity appears to be actively maintained in a system (May, 1984). The nature of the actual processes that actively maintain diversity (Hoffman, 1989; Signor, 1990) is not known.

In recent years it has been increasingly felt that a serious loss of species and a reduction in the genetic diversity of crops and wild species is taking place (Ehrlich and Ehrlich, 1981; Wilson and Peter, 1988). This feeling has arisen because of the greatly accelerated transformation of natural landscapes occurring around the world and particularly in the tropics (Wilcox and Murphy, 1985). An extensive reduction in species diversity may lead to loss of ecosystem stability and function. The danger to crop productivity from a reduction in the genetic variation needed to breed new varieties is also recognized.

Diversity occurs at all levels of biological organization, from molecules to entire biotas. Greater attention is usually given to the biological and ecological significance of biodiversity and its biogeographical implications (Solbrig, 1991). According to Diamond (1988), the great diversity in biological systems is a prerequisite for the proper functioning of the biosphere. Diversity seems to permit biological systems to cope with environmental stress and to recover from disturbance events.

Human-induced change reduces biodiversity but increases the complexity of human societies. People affect biodiversity both directly and indirectly. Use of renewable natural resources, especially in extractive industries such as forestry or fisheries, usually involves decreases in species diversity as stocks are depleted and unwanted species are destroyed. Agriculture and animal husbandry damage biological diversity by destroying or modifying the native biota. Since pre-agricultural times, the world's forests have declined from 5 to 4 billion hectares. Temperate forests have lost the highest percentage of their area (32-35%), followed by subtropical woody savannas and deciduous forests (24-25%), and old-growth tropical forests (15-20%). Indirect changes come from land-use, use of fossil and biomass energy, and from altering hydrological patterns. Introduction of exotic organisms, both international and accidental, has tended to reduce between-region biodiversity. The simplification of landscapes by the removal of hedges and forest fringes has reduced habitat diversity and has resulted in lowered biodiversity. A potent threat to biodiversity is the production and release of novel toxic chemical compounds into the environment.

Although diversity occurs at all levels of the biological hierarchy, it is the issue of species loss that has generated the greatest apprehension (Ehrlich and Ehrlich, 1981; Wilson, 1988). Crop scientists have been more apprehensive about reduction in the number of genes and genotypes (Plucknett, 1987). Conservation biologists concerned with endangered species are also concerned about diminished genetic diversity, with concomitant loss of adaptive potentials and increased breeding defects (Solbrig, 1991). Much less attention has been paid to diversity at the molecular level, even though it is here that biological diversity originates.

Biological diversity varies greatly according to the available habitat or living space. Over a wider range, diversity patterns can be established fairly easily, but this is not possible for a smaller range. There is no single process or theory that can help explain the complexity of the diversity phenomenon. With a view to understanding its occurrence, a knowledge of the ecological, evolutionary, and geological processes and also their interactions, is essential. Only by considering the various aspects of this highly delicate ecosystem can it be possible to preserve the genetic storehouse for future generations. In order to ensure effective protection, one must recognize the global value of our ecosystems ranging from natural reserves (where the legitimate interests of the indigenous population have to be preserved) to the planning of specific reafforestation efforts and appropriate systems of utilization.

Biological diversity is one of the most fascinating aspects of biology. Evolution by natural selection has produced, and is still producing, different species. Organisms generally have a tendency to multiply exponentially *ad infinitum* without intrinsic constraints. What does biological diversity mean for the evolutionary development of life? As genetic variation is pervasive and individual variation is the source of evolution by natural selection, the diversity at the species level has created a vast evolutionary potential (Wuketits, 1997). At least half a billion species have existed on our planet, probably many more. While the number of known extant species is approximately 1.5 million, and earlier estimations of the total number ranged between 5 and 10 million, the results of modern studies concerning biodiversity go far beyond these figures. Some estimates point to 20-80 million species now living on earth, many of them, however, already threatened by extinction (Price, 1996; Wilson, 1992). Wilson estimated an extinction rate of three species per hour. This evolutionary drama is being enacted particularly in the tropical forests which probably have the highest species density.

Each single species has a value in itself, and therefore we humans have no right to destroy the natural order; on the contrary, we must take care to save the other species. Indeed, humans are emotionally attracted by many species and find peace and satisfaction in observing them. According to Wilson (1984), 'to explore and affiliate with life is a deep and complicated process in mental development'. However, one trouble with this argument is that it can hardly be extended to all kinds of organisms. For vinstance, we do have positive feelings towards flowering plants, songbirds, dogs, elephants, the giant panda, and many other organisms but what are our emotions towards tapeworms, mosquitoes, rats and snakes?

The diversity of species is one of our most precious resources. Yet, we are far from knowing - or guessing - the economic value of many species since our agricultural policies are such that we depend on only a small number of species. A considerable number of now extinct plant and animal species could certainly have been utilized by humans for food.

We are losing biological diversity at an unprecedented rate. The disappearance of species in past eras has occurred through natural processes within the context of evolutionary time scales. But today, human activities are contributing more to the loss of biodiversity than any other factor. Biological resources are renewable resources, but they are being exploited at rates that exceed their sustainable yield. Human destruction of habitats, whether exploited for commercial or subsistence reasons, poses the greatest threat. The clearing of land for agriculture, overgrazing of grassland, cutting and buming of forests, unsustainable logging and fuelwood collection, indiscriminate use of fertilizers and pesticides, overexploitation of fisheries, draining and filling of wetlands, poor water management, and urbanization and pollution of air and water are prominent factors in the degradation of our biological resources. About seven million hectares of cultivated land are destroyed every year.

Genetic erosion –the reduction of diversity within and the main cause of extinction of a species- is a global threat to agriculture. The greatest loss of crop genetic resources results from the introduction of modern, uniform plant varieties in place of a mix of traditional ones. The Green Revolution introduced high-yielding varieties of wheat and rice in the developing world, but displaced traditional varieties and their wild relatives on a massive scale. In India, agronomists predict that just ten rice varieties will soon cover three-quarters of the total rice area where once over 30,000 different varieties were grown.

According to Wilson (1984), one in about ten plant species contains compounds with an anticancer effect. This underscores the tremendous relevance of biodiversity preservation. But this does not mean that a species should be evaluated only with regard to its utility. The ecological interdependence among all living beings, regardless of whether they may or may not be of some use in agriculture, pharmacy, or medicine, needs to be emphasized.

Each single species needs to be viewed as something unique that interacts with other species. Biodiversity research can go a long way in understanding what holds the biosphere together, and so demonstrates the social implications of biology.

The same is true of animal genetic resources. The introduction of a very few modern breeds that are better suited for the high input-output of industrial agriculture is displacing the diversity of indigenous livestock breeds. In Europe, half of all the breeds of domestic animals that existed at the beginning of the century have become extinct. A third of the remaining 770 breeds are in danger of disappearing within the coming two decades.

Loss of biodiversity is commonly regarded as an environmental problem, but the underlying causes are essentially social, economic and political. The excessive and unsustainable consumption of resources by a small but rich minority of the world's population, combined with the destructive impacts of the world's poor and hungry in a desperate bid for survival, have destroyed or overexploited habitats worldwide (FAO, 1993).

Fair and equitable sharing of the benefits accruing from the use of genetic resources is a principal objective of the *Convention on Biological Diversity* (CBD). Use of biodiversity has to be sustainable in order to meet the letter and spirit of the Convention. Lesser (1998) focused on economic incentives for sustainable use of genetic resources. According to him the sharing of benefits might be best ensured by contractual arrangements.

A good knowledge of the fluctuations in number of natural populations and the application of this understanding to the conservation of species must be based firmly on an understanding of the factors influencing spatial and temporal variability in population demography. All populations occur in heterogeneous landscapes, with different individuals experiencing different conditions depending on their location.

Environmental indicators reflect the condition of the environment at any given time. They reveal the pressures and stresses that humans place on the environment, and provide measures of the effectiveness of the responses to any problems. An indicator can be a microbe, plant or animal that is particularly sensitive to an environmental range; it can be a chemical or pollutant in water, soil or air; it can be a quantity measure of a natural resource; or it can even be a change in social or economic activities brought about by some environmental condition. Indicators can provide us with reliable, acceptable and understandable information about the health of the environment.

Without doubt, the best way to save biodiversity is to value it, whether as a resource for direct use on a sustainable basis, or for its indirect functions, such as maintaining freshwater supplies or providing a sink for greenhouse gases. Indeed, biodiversity is an index of how well we are doing in sustainable development. So long as we continue to loss it, our development will not be ecologically sustainable.

As in other Asian countries, in Indonesia also the Green Revolution led to the introduction of certain new high-yielding but genetically uniform varieties of rice that were highly vulnerable to attack by pathogens. Over two decades ago, a mysterious viral disease began to attack crops of the new rice varieties, and within a few years, over a million hectares of the crop were threatened, affecting millions of people throughout South East Asia. The International Rice Research Institute (IRRI), Manila, rapidly screened thousands of varieties in its collections for resistance to the virus and found that only one possessed it. This was a low-yielding, spindly species, collected from the wild in South India but believed to have no commercial value. The resultant new variety, IR36, has since been planted over millions of hectares in Asia.

Climate change is likely to radically alter the patterns of agriculture throughout the world. The question is : where will the genetic material come from to produce the new crop varieties? Gene banks may contain an odd variety that may sometimes turn out to be useful, but the greatest gene tank of all is nature-which is being destroyed at an alarming rate by humankind.

Today, the Earth's biological resources are under serious threat. Biodiversity is nothing but the myriad of genes, species and ecosystems that collectively make up Nature. Biodiversity may have taken a few billion years to evolve, but it may be largely destroyed in only a few human generations. Rates of species extinction are estimate to be 50 to 100 times the natural background rate: this could increase to 1000 to 10,000 times with the forest loss projected for the next 25 years. We must take direct action immediately to protect biodiversity, otherwise we will loss forever the opportunity of reaping its full potential benefit.

Tropical Biodiversity

Tropical forests have for long been regarded as the habitat of some of the richest variety of species on earth. This view has led to the idea that there is a virtually inexhaustible store of exploitable resources in these forests. This fallacy, coupled with man's unsatiable greed for profit, has been a chief reason for the widespread global destruction of these valuable ecosystems. Besides, the increasing overexploitation of resources by the ever-growing human population also contributes to this destruction.

The tropics appear to be at an ecological disadvantage because the tropical soils have only a limited yield capacity due to their low nutrient content. This nutrient paucity influences is the process of natural selection and seems to have a significant role in ensuring a wide variety of species. Thus, the wide biodiversity of the Amazonian lowland forests is quite in keeping with the soils in these ecosystems being notoriously low in nutrient. The tropical African rainforests also harbour a wide variety of species but, compared with the Amazonian forests, their biodiversity is lower, possibly due to a greater availability of nutrients. Interestingly, the conditions in SE Asia are totally different: here, despite the high nutrient content of the volcanic ash soils and fertile alluvial lands, an extremely high species diversity occurs, accompanied by a remarkably high density of some.

We tend to think of the tropics in terms of high diversity and yet a striking fact that impresses many travellers in the tropics is that all tropical gardens look exactly the same : *Bougainvillea, Hibiscus, Pyrostegia, Hedychium, Strelitzia* and not much else. A similar situation prevails with many tropical seminatural habitats : they have been infested by a motley assortment of escaped garden plants such as *Lantana camara* and *Rubus alceifolius,* and pernicious vines such as *Cryptostegia grandiflora* and *Passiflora mollissima.* These plants, and others like them, pose an enormous threat to the conservation of biodiversity in natural tropical habitats. Their impact can be devastating : *Miconia calvescens* dominates whole hillsides in Tahiti, travellers palm (*Ravenala madagascarensis*) towers above what was once a species-rich moist forest in Mauritius, or the vine *Hiptage benghalensis* literally buries vast areas of lowland dry forest in Reunion.

Whereas total eradication is seldom achieved, satisfactory management of alien species is achievable. Doubtless, it is far easier to nip problems of alien invasion in the bud, but it is also not wise to give up (as hopeless) cases involving fully naturalized aliens. Unfortunately, however, many of the tropical countries that suffer the greatest problems from alien invasions have neither the resources nor the management structures with which to solve them.

Biodiversity and Community Structure

Community structure includes the various ways in which organisms and population relate to and interact with one another. An important question is whether communities have both structure and properties not possessed by their component populations, referred to as 'emergent properties'. Possible examples include trophic structure, stability, guild structure, and successional stage (Solbrig, 1991).

Species diversity may be essential for the proper functioning of communities and for the emergence of new properties at the community level. Though species diversity is undoubtedly necessary for community structure, is any amount of diversity sufficient, or are there specific mixes of species that are necessary for the proper functioning of communities and ecosystems? There are two views on this issue. One maintains that a community is formed by those species that happened to arrive there (Gleason, 1926). Under this view the community would not possess any emergent properties. The other view is that of Elton (1933): 'In any fairly limited are only a fraction of the forms that could theoretically do so actually form a community at any one time.... the community really is an organized community in that it has "limited membership".

Every species in a community occupies a singular niche, defined as 'a region in a factor space whose axes represent the critical resources or environmental variables to which species in the community respond differentially' (Colwell, 1984). The axes of this hypervolume are the types and amounts of resources and the distribution of their use by species in space and time.

The inverse relationship of species richness with altitude has long been well known. High community and regional diversity in the tropics has been linked to available energy, species-specific predation, and delayed competitive exclusion (Holm-Nielsen *et al.*, 1989; Phillips *et al.*, 1994).

In some cases, direct interaction between previously isolted ecological communties occurs when geographical barriers break down, either through natural geological events or through man-made modifications of the landscape. Following the disappearance of a barrier, one community may swamp a second community, causing most or all of its species to become extinct.

Occasionally, two previously isolated communities may merge, following the removal of a barrier. This process is quite rare. Gilpin (1994) showed that two 'naive' competition communities mix randomly following the removal of a barrier. However, if the two communities have been 'assembled', or self-organized, through a history of competitive exclusion, the communities are likely to battle as co-ordinated armies, with one or the other side ultimately claiming the entire landscape.

Genetic Resources and Biodiversity

In the 1970s and 80s, much public interest in the crucial importance of genetic resources for agricultural development and food security generated debates among social scientists and NGs on control over genetic resources.

The genetic resources issue dominated public discussion on genetic resources from the beginning of the 1980s. The conservationist/environmental context gradually became part of the public concern as well and focused principally on the value of genetic resources for preserving the earth's biological diversity. The biodiversity angle differed from the genetic resources issue in that (at least initially) biodiversity aspects did not focus on the agricultural-economic value and use of genetic resources. It was only in the 1990s with the rise of initiatives such as biodiversity prospecting, that biodiversity issues became part of the North-South dichotomy (Pistorius, 1997).

Genetic Resources Issue and Its Critique

During the past several years, many controversies and debates relating to defining and assessing the pros and cons of genetic resources have created some confusion. Pistorius (1997) has attempted to rectify this by listing the criticism/counter-criticism as follows ('a' is a criticism, 'b' a counter-criticism) :

1a. The Northern countries are 'gene poor' while those in the South are 'gene rich'. The North is 'technology rich' while the South is 'technology poor'. The Green Revolution and current (biotechnology) research has widened this gap.

1b. To limit the negative consequences of this dichotomy, the North has created the global network of IARCs, including those of CGIAR. The Green Revolution crops which came out of the research of these networks have saved millions of people.

2a. The importation of advanced breeding lines from the North into the South replaces traditional landraces and contributes to a greater yield in the short run, if supplied with enough inputs, but causes unstable and/or lower yield in the long run.

2b. The importation of advanced breeding lines from the North into the South replaced traditional landraces but also contributed to long-term higher yields. Agricultural development has to result in the replacement of diverse ancestral crop varieties by improved varieties.

3a. The value of genetic resources in agricultural improvement in developing countries is misunderstood and underrated. Farmers have identified and classified valuable genetic material in landraces (often according to indigenous taxonomic systems), selected them, bred them and named them.

3b. Genetic resources used in the agroindustry in the North come from agricultural systems that are underdeveloped, agriculturally marginal and poorly controlled by farmers. The landraces have no direct value for crop improvement in agro-industry and, in most cases, are not consciously selected, bred or named.

4a. Farmers in developing countries are not rewarded for their contribution to the North's agricultural production since landraces are believed to be freely available. The patent system in the North leaves no room for reward for landraces.

4b. Crop improvement work will be seriously affected if the principle of free availability of genetic resources worldwide is abandoned. It is virtually impossible to trace what the specific contribution of landraces or related wild species has been to an advanced breeding line.

5a. The North controls not only the use of genetic resources, but also their collection and exchange. It favours centralized gen bank systems which limits the South's access to them.

5b. Genetic resources in the IARC, CGIAR Centres and most Western gene banks are freely available, while the South benefits from the know-how the technical assistance offered gratis through these institutions. When genetic resources are collected, duplicates are always left behind in the country of origin.

6a. The agroindustry is mostly dominated by the North and protects its products through property rights even though the basic material for these products usually originates from the South.

6b. The North's investments in research on genetic resources can only continue when intellectual property is recognized worldwide. Unimproved material from the South has no real value for agroindustry, but is freely available to the South.

The genetic resources issue is quiet distinct from the biodiversity issue. While the term 'genetic resources' refers to the genetic information contained in the genes, the term 'biological diversity'/ 'biodiversity' encompasses all species of plants, animals and micro-organisms (and the ecosystems and ecological processes of which they are parts). And while the genetic resources issue deals with 'use' and 'valorization' aspects, the biodiversity issue also covers non-use and non-value (such as ethical) aspects.

Biodiversity as a Renewable Resource

The biological diversity of our planet once seemed inexhaustible but today it is more like a finite, yet renewable, resource. If properly managed, it can support the world's population into the foreseeable future. The future of human civilization might well depend on man's ability to protect, conserve and make sustainable use of biological diversity.

It is biodiversity that nurtures people, contributing to long-term food security for all. Biodiversity forms the foundation for sustainable development. It is the basis for the environmental health of our planet and the source of economic and ecological security for future generations.

Since time immemorial, humans have been influencing the environment; even those ecosystems that appear most 'natural' have been altered directly or indirectly during the course of time. Starting some 12,000 years ago, our forebearers, as farmers, fishermen, hunters and foresters, have created a rich diversity of productive ecosystems. This heritage has spanned numerous generations but is now threatened by the recent rapid pace of change, undesirable side-effects of industrialization and increasing expansion of the world's population.

One notable weakness of the international economic system has been its inability to assign a value of exchange to biodiversity. Ascertaining how to incorporate the cost of conservation into that of production is a daunting task that must be met if we are to fulfill our obligations to future generations and halt the continuing impoverishment and misuse of biodiversity. It needs to be realized that the cost of conserving biodiversity is much less than the penalty of allowing its degradation. Once lost, this heritage cannot be recovered or restored at any cost.

The explanation of biodiversity as a sustainable resource involves the complex problem of how to use multispecies assemblages in a sustainable fashion. It requires that individual populations not be overharvested. It requires that sufficiently large areas of natural and seminatural ecosystems remain to sustain viable populations and ecosystems. Sustainable use also requires that we understand enough about the problems of community assembly, food-web dynamics, and systems of mutualism such that exploitation of part of the assemblage does not threaten the whole. Biodiversity cannot be a renewable resource if the world's human population continues to rise, and per capita consumption of resources continues to increase.

Contacts Between Biodiversity and Biotechnology

Biotechnology is likely to become a driving economic force in the 21st century, and will help provide practical solutions to global problems of food supply, health care, energy, waste treatment and industrial regeneration. It can also become a powerful agent for environmental change.

The practice of traditional (classical) biotechnology began some 10,000 years ago, with the first domestication of plants and animals, and the origin of agriculture. Six thousand years ago the Egyptians were using yeast to produce leavened bread. The plant and animal-breeding programmes throughout history and prehistory also constitute old forms of biotechnology. Modern biotechnology is largely associated with the frontiers of science, with medical biotechnology enjoying a particularly high public profile. Genetic engineering is a subdiscipline of biotechnology that allows gene components to be manipulated and transferred across species.

Biotechnological techniques are being used to introduce desirable traits in existing species more rapidly and more precisely than has so far been possible. In some cases, these advances cannot be achieved using other techniques. They are potent tools which are likely to be increasingly used to develop and produce novel medicines, agricultural products, foods, and other goods for everyday consumption and industrial use.

Biodiversity and biotechnology are mutually dependent on one another. Since biotechnology utilizes the capabilities of living organisms, it relies on biodiversity. The vast, but rapidly diminishing array of Earth's biota contains genetic resources that fuel biotechnology, and so provide opportunities for the future. This is necessary for the security of future generations, a major goal of sustainable development.

Biodiversity is undoubtedly the foundation from which biotechnology develops and upon which the industry is strongly anchored. Likewise biotechnology has much to offer to the conservation and sustainable use of biodiversity for the survival of species, in restoring or enhancing the resilience of ecosystems, and in reclaiming degraded lands.

The fact that the richest nations are home to the smallest pockets of biodiversity while the poorest are stewards of the richest reservoirs underlines the interdependency of all nations, and the urgency of formulating common strategies for sustaining biodiversity that share both responsibility and benefits.

One interesting aspect currently attracting attention is the extent to which there may exist redundancy in species functions in various ecosystems. Several species may perform similar roles but further studies may help distinguish them (Steinberg and Geller, 1993).

Cultural Diversity

During the course of evolution, biological diversity has increased because it arises mostly through adaptations to environmental heterogeneity, which itself has continually increased via the activities of living organisms. Biologists define culture as acquisition of behavioural traits from conspecifics through social learning; and man's close relatives, such as chimpanzees, have progressed further by introducing deliberate teaching. The capacity for tool use and symbolic communication has, however, made it possible for the human species to outcompete all others in the complexity of its culturally transmitted behaviour. This has generated a great variety of cultural behaviours in the different human populations across the world.

Human cultural behaviour includes many traits. Here, we are concerned only with patterns of behaviour relating to other living species. Any given human group may possess a wealth of culturally transmitted behaviour in this context; and just as we talk of alpha species diversity or ecological species packing, such within-group diversity of elements of cultural behaviour patterns may be termed alpha cultural diversity. These cultural patterns also differ from one human group to another, and this may be termed beta cultural diversity in analogy with beta species diversity or ecological species turnover (Gadgil, 1992).

However, certain patterns of human utilization of biological resources do not form part of an optimal foraging strategy. For instance many biological communities receive total protection (e.g. sacred groves or ponds) and function as refugia. Keystone species such as *Ficus* trees often receive total protection over a wide area and support a wide range of insects, birds, primates and other organisms. These practices suggest that human populations may indeed have developed traditions that specifically serve to conserve a diversity of biological resources of value to them.

Being a *K*-selected species with a long generation time, humans cannot quickly convert resources into increased population size. The well-being of a human group therefore requires the availability of resources, and possibly a wide diversity of resources, at a minimal level over periods of several years.

According to Gadgil (1991, 1992), cultural traditions of conservation would confer no advantage on human groups under the following conditions :

(a) When the resource level fluctuates independently of the extent to which the resources are harvested, for instance with populations of migratory prey species.

(b) When a conquering human group can move into new areas, displacing the former occupants.

(c) When humans have developed the technological abilities to deploy newer and newer resources to meet their various requirements, they may find little advantage to conservative use of any particular resource.

(d) When a human group cannot ensure the co-operative behaviour of its individuals.

Usually technological innovations favour non-conservative use of biological resources. Whenever a technologically superior group moves into a region occupied by groups at more primitive levels of technology, the dominant group has the option of moving on to fresh pastures as resources of any locality are exhausted, and thus would derive no advantage from traditions of sustainable, conservative use. Such a pattern of resource use has been termed sequential exploitation. This process would tend to deplete biological diversity. It also diminishes cultural diversity, for two reasons. First, a significant component of cultural diversity relating to fine-tuning of cultural behaviour to the local biological environment would lose its functionality. Secondly, subordinated groups may imitate the culture of a dominant group, thereby losing part of their cultural diversity.

Creation of genetically engineered organisms has the potential to revolutionize the world's biological diversity. The new, engineered organisms could also covert all lands and waters, however unproductive they may appear today, into highly profitable bases of resource production. This would tremendously increase the pressure for converting all habitats into sites for man-made production-perhaps, in the end, wiping out most natural biological communities and thereby a great deal of biological diversity. As the technological advances provide a competitive edge to technologically sophisticated societies, there is likely to be further erosion of cultural diversity also (see Gadgil, 1992).

Biological and cultural diversity are often linked. This is best seen in tropical areas of the world, where the greatest concentrations of species are found, and where people have the greatest cultural and linguistic diversity. Whereas geographical isolation by mountains and river systems favours biological speciation, it also favours diversification of human cultures. The cultural diversity found in Central Africa, Amazonia, and SE Asia, for instance, is one of the most valuable resources of human civilization, providing unique insights into philosophy, religion, music, art and resource management. Protection of these traditional cultures can contribute much to achieving the dual goals of protecting biological diversity and preserving cultural diversity.

Cultural diversity is also intimately connected with the genetic diversity of crop plant. In montane areas the inaccessible terrain leads to the evolution of various tribes that develop local plant varieties called landraces. These landraces are not only well adapted to the local climate, soils, and pests, but also satisfy the tastes of the local people. The genetic variation found in these landraces is highly valuable to modern plant breeding for the improvement of crop species.

Ecotourism

Tourism is one of the world's largest industries. It is potentially a powerful way to make biodiversity pay. In some countries (e.g. Kenya and Nepal) it is the major earner of foreign currency. In Kenya, land in some National Parks raises $40 per hectare each year, as opposed to the $0.80 it would raise if used for agricultural purposes. Coarse calculations suggest that each lion in the park raises about $20,000 per year and that the elephant herds raise about $610,000.

Most of the financial benefits of ecotourism accrue to the tour companies which are usually not based in the country the tourists have chosen to visit. Some criteria that have to be met before ecotourism can be regarded as sustainable or capable of further development are the following:

(1) Tourism should provide funds for purchase and maintenance of reserves.

(2) It should create opportunities for local people.

(3) It should also help to justify conservation to local governments. In Nepal, a project to build a dam near the Chitwan nature reserve was cancelled when an economic analysis showed that the Park raised more dollars than would the hydroelectric scheme.

Ecotourism is a large, rapidly growing enterprise that is currently estimated to be a $2 billion industry across the world (Herliczek, 1996). An explosive increase in travel is occurring internationally, especially in the tropics. In 1988, some 15 million tourists visited South and Central America, many of them going there to see the flora and fauna of natural areas. Thousands of tourists go to Argentina to view a huge colony of 200,000 Magellanic penguins. Visitors to Nepal had increased from 10,000 in 1965 to 250,000 by 1990. Ecotourism is the fastest-growing component of tourism worldwide (Herliczek, 1996). It can be a strong motivation to conserve natural sites. It has tremendous economic potential for local sustainable development.

Though less destructive than many other uses, ecotourism also involves some unavoidable costs, and a balance has to be struck between preservation and development. Even non-consumptive visitation, if done by large numbers of tourists, can subtly affect ecosystems. Animal behaviours may be altered, breeding grounds may be disturbed, and potential prey may be frightened away. Human intrusions can lead to changes in animal behaviour that may hamper, though not threaten, wildlife (Knight and Gutzwweiler, 1995) and some other intrusions may seriously affect populations by inhibiting reproduction.

Ecotourism often results in littering. Although tourists are encouraged to pack out what they bring in, certain objects may be left behind along tourist trails. Ecotourism can also lead to site degradation. Ecotourism reserves must be large enough to support the populations of animals and plants people want to see. Questions such as minimal numbers of breeding pairs and insufficient territory sizes become very critical in small reserves.

Initial development focused on unique natural sites supporting high-profile species such as lemurs, elephants, or grizzly bears, or ecosystems with obvious public appeal such as tropical rain-forests and marine coral reefs as being suitable. In general, rare places having some special attractions have been successfully developed for ecotourism. It appears that ecotourism is likely to result in isolated, specialized reserves scattered the world over, rather than vast protected tracts. Consequently, it will act as a large scale, sustainable development strategy only in concert with other protective measures. Scattered, specialized ecotourism reserves would be successful in protecting some of the larger and more charismatic vertebrates from extinction, but they would not be big enough to protect the broader diversity of plants and animals.

Role of Traditional Societies in Biological Conservation

A clear separation between lands in developing countries used by the natives to obtain natural resources and strictly protected national parks is often impossible. Local people sometimes live in protected areas and use their resources. In some cases, people are allowed to enter protected areas periodically to collect natural products. In Biosphere Reserves, natives are allowed to use resources from designated buffer zones. The protection of biological diversity, the customs of traditional societies, the genetic variation of traditional crops, and economic development can be meaningfully integrated through what are called Integrated Conservation-Development Projects (Caldecott, 1996; Maser, 1997), which are some of the best conservation strategies known today.

One of the objectives of UNESCO's Man and the Biosphere Program (MAB) is the maintenance of samples of varied and harmonious landscapes resulting from long-established land-use patterns (Batisse, 1997). Under this programme 327 Biosphere Reserves in 85 countries now cover over 200 million ha (Fig 1.2). The MAB Program hopes to find ways in which native people may sustainably use natural resources without degrading the environment.

In-situ Agricultural Conservation. In many areas farmers cultivating locally adapted crop varieties can preserve genetic variability in these species because local varieties often have unique genes for dealing with disease, nutrient deficiencies, pest resistance, drought tolerance and other

Fig. 1.2 : Locations of Biosphere Reserves (dots) in the world. Note the paucity of reserves in some parts, e.g. India, New Guinea and S. Africa.

environmental variations. These local varieties continue to evolve new genetic combinations, some of which may well be effective in dealing with looming global environmental threats. Unfortunately, however, farmers in many places are abandoning their traditional farming practices with local landraces and going in for new high-yielding varieties using capital-intensive methods, including fertilizers, pesticides and machinery. In some countries such as Indonesia, Sri Lanka and the Philippines, over three-fourths of the farmers have already adopted modern varieties which give them increased yields and income quickly. But this can have adverse effects in the long term in so far as the preservation of the genetic variability in local varieties is concerned. Therefore, some villages are now being identified as *in-situ* landrace custodians. In these the farmers would be encouraged to grow their traditional crops in the old, traditional way : the governments will adequately subsidize/compensate them for the reduction in yields and incomes generally associated with this practice.

Creative Conservation

The crisis of species extinctions is one of the greatest challenges facing mankind and calls for earnest, ingenious and creative means of tackling it.

Society is now struggling with unbearable pressures on land, water and their biota. Lengthening lists of endangered species show that biological diversity conservation can be best dealt with at the ecosystem and landscape level rather than at the species level, and that some form of ecosystem management approach is preferable. Innovative approaches are needed to manage landscapes with some degree of ecological cohesion in ways that provide flexibility, multiple options for human aspirations and opportunities for creative solutions (Lovejoy, 1997).

The best index of whether an ecosystem is being managed sustainably is whether it maintains its characteristic biodiversity and ecosystem processes. In other words, whther the native species list for some particular ecosystem a century hence should essentially be that which it has today, and that functions such as watershed protection, nutrient cycling, or pollination are well maintained. Achieving these goals will require considering every factor intrinsic to the ecosystem, as well as every factor extrinsic to the ecosystem. And it can work only if there is active involvement of local people.

Rather than viewing nature as a fenced-off patch in the middle of a human-dominated landscape where people go on doing what they like, people should learn to live and pursue their activities and aspirations within nature. The two important operational measures of sustainable ecosystem management, viz. characteristic biodiversity and ecosystem processes, should be well appreciated and respected by everybody.

Biodiversity Planning

The current rate of species extinction is some 100 times the natural background rate (Lawton and May, 1995; Stuart and Lawton, 1998). This rate is likely to increase in the coming decades. Only about 5% of the earth's land surface has protected reserves. When humankind destroys or modifies the remaining 95% of the land, only half the world's species will survive in the protected 5% the other half will be lost forever.

The most vulnerable species are those with the smallest geographical ranges; these are not distributed randomly. There are some natural hot spots where these rare, endemic species are concentrated. If the reserves are judiciously spread over these special places, there is a slight chance that a greater fraction of species might be saved. It is paradoxical that, globally, the larger reserves (high mountains, tundra, and the driest deserts) are areas not particularly species-rich. Hot spots such as Madagascar and the Philippines protect less than 2% of their land. The need is for setting up reserves in such a way as to protect the maximum number of species at the lowest possible cost.

Biodiversity and Development

Human beings as a species are rapidly altering the world in various ways. Unprecedented population growth and non-sustainable development are not only degrading the land, water and atmosphere, but also decimating a wide variety of the Earth's organisms and their habitats.

We humans consume well over a third of the total terrestrial photosynthetic productivity. Human activities threaten, over the next few decades, to eliminate about one-fourth of the world's species. Some of tehse species we may not use directly but our survival can depend on them in various other ways.

Although population growth is not the only cause of environmental degradation, it is an important exacerbating factor. It can undermine the capacity of many developing countries to conserve resources and meet basic human needs.

Depletion of the Earth's biological diversity has much more profound consequences than some other environmental threats. As the loss of biodiversity is irreversible, the potential impact on the Earth's living systems is enormous. The human species has evolved biologically and culturally in a highly diverse world.

One common myth about biodiversity is that its conservation in protected areas necessarily means totally prohibiting resource use. In keeping with this conventional misconception, land managements were proposed to exclude human use (sustainable or otherwise). Local people were sometimes displaced to make room for the protected area. Consequently, social conflicts arose and the area was threatened by surrounding land-use pressures. The fact is that local support for biodiversity conservation is critical and desirable. In integrated conservation, a management unit integrates core natural areas and surrounding land-uses, and managers solicit the support and participation of the local population, in part through research contributing to their social and economic development (UNESCO, 1994).

Local support for effective conservation may be ensured in the following ways:

1. Recognize that it is fundamental to the long-term conservation and sustainable use of biodiversity.
2. Ensure that local populations participate as true partners in designing and managing conservation programmes.
3. Allow local populations to identify their own socioeconomic needs.
4. Ensure that people who bear the costs of conservation projects (e.g. restrictions on fishing) also receive a high proportion of the benefits (e.g. tourist revenues).
5. Identify options for sustainable use of biodiversity.
6. Use indigenous knowledge to manage protected areas as far as practicable.
7. Local populations should have maximum stewardship over local resources.
8. Offer income-earning activities and/or service (e.g. improved access to markets, low-interest credit, controlled access to resources) to local populations as a *quid pro quo* for respecting protected area regulations (Brown and Wyckof-Baird, 1992).

Some conservation programmes focus only on augmenting protected area networks but this is generally not sufficient. Regional planning schemes are needed for conserving biodiversity outside natural areas (e.g. biodiversity in seminatural areas and in cultivated or domesticated species) and for minimizing the impact of human activities on natural areas. Only some 5% of the land area of the Earth lies within protected areas, and no marine protected areas have been established so far. However, even if all remaining wild areas were protected, much biodiversity would remain unprotected. Much of the Earth's biodiversity actually occurs in seminatural and rural areas, consisting of fields, forests, hedges, edges of fields and abandones areas at various stages of succession. Also much economically valuable biodiversity consists of cultivated plants and domesticated animals rather than the wild species of natural areas (UNESCO, 1994).

Some laymen believe that protected natural areas meet most conservation needs. This is not true because protected areas are not completely isolated; they are affected by activities in surrounding areas.

Mining and agriculture upstream of a protected area can affect water quality inside. Industrial activities, even from long distances, affect the air quality of a protected area downwind.

Protected areas surrounded by development effectively become islands vulnerable to environmental change (e.g. global climate change). Species in these area do not have escape pathways for migrating to new areas when environmental conditions change.

Integration of conservation and development planning can maintain the integrity of ecological processes providing key services (e.g. water quality). Conservation and development planning need to be linked through legislation, regulations and institutional mechanisms. Habitat protection is the single most effective means of protecting biodiversity at all levels of organization (genetic, species, ecosystem).

2
Importance of Biodiversity

Introduction

Biodiversity is the vast repertoire of non-human organisms of our planet. It is to be valued for four reasons. First, as the dominant species on Earth, *Homo sapiens* has an ethical leadership responsibility towards humanity. Second, activities as diverse as gardening, making nature films and ecotourism, attest to the aesthetic values of biodiversity. Third, humanity has derived many direct economic values from biodiversity, including all of its food and many of this medicines and industrial products. Fourth, and most important from an anthropocentric perspective, plants, animals, and micro-organisms help to supply human beings with an array of free ecosystem services, without which civilization could not persist.

A major problem facing us today is how to measure the value of environmental goods whose destruction generates vast externalities. A good example of these goods is biodiversity -the variety of genetically distinct populations and species of plants, animals and micro-organisms with which *Homo sapiens* shares the Earth, and the variety of ecosystems of which they are functioning parts (Ehrlich and Ehrlich, 1992). It is generally agreed that biodiversity has value to humanity, but whether or not it has value independent of human needs is not clear. The value of biodiversity to humanity has both use and non-use components. Biodiversity supplies us with food (fishing and hunting), direct enjoyment (scenic value, bird-watching), or ecosystem services (recycling of nutrients). Its non-use values are, for example, the pleasure a non-African who will never travel to Africa may derive from knowing that free-living black rhinos exist there. Interface ecosystems such as estuaries are rich in biological diversity and serve as important life-support systems also for human beings. The economic values of biodiversity can be divided into four categories: ethical, aesthetic, direct economic and indirect economic (Ehrlich and Ehrlich, 1992).

Natural ecosystems provide civilization with a variety of essential services *gratis* and on a large scale. The strongest anthropocentric reason for the preservation of Earth's life-forms is their intimate involvement in the delivery of these services. One very essential ecosystem service is maintenance of the gaseous composition of the atmosphere. Ecosystems prevent changes in the mix of gases and particulate matter from being too rapid. They act as a sort of buffer.

Sea-birds regulate fish populations and transport nutrients and organic materials from the sea to the land. They are important nodes in the structuring of both aquatic and terrestrial ecosystems, and therefore contribute to the generation of indispensable environmentally produced goods and services.

Value of Species

Although the value or usefulness of many species of plants, animals and microbes is very well known, there are countless other species to which no known use or value can be assigned. So what, if any, may be the justification or reason for preserving these species? Organisms have direct value as resources for humanity, both now and in the future. Also, species provide goods and services indirectly as important components of the web of life. Some workers question economic arguments for protecting species. Ehrenfeld (1988) stated that the value of biodiversity does not depend on the uses to which particular species are put, or their role, if any, in global ecosystems.

Some conservationists believe that even if the *in-situ* protection of an endangered species cannot be justified by its potential as a source of pharmaceutical, agricultural, or other raw materials, the species nevertheless plays some role in supporting the ecosystem services and processes on which all life depends. But this belief is not shared by others (e.g. Lawton and Brown, 1993) who feel that the functional redundancy of species is so great that ecological processes would function quite well even if all the creatures now threatened and a great many more besides become extinct (Lawton and Brown, 1993) because the functions of the lost species would be taken over by some other living species. The Global Biodiversity Assessment (Perrings, 1995) makes this point, as does Holling (1992) : 'Although any ecosystem contains hundreds to thousands of species interacting among themselves and their physical environment, the system seems to be driven by a small number of biotic and abiotic variables on whose interactions the balance of species is carried along' (Holling, 1992; Perrings, 1995).

Even though both diversity and the ecosystem have long been central concepts in ecology, the merger of the two concepts is quite recent. Research on food webs and energy flow (Hairston *et al.*, 1960; deAngelis, 1992) conceptually linked the concerns for declining biodiversity and for ecosystem function. Tilman *et al.*, (1996) showed that ecosystem productivity may be affected by diversity but many workers feel that this issue is highly complex (Bengtsson *et al.*, 1997).

The complexity of ecosystem responses to diversity has been further borne out by a Swedish study on aquatic systems where zooplankton grazer diversity did affect primary productivity, and phosphorus levels, size-and species-related effects were more important for plant biomass and nitrogen levels. It was also observed that different ecosystem processes can respond in different ways to variation in diversity. It appear that ecosystem processes often respond to variation in diversity in a system-specific species-specific and process-specific manner, and that diversity may sometimes not have much effect at all.

Diversity at the Landscape Level

Several current definitions of biodiversity have lacked the ecosystem and landscape perspectives. According to Gaston (1996), greater consideration has been given to the species-diversity component of the individual landscapes, in most cases, even when landscapes themselves were purported to be the level of study. What role do landscape-scale processes play in our understanding of diversity? What is the 'minimum diversity' necessary for riparian habitats to provide important ecosystem services such as nutrient retention in agricultural landscapes? There is need to intensify work on how to use biodiversity to manage landscapes sustainably and restore habitats.

Many ecologists dislike the idea of judging biodiversity and the processes that species perform in ecosystems in economic terms. The value-criterion of welfare economics is that human preferences should count and constitute all values; the words 'ecosystem services' contributing to the satisfaction of human preferences have an economic value. Economic values of ecosystems can be use-motivated (e.g. production of food and recreation) as well as non-use-motivated.

But what is the role of biodiversity in providing ecosystem services? Biodiversity may be needed to maintain ecosystem functions at current levels and also, it may be vital as an insurance for rare events or environmental changes that may impair ecosystem functions (Folke *et al.*, 1996).

In the first case, biodiversity can be directly important for managing ecosystems, in habitat restoration and in landscape planning. It will then have an instrumental value that might possibly be evaluated from existing market prices. Table 2.1 is a tentative list of ecosystem services that may be needed for managed landscapes to function and provide values for humans, and ascertaining for whether present knowledge can assign some importance to biodiversity for these services. A diversity of primary producers is important, for example, for food production, production of materials and energy, environmental protection, recreation and culture. Decomposer diversity is needed for many functions.

Table 2.1 : Some selected ecological functional groups for a landscape to provide ecosystem services for humans, and the possible importance of biodiversity for these functions (modified from Bengtsson *et al.*, 1997).

Ecosystem service/ landscape function	Primary producers	Herbivores	Predators	Decomposers	Pollinators	Bioturbators
Plant production (cereals, vegetables, fruit)	+ +	–	–	+ +	+ + +	–
Meat production	+ +	+	+	+ +	+	+
Fuel (bioenergy)	+	–	–	+ +	–	+
Building materials	+ +	–	–	+ +	–	+
Fibre production	+ +	–	–	+	+	+
Erosion control/flood regulation	+	–	–	+	–	–
CO_2 fixation	+	–	–	+	–	–
Drinking water	+	+	+	+	–	+
Waste treatment	+	–	–	+ +	–	+
Landscape heterogeneity	+ +	+ +	+ +	+ +	–	+ +
Recreation and natural history	+ +	+ +	+ +	+ +	+ +	(– to + +)
Urban environment (aesthetics)	+ +	–	–	+	+	+

Key : + +, biodiversity is important; +, some biodiversity is needed; –, no obvious relation to biodiversity, or no information available.

Note : Some biodiversity is needed for functioning of dispersers for landscape heterogeneity and for urban cultural values.

Economics and Agriculture

Biological resources, especially wild species, contribute significantly to economic activity and if well managed, also to sustainable development. For instance, Brazil nuts and wild rubber in Brazil and the *Babassu* palm forests of Amazonia are potentially promising as extractive reserves. In South East Asia, the various wild species of the climbing palms known as rattans are of great economic value.

Another way in which biodiversity contributes to economic development is via nature tourism (Boo, 1990). In Kenya and Costs Rica, nature-oriented tourism is the most important foreign exchange earner.

Agricultural uses of biodiversity for economic activity include the use of genetic traits from wild relatives of domestic crop species. The international centres for such crops as rice and wheat look for wild relatives for disease and pest resistance. Biodiversity also contributes to economic activity in the area of integrated pest management. Thus, the fungi *Metarhizium flavoviride* and *Beauveria bassiana* are effective in field testing against migratory locusts and agriculturally problematical grasshoppers (Lovejoy, 1994). In Australia the spread of two species of *Opunita* (cactus) has been effectively controlled by the introduction of the moth *Cactoblastis cactorum.*

Two kinds of ecosystems are crucial to the functioning of human society today. (1) Agricultural, which are simplified versions of natural ecosystems, artificially maintained by humanity to increase the production of commodities people need and desire and (2) Natural ecosystems, whose importance is less widely appreciated, but society depends upon them as much as it depends upon agricultural ecosystems. That is because agricultural ecosystems are embedded in natural ones and depend on the natural components for their sustained productivity (Ehrlich and Ehrlich, 1992). Natural ecosystem directly provide us with food and material of all sorts. A crucial portion of the protein in our diets comes straight from nature in the form of fishes and other animals harvested from the seas. This service is provided by the oceans in conjuction with coastal wetland habitats, which are nurseries for marine life that is either harvested directly or serves as a food supply for the sea-life that we eat. Forests also provide timber and other wood products. Rubber, oils and organic chemicals, spices and herbs, and game are provided by natural ecosystem. The active ingredients in at least a third of the prescription drugs used by civilization come directly from or were derived from chemical compounds found in wild plants, fungi, or other organisms, especially in tropical forests; digitalis, morphine, quinine, and antibiotics are familiar examples.

Biodiversity has a strong relationship with sustainable development, including direct harvest, nature tourism, wild genes improving domestic crops, wild species contributing to crop productivity, pest management, sources of medicine and bioremediation (Lovejoy, 1994). Biodiversity relates through services, individual species indicating environmental, change or stress, and through wealth generation.

A major weakness of the international economic system has been its inability to assign a value of exchange to biodiversity and other environmental components. Determination of how to incorporate the cost of conservation into that of production is a challenge we must meet with a view to fulfilling our obligations to future generations by halting the continuing impoverishment and misuse of biodiversity.

Genetic diversity in agriculture enables crops and animals to adapt to different environments and growing conditions. The ability of a particular variety to withstand drought or flooding, or resist insect pests or diseases, give higher protein yields or produce a better-tasting food, are traits passed on naturally by its genes. This genetic material constitutes the raw material that plant and animal breeders and biotechnologists use to produce new varieties and breeds. Without this diversity we would lose the ability to adapt to ever-changing needs and conditions. Sustainable agriculture could not then be achieved in many of the world's different food production environments. The rural poor depend upon biological resources for an estimated 90% of their needs.

Diversity among individual plants and animals, species and ecosystems provides the raw material that enables human communities to adapt to change. Deprived of biodiversity, the ability of humankind to meet environmental challenges would be severely limited. Through modern biotechnologies, wild diversity can also be incorporated into crops and contribute to world agricultural development.

Pest Control and Pollination

Natural ecosystems help control an estimated 99% of pests and disease that can potentially attack crops or domestic animals. Most of the potential pests are herbivorous insects, which are controlled primarily by predacious insects that consume them. This service has been disrupted by the misuse of artificial insecticides, because insect pests are generally less susceptible to pesticides than are their predators. Populations of pests tend to be large and so have a better chance of evolving resistance to pesticides. Herbivorous insects also have long been engaged in a coevolutionary race with plants. Plants have evolved many deadly compounds in attempts to poison their attackers; compounds familiar as the active ingredients of certain species, drugs, and medicines. In turn, the insects have evolved resistance to these poisons; this ability to evolve resistance preadapts them for dealing with the insecticides developed to poison them.

Natural ecosystems provide crop plants with stable climates, water, soils and nutrients, and protect them from pests.

Pharmaceuticals

Ever since hippocrates prescribed infusions of willow bark (the precursor of aspirin) as an analgesic, wild species (mostly of plants) have yielded valuable medicines for humans.

More recently, taxol has been isolated from the Pacific yew tree (*Taxus brevifolia*) and is being used in the treatment of cancer. The vine *Ancistrocladus korupenis*, collected from Cameroon, has yielded the compound michellamine B, which protects human cells from the HIV virus *in vitro;* it is the first natural compound to move to animal testing for AIDS research (Stix, 1993). Squalamine is an aminosterol from the dogfish (*Squalus acanthias*) that shows antiparasitic, antifungal, antiprotozoan and antibacterial activity (Moore *et al.,* 1993). The Ecuadorian arrow poison frog (*Epipedobates tricolor*) secretes a substance which *blocks pain* far more effectively than morphine but is not an opioid (Spande *et al.,* 1992).

Certain constituents of marine organisms have been examined for various uses : discodermolide from a Bahamian sponge (*Discodermia* sp.) appears to be useful in suppressing organ rejection after transplant operations; macrolectins from the Bahamas sea-floor mud bacteria hold promise in inhibiting growth of melanoma and colon cancer cells as well as the herpes simplex virus and HIV (Fenical, 1993).

The potential for biodiversity to supply new foods and medicines alone is enormous (Ehrlich and Ehrlich, 1981; Myers, 1979). Mankind depends for its very existence on other organisms in all their extraordinary variety. We are completely dependent on the totality of life-support systems to maintain the habitability of this planet. The basic dependence of mankind on other organisms is through the process of photosynthesis.

Ethical and Aesthetic Values

These are based on the religious or quasi-religious feelings of many people in many cultures that other-life-forms have intrinsic value and deserve to be protected from destruction by humanity. Buddha preached that man has no right to kill other animals at all.

The notion of *rights* now applies to animals also. Today, both domestic and non-domestic animals are being legally protected against abuse as is suggested by increasing opposition to hunting, laws to protect birds, the movement to protect whales and dolphins, and the general revulsion at the slaughter of baby seals.

As the dominant species on the planet, *Homo sapiens* carries an ethical responsibility to preserve biodiversity. This means opposing intentional exterminations of other species and supporting conservation efforts.

The beauty of birds, tropical fishes, butterflies and flowering plants is widely enjoyed and supports extensive economic activity, including bird-watching and feeding, scuba diving, butterfly collecting, photography and the making of nature films. Some tiny wasps and flies, when seen under the microscope, appear to be made out of solid gold. Diatoms have silica shells that are as exquisite and varied as snowflakes. Most organisms at least exhibit the beauty of design. Even the tiniest beetles have complete external skeletons, nervous and digestive systems, and complex musculature. Such insects show a degree of sophisticated miniaturization beyond the reach of human engineers. Many insects show the type of beauty, intricacy and diversity that captivates gun collectors, philatelists, computer hackers, bibliophiles and others.

Indigenous Knowledge

Human cultures have adapted to many diverse habitats and have used, changed and nurtured biological resources to meet innumerable needs. As a result of plant and animal domestication, and resource harvesting, a strong interdependence has evolved between 'natural' and 'human-induced' biodiversity. Rural people have managed genetic resources for as long as they have cultivated crops. For over 10 centuries they have selected varieties of crops and livestock breeds to meet environmental conditions and diverse nutritional and social needs. The immense genetic diversity of traditional farming systems is the product of human innovation and experimentation.

For many developing countries, self-reliance in food production will depend on improving low-input agriculture in difficult environmental conditions. The raw materials for these improvements are the biological resources sustained in forests, rangelands, fields and farms. The accumulated knowledge of farmers, coupled with access to modern technologies, provides vital clues to developing sustainable agricultural systems.

In many parts of the world, wild species and natural habitats still help support household food security – access by all people at all times to the food they need for a healthy life. In Nepal, for example, 135 tree species are used as fodder. In Ghana, most people depend on wildlife for their requirement of animal protein. There is now increasing recognition of the value of indigenous knowledge to address global agricultural, health and environmental problems. Conservation and use of biodiversity must be concerned not only with genes, genotypes, species and ecosystems, but also with the traditional knowledge that has helped to produce and maintain this diversity (FAO, 1993).

Traditional Medicines

An estimated three-quarters of prescription drugs derived from plants were discovered because of their prior use in indigenous medicine. Forest-dwelling indigenous people employ at least 1300 plant species for medicines and related purposes.

Traditional Food Plants

Subsistence farmers have been producing or gathering plants from the wild or semi wild that have long been accepted as desirable sources of food. Millions of people use such traditional plants to satisfy their food needs. They are essential to the diets of rural subsistence households throughout the developing world, providing soruces of energy, vitamins and minerals. In recent times, however, many countries have seen a shift away from traditional foods, resulting in a narrowing of the food base. Failure to appreciate their benefits, together with a growing demand for imported foods, is reducing availability and consumption of these foods.

In Nagaland, an agricultural system has evolved that centres on the alder tree (*Alnus nepalensis*) which is a source of fuelwood, timber and mulch. It is a nitrogen-fixing nodulated tree that enriches the soil nitrogen status. In the fields, alongside the alder trees, farmers grow main crops of maize, millet, potato, barley and wheat and secondry crops fo chilli, pumpkin and taro.

Endod (*Phytolacca dedecandra*), commonly known as the African soap berry, is a pernnial plant that has been cultivated for cencturies in many parts of Africa where its fruits are used as a laundry soap and shampoo. In 1964, the Ethiopian biologist Aklilu Lemma observed that downstream from where people were washign clothes with endod barries, dead snails were found floating in the water. Further research revealed that sundried and crushed endod berries, are lethal to all major species of snails, but do not harm animals or people, and are completely biodegradable.

For Africa, where one of the most serious diseases, schistosomiasis, is transmitted by freshwater snails, discovery for a low-cost and biodegradable lumicide is a major breakthrough (FAO, 1993).

Indian farmers have been using seeds from the neem tree (*Azadirachta indica*) as a natural insecticide to protect crops and stored grain. Scientists have isolated compounds that are extremely effective against insects, even in minute quantities. They control more than 200 species of insects, mites and nematodes, including major pests such as locusts, rice and maize borers, pulse beetles and rice weevils. Yet neem extracts do not harm birds, mammals and beneficial insects such as bees. Unlike most synthetic pesticides, insects apparently do not develop resistance to neem extracts because they contain several biologically active ingredients (FAO, 1993). In 1993, the world's first commercial-scale facility designed and built for neem-based natural biopesticide production opened in India, capable of processing 20 tons of neem seed per day.

Human Health

Growing human populations and their plundering and blatant exploitation of natural resources deplete or destroy many natural habitats and not only create a less healthy environment, but become exposed to a variety of new pathogens. This has been occurring at a time when global climate change may lead to expansions in the range of several infectious diseases when resistance can evolve to the antibiotics used to cure many common diseases. There are potential interactions between human health, the destruction of tropical forests, the emergence of new pathogens and the loss of natural products with therapeutic potential that could be used to treat new and existing ailments. The potential dangers to human health of the continued assault on the environment, particularly the destruction of tropical forests, should be realized.

The potential for tropical (and temperate forests to supply agents with medicinal value that might be used to treat infectious and congenital diseases has been known for some time. So far, one of every 125 plant species studied has produced a major drug. If we lose one tree species a day, then we lose three to four potentially valuable new drugs every year. If we contrast this with the synthesis of new drugs from chemicals, the potential for finding major new drugs is in the order of one in 10,000 for each compound tested. The difference in the success rate between screening programmes and searching for natural products from plants is patentability. Phamaceutical companies face a much greater difficulty in patenting natural products than ones they have synthesized.

Despite significant trade and legislative barriers, the overwhelming majority of prescription drugs used at present in the US are based on natural products. In the USA 118 of the top 150 prescription drug (74%) are based on plants, 18% on fungi, 5% on bacteria and 3% on vertebrates (indeed, all of the latter are from one species of snake, *Bothrops asper*). If we consider the top ten prescription drugs in the US, nine of the ten are based on natural plant products. And the USA is conservative in terms of its phamaceutical diversity! According to the World Health Organization, more than 80% of the world's population relies upon traditional plant medicine for primary health care.

In Europe, compounds distilled from leaves of the *Ginkgo* tree increase rate of cerebral blood flow and help prevent senile dementia. Other compounds widely available in Europe, but not available in the S, include (1) an extract of mistletoe, which increases the survival time of women suffering from breast cancer, (2) Theokal, a hawthorn extract widely used as an antiarrhythmic that is safer to use than digitalis, having fewer side effects on people with hearth problems, and (3) Thisilyn, a derivative of milk thistles, that all European hospitals stock and which is used for liver damage from alcoholism-also relieves sufferers from hepatitis. Garlic-based compounds are being used all over the world. Garlic reduces cholesterol through its action as an antioxidant

Only about 1100 of the world's 365,000 known species of plants have been examined for their medicinal properties thus far. Indigenous cultivators are the *in-situ* protectors of 90% of the world's remaining diversity. Their knowledge of the medicinal plants and their uses is of great value to human health. Currently, one indigenous cultivation goes extinct in the Amazon basin every year. Every time an old man of these tribes dies, knowledge of some tropical plants and their medicinal properties is lost. If tropical rainforest's products are used to put a value on preserving biodiversity, then intellectual property rights agreements must ensure that some proportion of the financial benefits goes back into protecting the concerned habitat and the indigenous tribes.

We need these new drugs for four main reasons (Chivian *et al.*, 1994; Rosenthal and Grifo, 1995) : (1) increasing large-scale development of resistance to many antibiotic and anthelminthic compounds currently in use, (2) emergence of new human diseases, particularly HIV/AIDS; (3) the resurgence of diseases such as TB, and (4) changes in the geographical distribution of older diseases owing to increased human movement and global climate change. Conservation biologists are desperately searching for examples of human benefits that might come from conserving biodiversity. Pharmaceutical products from rain forest plants can be a good justification for rain forest protection in a world where we are beginning to lose our battle with infectious diseases.

New pathogens continue to emerge. The larger the number of humans sampling the environment, the greater the chance that some of them may pick up a new pathogen and transport it to a sufficiently populous area to cause an epidemic outbreak. However, parasites and pathogens that have been with us for a long time still pose the greatest threat to the welfare of human populations, particularly in a world where a changing climate allows pathogens to establish either in areas where they have previously been eradicated, or in areas that have become noticeably warmer and damper. Resistance has already evolved to many of the manufactured compounds used to control these pests and their vectors; hence the search for new compounds will intensify. Drugs derived from natural products are likely to have novel properties to which the pathogens are unlikely to show much resistance. Furthermore, as many natural products contain not one but several active compounds, this would be like using more than one drug to treat an outbreak and could considerably reduce the rate at which resistance evolves.

Plant-derived Drugs

Over 110 pure chemical substances extracted from higher plants are used in medicine throughout the world (Table : 2.2). Discovery of the drugs mostly emanates from knowledge that their extracts can be used to treat one or more diseases in humans.

A large majority of the poeple in developing countries depend on traditional medicine for their primary health care needs, and most of traditional medicine involves the use of plant extracts.

Aloe

Extracts from the *Aloe vera* plant (related to the onion and garlic plants) are now extensively used for softening the skin and warding off wrinkles. It is used as an ingredient in many skin creams and other cosmetics. The enormous medicinal importance of aloe has earned it such names as the Burn plant, Medicine plant, First Aid plant, Miracle plant, Wand of Heaven, and Plant of Life. *Aloe vera* has been claimed to kill *Mycobacterium tuberculosis* which causes TB as well as the herpes virus responsible for herpes genitalis. It inhibits growth of many common organisms, such as yeasts, fungi and the bacteria associated with wound infections. The genus *Aloe* has over 300 species, of which the following *four* have been proven to have medicinal properties : *Aloe babadensis, Aloe perryi, Aloe ferox* and *Aloe arborescens* (Atherton, 1998). It is quite possible, however, that some other species may prove medically important.

Most aloe plants are non-toxic but a few are poisonous. *Aloe barbadensis* may be the most potent medicinally. The resinous mucilage from the cut leaves of aloe is used as a wound healer to treat cuts, stings, abrasions, incised wounds and ulcers the plant contains the following ingredients :

Vitamins - including the important antioxidant vitamins A,C,E,B (thiamine), B_3 (riboflavin), as well as choline, folic acid and vitamin B_{12}.

Table 2.2 : Source and uses of some secondary plant constituents used as drugs (after Farnsworth, 1988).

Name	Therapeutic category	Plant source	Plant use in traditional medicine	Correlation between two uses
Acetyldigitoxin	Cardiotonic	*Digitalis lanata*	-	Indirect
Adoniside	Cardiotonic	*Adonis vernalis*	Heart conditions	+
Aescin	Anti-inflammatory	*Aesculus hippocastanum*	Inflammations	+
Aesculetin	Antidysentery	*Fraxinus rhynchophylla*	Dysentery	+
Ajmalicine	Circulatory stimulant	*Rauvolfia serpentina*	Tranquilizer	Indirect
Anabasine	Skeletal muscle relaxant	*Anabasis aphylla*	-	-
Andrographolide	Antibacterial	*Andrographis paniculata*	Dysentery	+
Arecoline	Anthelminthic	*Areca catechu*	Anthelminthic	+
Asiaticoside	Vulnerary	*Centella asiatica*	Vulnerary	+
Atropine	Anticholinergic	*Atropa belladonna*	Dilation of eye pupil	+
Caffeine	Central nervous system stimulant	*Camellia sinensis*	Stimulant	+
Chymopapain	Proteolytic; mucolytic	*Carica papaya*	Digestant	+
Codeine	Analgesic; antitussive	*Papaver somniferum*	Analgesic; sedative	+
Colchicine amide	Antitumour agent	*Colchicum autumnale*	Gout	-
Curcumin	Choleretic	*Curcuma longa*	Choleretic	+
Danthron	Laxative	*Cassia* species	Laxative	+
Digitalin, Digitoxin	Cardiotonic	*Digitalis purpurea*	Cardiotonic	+
Digoxin	Cardiotonic	*Digitalis lanata*	-	Indirect
L-Dopa	Antiparkinsonism	*Mucuna deeringiana*	-	-
Emetine	Amoebicide; emetic	*Cephaelis ipecacuanha*	Amoebicide; emetic	+
Ephedrine	Sympathomimetic	*Ephedra sinica*	Chronic bronchitis	+
Glycyrrhizin	Sweetener	*Glycyrrhiza glabra*	Sweetener	+
Gossypol	Male contraceptive	*Gossypium* species	Decreased fertility	+
Hyoscyamine	Anticholinergic	*Hyoscyamus niger*	Sedative	+
Kainic acid	Ascaricide	*Digenea simplex*	Anthelminthic	+
Khellin	Bronchodilator	*Ammi visnaga* (L.) Lamk. (toothpick plant)	Asthma	+
Lobeline	Respiratory stimulant	*Lobelia inflata*	Expectorant	+
Menthol	Rubefacient	*Mentha species*	Carminative	-
Monocrotaline	Antitumour agent (topical)	*Crotalaria spectabilis*	Skin cancer	+
Morphine	Analgesic	*Papaver somniferum*	Analgesic; sedative	+
Nicotine	Insecticide	*Nicotiana tabacm*	Narcotic	-
Noscapine (narcotine)	Antitussive	*Papaver somniferum*	Analgesic; sedative	+
Papain	Proteolytic; mucolytic	*Carica papaya*	Digestant	+

Contd...

Table 2.2 – *Contd...*

Name	Therapeutic category	Plant source	Plant use in traditional medicine	Correlation between two uses
Protoveratrines	Antihypertensive	*Veratrum album*	Hypertension	+
Pseudoephedrine	Bronchodilator	*Ephedra sinica*	Chronic bronchitis	+
Quinidine	Antiarrhythmic	*Cinchona ledgeriana*	Malaria	-
Quinine	Antimalarial; antipyretic	*Cinchona ledgeriana* Moens ex Trimen (yellow cinchona)	Malaria	+
Quisqualic acid	Anthelminthic	*Quisqualis indica* L.	Anthelminthic	+
Scopolamine	Sedative	*Datura metel*	Sedative	+
Silymarin	Antihepatotoxic	*Silybum marianum* (L.) Gaertn.	Liver disorders	+
Strychnine	Central nervous system	*Strychnos nux-vomica*	Toxic stimulant	+
Theobromine	Diuretic; vasodilator	*Theobroma caco*	Diuretic	+
Theophylline	Diuertic; broncholdilator	*Camellia sinensis*	Diuretic stimulant	+
Trichosanthin	Abortifacient	*Trichosanthes kirilowii*	Abortifacient	+
Vincamine	Cerebral stimulant	*Vinca minor*	Cardiovascular disorders	+
Vinblastine, Vincristine	Antitumor agent	*Catharanths roseus*	-	-
Yohimbine	Adrenergic blocker; aphrodiasiac	*Pausinystalia johimbe*	Aphrodisiac	+

Note : + means yes, - means no; a few of the compounds are currently being synthesized commercially

Enzymes - several different types incl. amylases, lipases and carboxypeptidase which has an anti-inflammatory effect.

Minerals - sodium, potassium, calcium, magnesim, manganese, copper, chromium and iron. Magnesium lactate inhibits histamine decarboxylase, preventer of formation of histamine from the amino acid histidine. Mg lactate has powerful anti-itching effect.

Sugars - the most important are the long-chain polysaccharides, e.g. glcose and mannose. If taken orally, some of these polysaccharides bind to receptor sites lining the gut and help prevent 'leaky gut syndrome'. Unlike the other sugars which are broken down before absorption, the aloe polysaccharides are fully absorbed and appear in the bloodstream unchanged where they act as immunomodulators, both enhancing and retarding the immune response.

Anthraquinones - these phenolic compounds found in the 'bitter aloes' exert a powerful purgative effect.

Lignin - facilitates penetration of other active ingredients deep into the skin.

Saponins - These soapy substances form 3% of the gel and are general cleansers with antiseptic properties.

Sterols - three important anti-inflammatory plant steroids present (cholesterol, campesterol, β-sitosterol.

Salicylic acid - An aspirin-like compound with anti-inflammatory and antibacterial properties.

Amino acids - *Aloe vera* gel provides most of the essential amino acids.

Aloe vera gel si particularly effective on burns, soothing, relieving pain, reducing inflammation and promoting healing with minimal scar formation. Its anti-inflammatory action probably depends on its ability to inhibit the blood-clotting agent thromboxane and prostaglandin production, causing increased blood flow to the area and reduced platelet aggregation (Atherton, 1998).

3

Ecosystems, Environment and Biodiversity

Introduction

The concept of 'diversity' and its relationship to other aspects of community organization and dynamics is of great significance in the development of plant ecology, but a precise meaning of diversity is lacking. Two individuals are not diverse if they are of different sexes; they are diverse only if they are of different species. Similarly, subspecies do not count as diverse only if they are of different species. Similarly, subspecies do not count as diverse, nor are larvae from adults. There has been little critical evaluation of whether species are the most appropriate units of currency of diversity for understanding ecological phenomena. Cousins (1991) discussed the limitations of treating all species equally, and suggested that ranking or weighting species with respect of function or size or systematic relationships provides a better and more relevant measure of diversity.

Two active areas in community ecology concern the relationship between species diversity and both disturbance and non-competitive biotic interactions such as mycorrhizal infections (Wayne and Bazzaz, 1991). The rapid loss of natural areas and high rates of extinction of species and subspecies in the tropics has led to resurgence of interest in cataloging species in unexplored areas and trying to understand the factors generating and maintaining this bio-diversity (Wilson, 1988).

The ecologist's definition of a diverse community implies that it is biotically and/or abiotically heterogeneous in space or time. More often it is the biotic components of communities that are described with measures of diversity. It is not clear, however, why ecologists have mostly tended to describe biotic diversity using species classes.

In reality, the term 'diversity' biologically denotes little on its own; it requires an adjective such as 'species', 'genetic' or 'chemical' to be meaningful. The potential number of abiotic and biotic classes of diversity in natural ecosystems can be quite high. The adjective most appropriate to describe ecological phenomena will depend on the patterns and processes one hopes to study. For systematic inventory, species or generic diversity is the most appropriate unit of study. To understand the influence of vegetation on the diversity of bird communities, one could study canopy structural diversity or fruit size diversity. For pollination biology, floral-colour diversity is the appropriate parameter.

Use and Abuse of Species

Ecologists have mainly focused on 'species' diversity. However, the ecological use of 'species' has certain drawbacks. Wayne and Bazzaz (1991) discussed five aspects of within-species variation that reduce the reliability and usefulness of species as ecological units of diversity in studies of plant communities : genetic variability, phenotypic plasticity, ontogenetic diversity, gender diversity and environmental maternal effects. The species diversity of a region can be partly a function of the taxonomic flora one uses.

Simple numbers or lists of species, or indices based on the algorithmic manipulation of species relative abundances can underestimate the complexity of systems, and signify very different ecological circumstances, depending on the species within a particular system, their population sizes, life stages and environments (Wayne and Bazzaz 1991). Some species-rich communities may be less ecologically diverse with respect to certain features than species-poor communities, yet this point is often overlooked by ecologists and conservationists. According to Wayne and Bazzaz, in studying the ecological and evolutionary processes regulating communities, such as disturbance, competition and mutualisms, it is desirable to explore and employ alternative classes of diversity.

The ecosystem is the 'life-support system' for the community and has a strong relationship to biodiversity.

Species Interactions

All species in a community do not contribute equally to community structure and processes. Some special (called 'drivers') species interactions are quite critical because they effect many other species and process materials out of proportion to their numbers or biomass, or show particularly strong linkage to other species. In contrast, the loss of some species (called 'passengers') from a community has little or no effect on its structure or function. For conservation the focus should be more on the drivers rather than the passengers (Holling, 1992, 1995).

Mutualisms

Mutualisms are relationships in which two or more species benefit from a specific and often highly coevolved interaction. They can be very important in conservation. Good examples are seen in flowering plants, some of which have evolved elegant mechanisms to attract the services of their pollinator, while others have evolved similar mechanisms to attract the animals that disperse their seeds. Legumes and their nitrogen-fixing Rhizobia in root nodules also exemplify mutualism.

Some animals also form mutualistic relationships with other animals. Parasite pickers remove parasites from other species. Some birds clean parasites from rhinoceroses or other large mammals. Some small marine fishes establish 'cleaning stations' where large, often predacious fishes come for removal of their parasites (Feder, 1966).

Since they affect more than one species, and often incorporate a critical function with far-reaching community effects, mutualisms need to be identified and maintained. Loss of a specialist pollinator can lead to local extinction of its host flowering annual which in turn may lead to loss of other insects associated with that annual.

Indirect Effects and Diffuse Interactions

The indirect interactions of keystone predators with non-prey species can be very strong and their presence in a community often has far-reaching effects, even on other species not consumed. These indirect effects can be critical to community structure and hence are of conservation relevance.

Good examples are known from some birds, e.g. the peregrine falcon (*Falco peregrinus*) and fishes (e.g. *Campostoma anomalum*).

Keystone Species

Some species, because of their trophic position, their production of food resources, or other interactions, have a disproportionately great role in community structure. These species are called keystone species. One good example is the sea other (*Enhydra lutris*) which preys heavily on sea–urchins (*Strongylocentrotus droebachiensis*). When sea-otter populations declined in the 20th century due to fur trapping and removal by fishermen, sea-urchin populations increased greatly and grazed heavily on seaweed, occasionally destroying entire kelp forests, which in turn lowered the diversity of the other plants and animals in that habitat (Estes and Palmisano, 1974). The situation could be successfully reversed by reintroducing sea-otters.

Another type of keystone parameter is a keystone food resource. Best examples in tropical forests include figs, nectar and certain fruits, representing less than 1% of plant species diversity; these resources sustain nearly the entire frugivore community at times when other food sources are scarce.

A habitat modifier (an ecosystem engineer) (Jones *et al.*, 1994) exemplifies yet another type of keystone species, e.g. the Canadian beaver and the African elephant (*Loxodanta africana*). The African elephant (like the bison) is herbivorous, browsing on a variety of woody plants and grasses. When feeding on shrubs or small to large trees, the elephants strip bark and branches, and occasionally uproot entire trees. Dense woodlands are often transformed by elephant-feeding into woodland-grasslands or even open grasslands. These areas support more grazing ungulates but are more susceptible to fire, which further favours grasses over woody plants. This animal can therefore change whole landscapes.

Several microbes (bacteria, algae, fungi, protozoa) are found in soils; they break down and decompose dead matter that would otherwise accumulate in a form that is unavailable to living species. Natural cycles would shut down were it not for these inconspicuous but keystone species.

A true keystone species is less abundant in a community but has a marked influence on other species and on community structure. Removal, addition, or local changes in population size of keystone species have wide-ranging effects on other species, on processes and interactions, and even on landforms.

Ecosystem Processes

Although a substantial fraction of the Earth's species are likely to become extinct as a result of human activities before we even document their existence, we cannot predict the consequences of this loss. Nonetheless, certain patterns of biodiversity are emerging. There has been a general increase in species diversity throughout the fossil record, despite periods of mass extinction.

The concept of functional groups involves the grouping of species that have 'similar' effects on ecosystem processes. Functional groups are a practical necessity because it cannot be determined how each species affects ecosystem processes. Such groups are generally useful in organizing our ecological understanding of diversity.

The functional-group concept implies that species within a functional group are equivalent or 'redundant' in their impact on ecosystem processes and that ecosystems could function equally well with fewer species. There are two extreme viewpoints on redundancy : (1) Each species in an ecosystem plays a fundamental role, such that removal of each species incrementally weakens the integrity of the system. Alternatively, (2) a community is composed of a few functional groups, each with several ecologically equivalent species, such that species can be lost from the community with little effect on ecosystem processes, as long as each functional group is represented (Chapin *et al.*, 1992). The truth may lie between these extremes.

A keystone species is a functional group without redundancy. Addition or removal of a keystone species causes massive changes in community structure and ecosystem process. To predict which species in a community are keystone species or could become so under reasonable scenarios of environmental change is a daunting task.

Manipulation of plant diversity can alter the stability of plant communities. This occurs partly because the high phenological and physiological diversity associated with high species diversity enables a community to capture resources and grow over a wide range of unpredictable and constantly changing conditions. In several ecosystems, long-term population stability – which is often a product of community diversity – may be at least as important as ecosystem measures as an index for the long-term maintenance of communities (see Chapin *et al.*, 1992).

Despite the predicted increase in world economic activity, the reality in many developing countries is bleaker than ever. The economic growth experienced on a global level during the 1980s was accompanied by greatly increased world poverty in most developing countries (Abramovitz, 1991). A major factor in the global economic crisis is that during the 1980s the net flow of financial resources was reversed : the countries of the South began exporting more money in the form of debt service than they received in development assistance and private investments (Schmidt, 1990).

Ecosystem Productivity

Many ecologists once believed that diverse, complex ecological communities are the most stable but during the last three decades the universality of this theory came to be questioned when it was observed that the nature of species interactions, rather than species number alone, determines the stability of ecological systems. Recently, expanding concern over loss of biodiversity has led to a reappraisal of the idea that species diversity enhances the productivity and stability of ecosystems (Baskin, 1995).

According to the diversity-stability hypothesis (see Kumar, 1995), increasing the number of trophically interacting species in an ecological community should increase the collective ability of member populations to maintain their abundances after disturbance. Energy is transferred from one trophic level to another as a quintessential ecosystem function. MacArthur's idea that species diversity is crucial for maintaining ecosystem functions received support from the results of a microcosm experiment in a controlled laboratory facility called the Ecotron, and from plant community responses in field experiments conducted in herb-dominated systems in seasonal environments (Tilman and Downing, 1994; Johnson *et al.*, 1996).

Generally, species diversity increases and biomass production decreases with successional time. In some cases it was found that primary productivity increased with plant species diversity (see Johnson *et al.*, 1996).

The quintessential issue of how important species diversity is to ecosystem function boils down to the way the species in that system interact (Walker, 1992). The inherent properties of an ecosystem affect the feedback interactions among its resident species.

Besides disturbance, some other environmental factors also can influence the functional redundancy of various species.

In the context of the role of species diversity in maintaining ecosystem function, the critical point is whether or not species designations best distinguish functional groupings. Where in taxonomic hierarchy do functionally distinct phenotypes exist? Phenotypes have a strong genetic determinant, and it is the phenotype that translates into functional role. Taxonomy itself has continually envolved to embrace new technology and conventions in classifying biotas. With recent advances in protein, chromosome and DNA isolation, genetics has become increasingly important in species determination. However, explicit functional distinction has not yet become an important taxonomic criterion for species identification. As some species designations are controversial and dynamic, species names seem unreliable as a basis for differentiating functional groups (Johnson *et al.*, 1996). The results from some recent studies have supported the hypothesis that species diversity enhances productivity in some ecosystems but not in others (Johnson *et al.*, 1996).

Environmental Services of Biodiversity

Environmental or ecosystem services are defined as any functional attribute of natural ecosystems which is beneficial to humankind. They comprise the main indirect values of biodiversity, as opposed to direct values in the form of material goods such as timber, fish, plant-based pharmaceuticals and germplasm infusions for major crops. They include generating and maintaining soils, converting solar energy into plant tissue, sustaining hydrological cycles, storing and cycling essential nutrients, notably through nitrogen fixation, supplying clean air and water, absorbing and detoxifying pollutants, decomposing wastes, pollinating crops and other plants, controlling pests, running biogeochemical cycles of vital elements, controlling the gaseous mixture of the atmosphere and relating weather and climate (Myers, 1996, 1997). They basically include three forms of processing, namely, of minerals, energy and water (Perrings, 1987; Perrings *et al.*, 1992, 1995).

The services supplied by one form of biodiversity in one locality may not necessarily be supplied by a similar form of biodiversity in another locality. Services are often site-specific.

Regulation of climate and biogeochemical cycles, hydrological functions, soil protection, crop pollination, pest control, recreation and ecotourism are some of the environmental services of biotas and ecosystems. The services are indeed significant, both in an ecological and economic sense. Ecosystem resilience underpins many of the services. It should not be inferred however, that environmental services stem necessarily and exclusively from biodiversity. While biodiversity often plays a key role, the services can also come from biomass and other attributes to biotas.

Conservation biologists emphasize the numerous contributions of biodiversity to the human cause. These contributions fall into two classes : material goods and environmental services. The first is widely documented (Schulze and Mooney, 1994; Myers, 1996) in the form of new and improved foods, medicines and drugs, raw materials for industry and sources of bioenergy. The second is less well known even though it may be even more valuable than the first. Whereas the benefits of material goods tend to accrue to individuals, often as producers or consumers in the marketplace, the values of environmental services generally pertain to society and therefore they mostly remain unmarketed (Brown *et al.*, 1993; Barbier, 1994).

Interest in below-ground phenomena has increased greatly in recent years, largely due to the fact that, globally, the majority of primary production goes belowground. In the global carbon cycle, below-ground processes play important roles. A popular current view of below-ground parts of ecosystems is based on the realization that organisms exist as functional groups. According to this idea, below-ground biotas can largely be treated as biomass and their taxonomic composition may be ignored because our understanding of their taxonomy is very poor.

Below-ground and above-ground ecology has diverged. The properties of communities such as resilience, stability and persistence are not being seriously tackled in below-ground ecology probably because soil ecologists has become distracted by the great complexity of the system and the various kinds of technical difficulties it poses (Fitter, 1995).

Climate Change

Thousands of species are believed to become extinct each year due to habitat destruction (Peters and Myers, 1992). Now, an even more dangerous and literally invisible threat looms-global climate change caused by the build-up of greenhouse gases in the atmosphere. Greenhouse warming is likely to trigger a massive disruption of natural environments and may set off a wave of mass extinctions. According to Peters and Myers, most scientific reports downplay the dramatic changes global warming could wreak on the world's biotas.

If global warming occurs, the ensuing ecological reorganization might well be as great as the one at the end of the last ice age, a time of mass extinction of species and reshuffling of biological communities. Many scientists have speculated about the shape of things to come in the future : they describe a world filled with biological refugees, species forced to move as local climates become too hot or dry, and one in which more frequent storms and fires will accelerate the disappearance of several ecosystems. Vegetation belts will move hundreds of kilometers towards the poles and hundreds of metres up mountains. Rising seas will inundate coastal marshes and melting polar ice will disrupt the Arctic food chain. Coral reefs will die, migratory birds will find it difficult to locate their food sources or bred, and fish restricted to streams and rivers in drying regions will perish. The future world may turn out to be one in which large areas are biologically impoverished, inhabited primarily by weedy plants and species such as raccoons and deer that reproduce rapidly, colonize well, and do not need mature biological communities thrive. Of course, the picture isn't uniformly bleak. New coral reef would grow in some areas and migrating trees would form new colonies. But overall, the speed and magnitude of change would cause the majority of species to decrease in abundance and distribution, and many will be lost entirely (Peters and Myers, 1992).

Massive changes in vegetation would dramatically affect the lives of many people. Deciduous forests will decline (Peters, 1990), damaging economies, particularly those dependent on the recreation and lumber industries. The disturbances would also reduce other benefits of natural ecosystems are critical for the quality of human life yet difficult to quantify economically and thus often undervalued by economists when they assess climate change impacts (Peters and Myers, 1992).

Since climate-caused disruption is likely to be a severe ecological disaster, action must begin now to ward off some of the consequences. Efforts currently underway at slowing global warming may not be sufficient to preserve biodiversity.

Any efforts to conserve biodiversity should not only ultimately reach beyond the species level, but also include communities and ecosystems. Understanding a community's critical components and strong interactions, such as keystone species and mutualisms, can aid and guide conservation efforts.

Though a variety of changes occur in communities over ecological time, natural disturbance regimes and introductions of exotic species are particularly important in relation to conservation. Most or all natural communities suffer some normal disturbance regime, such as fire, flooding, drought and herbivory. Understanding the scale, frequency and intensity of natural disturbances is critical to conservation at the community level. Continuance of normal disturbance regimes at appropriate scales needs to be encouraged and new artificial disturbances may be avoided. New disturbance through invasion by exotic species can be highly damaging. But not all species introductions to and losses from communities are equal with respect to their community effects - some are benign whereas others are devastating.

Evidence of biotic impoverishment has been gathered from tundra, forest, grassland, wetland, and freshwater and marine ecosystems. Biotic impoverishment appears to be a multifactorial intrigue. Air pollution alone may not kill the ecosystems but multiple and chronic effects can. Until recently most ecologists used to study ecosystems to satiate scientific curiosity. Only recently have they begun to relate ecosystem processes to applied issues. 'To arrest the trend of biotic impoverishment by legislation, the following vital attributes need to be considered : economic prosperity and environmental richness, realistic and workable policies, and the political and social will to overcome degradation' (Wali, 1992).

Global environmental change demands detailed understanding of how biological diversity works but we do not yet have that understanding. It is not enough to know which plant and animal species are present, locally or globally. We also need to understand how individual populations and communities of interacting populations are affected by habitat fragmentation, and by other changes in their physical environment such as long-term changes in temperature or soil moisture, resulting from global climate change. Policy choices require a better knowledge of the degree to which particular ecosystems continue to function as constituent species become extinct or substituted, in other words, the relation between diversity and community stability. We need to know much more about the mutual influence of physical and ecological systems on each other, For instance, to what extent do tropical forests generate cloud cover or buffer carbon dioxide inputs? It should be remembered that our oxygen-rich atmosphere was produced by living organisms. We need to know how, ultimately, biological diversity may influence climate.

Irreversibility

As against many uncertainties relating to biodiversity, there is one certainty-once lost, species cannot be brought back from extinction. Changes which are irreversible assume a special status in environmental policy. They prompt us to look very carefully at the need to be precautionary. If by reversibility we mean the complete reconstruction of a species lost from the gener pool altogether, then the loss of a species is indeed totally irreversible within the bounds of our present knowledge.

Of course some species losses have been occurring naturally during the course of evolution. Darvin's theory of natural selection was based on the idea that species become extinct if they are unable to adapt to a change in conditions. But, the rate of species lost through man-made effects is much greater than those lost through the evolutionary process of 'natural wastage'.

Tropical Forests

Tropical forests cover about 7% of the earth's land area, but they contain more than half the world's known plant and animal species. Human population increase dictates that the forests provide revenue or resources for people living in rain-forest areas. In the absence of sustainable means of economic use, destructive exploitation for timber and mineral resources can eventually lead to the disappearance of rain forests from most parts of the world (Connell, 1975; 1978).

Forestry, fisheries, mineral resources and pharmaceutical and tourist opportunities are the ways by which tropical rain forests have been used on a global scale.

The economies of most rain-forest countries need strengthening. However, there is a difference between a stable economy that allows a low-density human population to maintain itself and a developing economy that allows a country to compete internationally.

Unfortunately, most exploited systems are susceptible to capitalization-the tendency towards over-exploitations that occurs when the growth rate of the managed resource is less than that of capital invested in alternative ventures.

Forests provide local communities with food, shelter, medicinal plants, firewood, lumber, and other goods and services including employment opportunities, income, and soil and water system protection.

During the last few decades concern about the fate of tropical forests and, more recently, about all types of forests has grown. Strong arguments have been raised about protecting tropical and non-tropical forests from exploitation and misuse; saving old-growth forests and reducing consumption of firewood by local populations.

The forces changing the extent, distribution and quality of forests in the past are largely still at work today, although not always in the same area (Zentilli, 1992). The needs for energy, for agricultural land, and for urban settlements and infrastructures have been the basic factors of deforestation, which now happens much faster and dramatically affects the livelihood of rural and urban people.

Forests are traditionally considered as wastelands or hunting grounds rather than a comprehensive system of physical and biological components that benefit society by producing goods and services and moderating climate and water cycles (see Ewel, 1977).

The world forests are subject to natural and anthropogenic pressures : climate changes; air, water and soil pollution; pests and fires; recreational needs; needs for more farmlands to produce food, fodder, or raw materials; and needs for timber, fuel, or other goods. Indirect pressures derive from the burden of the foreign debt on developing countries, the deleterious effects of some foreign loans and investments, and conditions of aid and trade that promote unsustainable exploitation of natural renewable resources. All these pressures have resulted in substantial forest loss.

Probably the only new factors are air pollution, the effect of population increase, and the much shorter time frame in which the process is developing.

Precise and reliable data on the total area of world forests are lacking; the figures used are an aggregation of figures from different sources based on various definitions of trees, forests, forestland and forest cover.

Forest inventory plots provide a good approach to documentation and monitoring of plant diversity. They are a means for obtaining long-term data on the growth, mortality, regeneration and dynamics of forests trees. Many forests plots have been set up over the last few decades in some countries. In recent years, modern computer-based technologies have created novel possibilities for surveying and establishing forests plots, for inventories and monitoring of plants growing in those plots, and for using the inventoried plots in resource management efforts.

To conserve tropical forest diversity and promote sustainable use of forest resources, it is necessary to understand how forests change. The permanent plot inventory system provides a baseline distribution of trees species and describes habitats within a particular sites. Monitoring any changes in these plots over time enables an understanding of the impact of natural and man-made disturbances on the composition of species and communities. Data produced through monitoring may be used to predict future changes. Small plots up to two ha in size and containing 1200 to 1400 trees within a protected area can reveal much about species diversity and allow good characterization of the larger forest (Dallmeier, 1992).

Long-term forest inventories provide important information needed to establish conservation priorities. They are also used in integrating management of protected areas with compatible development on surrounding lands and timber lands, information that is critical in planning for sustainable logging. Similar information is needed to ensure that logging in tropical forests proceeds in a sustainable fashion (Dallmeier, 1992).

Medicinal Plants

An important reason to save the world's tropical rain forests is the potential medicinal value of their plants. The World Health Organization estimates that 80% of the world's population relies on traditional plant-based medicines. Products of rain-forest plants have been used to treat many devastating diseases including childhood leukemia, hodgkins disease, glaucoma and hearth disease. Unfortunately, however, a mere 1100 of about 3,65,000 known species of plants have been examined for medicinal properties (Dobson and Absher, 1991).

In general, there seems to be greater success with plants that are used to treat pathogens that only require small populations to sustain them, such as fungal infections. Nevertheless, the potential for drugs to assist with the world's biggest problem (i.e., birth control) may lurk in the rain forest;

a number of rain-forest tribes in different parts of the world have discovered plants that may be used for this purpose.

Developing local supply and extraction methods is important and it may be wiser to concentrate on weedy species. The chances of success seem to be greater if searches are directed not to miracle cures for AIDS or cancer, but to drugs effective against fungal infections, cold sores and herpes; these are common both in temperate and tropical countries and there is a large market for drugs used to treat them.

Aquatic Biodiversity

There are at least 20,000 species of fish in a variety of marine aquatic ecosystems worldwide, and in freshwater environments many new fish species continue to be discovered by science. Yet, 40% of the world's fish catch comprises only 20 species. This attests to the potential of other types of fish in providing food for humankind but masks the fact that complex, face-reaching interdependencies occur between fish species and their environment that are essential to sustaining global fisheries production.

Over 90% of the marine fish catch comes from the 9.9% of the ocean that lies over the continental shelf, especially in key habitats in the nearshore waters, in the intertidal zone and in the land areas immediately adjacent to the coast, e.g. coastal rivers, bays, wetlands, estuaries, mangroves, salt-marshes, mudflats, sea-grass and seaweed beds, and coral reefs. About two-thirds of all commercial valuable fish species spend the first-and most vulnerable-stages of their life in these waters (Weber, 1994). About 90% of the remaining open oceans yield relatively few fish but are essential for providing resources needed to maintain fish populations in the productive 10% of the seas. Fisheries ultimately depend on the quality and integrity of the whole ecosystem and the biodiversity within it. Any severe loss or degradation of biodiversity in an aquatic ecosystem causes irreversible changes in that system. Therefore, it is essential to conserve habitats and within-habitat diversity, providing varied environmental niches to which particular species are genetically suited.

All the world's 17 main fishing grounds are now being fished at or above their sustainable limits.

The following are four principal and interrelated causes of the loss of aquatic biodiversity :

1. **Habitat destruction** : reduces the number of species of all aquatic organisms.
2. **Overfishing** : affects whole ecosystem because of disruptions to the food web caused by the loss of the targeted species that is overfished.
3. **Wasting of fish** : the losses of untargeted species caught unintentionally and simply thrown away.
4. **Introduction of exotic species** : deliberate or accidental introduction disturbs the ecological balance between aquatic species.

Capture Versus Nurture Fisheries

Capture fishery strategies consider fishing a hunting activity, targeting selected species for mass markets, where the range is open access and the fish stocks are common property. Responsibility for managing fisheries is ill defined leading to the socalled 'tragedy of the commons' where what is left by one user is taken by another. So, all the world's main fishing grounds are being fished at or above their sustainable limits. Some 70% of global fish stocks are now regarded as fully exploited, overexploited, depleted or recovering (Garcai, 1994).

In contrast, nurture fishery strategies give due importance to the time needed for stocks to replenish themselves and the need to conserve species diversity. The nurture fishery strategies of coastal fishing communities throughout the world have evolved several practices to regulate their fisheries. Some govern who may fish in which season and where; other stipulate the sort of fish that may be caught; others relate to the kind of fishing gear that may be used; and still others govern onshore activities such as processing, net-making, and marketing (Fairlie *et al,,* 1995).

Tropical Rivers and Their Fisheries

Tropical rivers contain some of the world's most complex fisheries. Fish species diversity in Amazon exceeds the total number of known mammal species. Each year, 80,000 tonnes of fish are removed from the Amazon River. On any day there may be 50 different species for sale in the local markets. Many of the fish species are frugivores or detritivores; they eat the products of the forest that humans don't. Unfortunately, the fisheries of many rivers are being perturbed in various ways, e.g. hydroelectric projects which may affect both fish survival and migration patterns through their effects on temperature and turbidity. Siltation due to deforestation and mineral extraction is sure to cause the collapse of some populations. Changes in the habitat are compounded by changes in fishing technology : most fish are now caught either by gill nets or seine nets; both of these have only been introduced in the last few decades (Dobson and Absher, 1991). For the Amazon fishery a serious threat is the mercury poisoning which occurs as a by-product of gold mining. Large amounts of mercury are accumulating in the Amazonian food chains, not only disrupting animal and plant communities, but also causing widespread human debilitation (Dobson and Absher, 1991).

Anthropogenic eutrophication (the human addition of nutrients above natural values) is a major environmental problem in many water bodies. Good understanding of the mechanisms of nutrient dynamics in lakes can help design management strategies for tackling such eutrophication issues as nuisance algae.

In temperate lakes planktivorous fish play a fairly minor role. It is commonly assumed that phosphorus (P) is the primary limiting nutrient of primary production. A link exists among P loading rate, hydraulic loading rate, and P retention in the lake with total P concentration. Purely physical principles can help one to estimate total P; but biotic interactions create interesting deviations from this value. The biotas are aggregated into a few key functional components : algae biomass, algal P and zooplankton. The key biological processes of nutrient uptake, algal growth and zooplankton consumption affect algal biomass and P retention (DeAngelis, 1997). In temperate lakes where there is little influence by planktivorous fish, we may assume that large zooplankton such as daphnids constitute the herbivore component.

Riparian zones have been found to have an unusually high diversity of species and environmental processes. The ecological diversity is related to variable flood regimes, geographically unique channel processes, altitudinal climate shifts and upland influences on the fluvial corridor. The resultant dynamic environment supports a variety of life-history strategies, biogeochemical cycles and rates, and organisms adapted to disturbance regimes over broad spatial and temporal scales. Innovations in riparian zone management have been effective in ameliorating many ecological issues related to land-use and environmental quality. Riparian zones play essential roles in water and landscape planning, in restoration of aquatic system, and in catalyzing institutional and societal co-operation for these efforts (Naiman and Decamps, 1997).

Biodiversity and Small Islands

Many small islands are rich reservoirs of flora and fauna. Almost 900 bird species-10% of the world's total-are endemic to one or other of the world's islands. About 2470 of New Caledonia's 3250 plant species are endemic. So are 50% of Cuba's 6-7,000 plant species. A single volcanic island in the Philippines, Mount Makiliang, has more types of woody plants than the entire USA (Cropper, 1994).

About 75% of all the animals and birds known to have become extinct in recent history used to live on islands. The dodo, the most famous of all extinct species, lived on Mauritius until the last bird was killed in 1662. About one-half of the island's endemic bird, mammal and reptile species have now followed it to entinction. About 60% of its endemic plant species are rare, threatened or extinct. Of the bird species, 108 are known to have become extinct over the last four centuries; 97 of them were island dwellers. Of the birds listed in the Red Data Books, 60% are island birds. Of the endemic plant species of Lord Howe Island 97% those of St. Helena, rodrigues and Norfolk Islands, 96% of those on Ascension Island 91% and of those of Juan Fernandez 81% are rare, threatened or extinct. So are 81% of those in the whole of the Seychelles, 67% of those in the Canaries and 66% of those of the Galapagos.

The Cafe Marron tree on the island of Rodrigues is the rarest in the world-just one tree survives today (cropper, 1994).

Marine Biodiversity

In recent years, the conditions of the ocean environment, its resources, biodiversity and the livelihood of coastal people and economies of maritime developing states have been declining at an unprecedented rate, and there are no signs of reversal in these negative trends. Added to this, there will be impact to the ocean ecosystem from global warming and ozone depletion. The very survival of some coastal communities which include some of the poorest people in the world who greatly depend on coastal resources for their food and livelihood, is at stake. In the majority of cases the ocean is the only source of protein and income. Over 90% of the fish production comes from ocean waters over continental shelves and from major ocean currents.

According to the United Nations Food and Agricultural Organization (FAO), the maximum production from world marine fisheries is about 82 million tonnes. Therefore, the current annual fish production of about 84 Mt means that we are harvesting the full potential of wild captured fisheries. Further, some 27 Mt, consisting of low value and juvenile fish as well as by-catch and incidentals (sea-birds, sea-turtles, marine mammals and other ocean life), are discarded annually, raising the average annual production to 110 million metric tonns. Also, fish pirating is a major factor in fisheries management, especially in developing countries. In 1993, the top five captured fishes were anchoveta, Alaskan pollock, Chilean jack mackerel, silver carp and Japanese pilchard, accounting for 20% of the total catch. China, Peru, Japan, Chile, and the United States landed most of the fish (Ngoile, 1997).

Overfishing is the main problem plaguing our oceans. Fish have never been in greater demand as seafood, nor more threatened as marine wildlife. Fishing is the last major industry exploiting wild animals for food. Yet, the worldwide marine fish catch has been declining since 1989, following a 500% increase since the end of the Second World War. Excessive fishing in some areas has driven staple species such as Atlantic cod to commerical extinction.

Fishing activities significantly affect the productivity and biodiversity of the oceans. Fishing gear such as bottom trawlers, longlines and driftnets damage ocean ecosystems, reducing the oceanic biodiversity. Shipping adds hydrocarbons which polute waters, harming biotas. Ships also introduce several alien species when emptying their ballast waters.

Since Rio, some intergovernmental agreements have been concluded. Integrated Coastal Management (ICM) has been identified as the key process for conservation and sustainable development, but the chief constraint has been in implementing these agreements and programmes.

There are not enough Marine Protected Areas - as yet 0.25% of the ocean surface to realize the benefits of these in maintaining marine biodiversity (Ngoile, 1997).

Marine biodiversity is higher in benthic rather than pelagic systems, and in coasts rather than the open ocean since there is a greater range of habitats near the coast. The highest species diversity occurs in the Indonesian archipelago and decreases radially from there (Gray, 1997). The terrestrial pattern of increasing diversity from poles to tropics occurs from the Arctic to the tropics but does not appear to occur in the Southern Hemisphere where diversity is high at high latitudes. Highest losses of marine diversity occur in coastal areas largely as a result of conflicting uses of coastal habitats. The best way to conserve marine diversity is to conserve habit and landscape diversity in coastal areas. Marine protected areas form only a part of the conservation strategy needed. A suitable framework for coastal conservation is integrated coastal area management where one of the primary goals is sustainable use of coastal biodiversity (Gray, 1997).

There are genetic differences both among individuals and among populations. Populations with higher genetic diversity are more likely to have some individuals that can withstand environmental change and thereby pass on their genes to the next generation. On an evolutionary time scale, genetic diversity is higher in species found in unstable, stressed environments compared with counterparts from more stable environments. In contrast, on an ecological time scale (only a few generations), stress reduces genetic diversity; long-term exposure to contaminants can decrease genetic diversity, the remaining population becoming more vulnerable to extinction. Commercial fishing, which concentrates on specific size ranges, has significantly altered the genetic composition of populations (Elliott and War, 1992). In general, marine species have higher genetic diversity than freshwater and terrestrial species.

Most marine species diversity is benthic rather than pelagic (Angel, 1993), probably because the marine fauna originated in benthic sediments. The pelagic realm has an enormous volume compared with the inhabitable part of the benthic realm. Yet there are only 3500-4500 species of phytoplankton (Sournia and Chretiennot-Dinet, 1991) compared to the 2,50,000 species of flowering plants on land. There are probably only 1200 oceanic fish species against 13,000 coastal species

(Angle, 1993). In the pelagic realm, diversity is higher in coastal rather than oceanic areas and, therefore, conservation efforts should be concentrated in coastal areas.

According to Gray (1997), in the marine domain there are more animal phyla than on land. Thirty-five phyla are marine and of these 14 are endemic, whereas only 14 occur in fresh water, where none are endemic. Eleven are terrestrial with one phylum being endemic and 15 phyla symbiotic with 4 endemic (Briggs, 1994). This figure includes the newly described phylum Cycliophora found in the gills of the Norway lobster (Funch and Kristensen, 1995). Thus phyletic diversity is also highest in the sea. Of the 35 marine phyla, only 11 are represented in the pelagic realm; most phyla occur in the benthos which is the archetypal habitat.

The concept of functional redundancy suggested that there are more species present in communities than are needed for efficient biogeochemical and trophic functions, but recent researches have shown that this is not the case-the higher the number of species in a community, the greater the efficiency of biogeochemical processes (Naeem *et al.,* 1994; Tilman and Downing, 1994). Unfortunately, however, no such work has been done in the marine environment.

The most frequently used quantitative measure of biodiversity is for a given area rather than for a given biological community. In ecological terms, physical areas and the biotic components they contain are termed habitats. Habitat diversity is a more useful term than 'ecosystem diversity' since habits are easy to envisage (e.g. a mangrove forest, a coral reef, an estuary). Furthermore, habitats often have clear boundaries. Indeed habitats have been termed 'templets' for ecology (Southwood, 1977). Within coastal area, a wide variety of habitats are known to have high species diversity. Examples are sea-grass beds and coastal sedimentary habitats.

Patterns

Whereas in terrestrial systems, species diversity of genera and families commonly increases from poles to tropics, in the marine domain there is an apparent increase in species diversity of hard substratum epifauna from the Arctic to the tropics. Whereas the Arctic is dominated by many commercial fish species, the Antarctic is characterized by invertebrates (krill and squid) which support birds and mammals but only a small fishery (Gray, 1997).

The best known marine diversity pattern is that of corals which show highest values in the Indonesian archipelago and falling values radiating westwards across the Pacific Ocean (Stehli and Wells, 1971). The Indonesian archipelago may well be the 'epicentre' for evolution of marine tropical biodiversity (Veron, 1995).

Another notable pattern is that soft sediments harbour increasing diversity from shallow areas to the deep sea. There may possibly be a general pattern of low species diversity in shallow coastal areas.

Threats

Two important human activities that pose major threats to open ocean biodiversity are ocean dumping and UV-B radiation. Long-transported materials enter the open ocean and concerns have arisen about effects of organochlorine compounds on marine planktonic and benthic systems.

Some important threats to coastal systems are : habitat loss; global climate change; overexploitation of fishing; pollution, eutrophication, algal toxins; radionuclides); species introduction/invasions; watershed alteration and physical alterations of coasts; tourism, marine litter; and the fact that humans have very little perception of the oceans and their marine life (Norse, 1993; Sebens, 1994). All these are frequently interlinked. The most critical threat is undoubtedly habitat loss.

Another grave problem is that habitats become divided into small fragments. The resultant small habitat islands experience higher rats of species extinctions and lower immigration rates than large 'habitat islands'. Fragmentation of habitats increases losses of species diversity.

The warming of the coastal ocean produces strong effects on corals. As a consequence of ozone depletion in the stratosphere, UV-B radiation flux on the earth is increasing. Consequently productivity of phytoplankton in surface waters is being reduced (Hader *et al.*, 1995). Effects on the symbiotic zoozanthellae in corals are likely. There are also concerns about impacts on diatoms on sand and mud flats. Dunlap and Yamamoto (1995) showed that several marine organisms contain mycosporine-glycine which functions as a biological antioxidant, thereby protecting the marine organisms from the adverse effects of UV-B radiation. More research is needed before reliable predictions can be made of effects on marine bioidiversity.

In summary, there are few threats to the open ocean : they are concentrated in coastal areas. Habitat destruction is particularly severe in tropical areas where mangroves, coral reefs and wetlands are being destroyed at alarming rates. In temperate areas there are severe threats to wetlands areas and estuaries; also conflicts between industrial and tourism development and conservation are universal (Gray, 1997).

Marine Protected Areas

In 1995, Kelleher *et al.*, reported over 1300 marine protected areas (MPAs) throughout the world, convering more than 80 million hectares of coastal seas. Many of these suffer from the lack of good management. According to Agardy (1997), many of these subtidal protected areas are really paper parks, with no sound management plan.

The concept of an MPA is simple - a marine area with defined boundaries within which there is some planned protection of species, system, and/or processes. They range in size from a couple of hectares to the Great Barrier Reef Marine Park, which extends over 35,000 km^2 of Australia's continental shelf. For conservationists, MPAs represent tangible steps to protect natural systems from our overuse; for managers they provide a better alternative to control human use of coastal seas.

Sustainable use requires basic knowledge about the underpinnings of the system. From a fisheries management perspective, an important consideration in the identification of refugia is where the recruits come from. In this context, Agardy (1997) suggested the following to identify ecologically important areas : high species diversity, high endemism, significant productivity, concentration of important ecological processes, habitat for keystone or commercially important species, and high habitat diversity (beta diversity). The multistep process is a complex one and ecologists, managers and conservationists need to develop the science to guide these decisions (Sale, 1997).

Deep-sea Biodiversity

There has been a common misconception that organisms living at the bottom of deep seas are immune from large-scale shifts in climate. Recent studies have demolished this myth : there are links between benthic biodiversity, glaciation and the Earth's obliquity (Rex, 1997). It used to be assumed that the sea-floor is spatially and temporally uniform, and therefore deep-sea biodiversity is buffered from climatic change at the surface by the overlying water column. That changed with the finding of surprisingly high species diversity in the deep sea. It is now known that deep-sea biodiversity also varies on geological timescales of 10^3-10^4 years. These long-term fluctuations in diversity correspond to climatic shifts at the surface associated with glaciation and are linked to changes in solar insolation (Rex, 1997).

Recent researchers suggest that some features of nutrient input (rate, spatiotemporal variation, quality) may be involved in shaping patterns of past and present deep-sea biodiversity. In the deep sea, species diversity varies geographically and bathymetrically, and can reach levels that rival tropical communities. The nature of sediments may be important in structuring deep-sea communities because deposit feeders rely on the sediments for nutrition and comprise most of the organisms in the deep sea. The composition of soft sediment communities is influenced by sediment particle size. Shallow-water deposit feeders selectively ingest particular size fractions of the sediments and there are interspecific differences in particle size preference (Etter and Grassle, 1992). Partitioning of sediments with respect to size may be more likely in the deep sea if there is strong selection for macrophagy as a result of reduced food supply and digestive constraints imposed by feeding on deposits; macrophagy world enable species to ingest selectively the more labile components of the sediments. If deposit feeders in the deep sea partition the sediments with respect to size, species diversity may at least partly be a function of sediment particle size diversity. Also, sediment particle size diversity may reflect habitat complexity because the organisms live on or within the sediments. Etter and Grassle (1992) have shown that species diversity is a significant positive function of sediment particle size diversity. The relationship seems to be scale invariant, accounting for a similar proportion of the variance at interregional, regional and local scales. Bathymetric patterns of species diversity also seem largely attributable to changes in sediment characteristics with depth. Sediment diversity may have an important role in determining the number of species within a community.

In the face of the various impediments to marine conservation, there is an intrinsic need to manage marine biological resources (e.g., commercial fisheries) at scales which cross geographical and political boundaries. Yet, many of the effective management schemes so far have been effective because they were local, grass-root type efforts motivated by a strong commitment to protect and sustain resources in one's own backyard. The recommendations for implementing the conservation efforts range from managing particular species to broad mandates for entire organizations, governments and institutions.

Species Introductions in Marine Environments

Rather little is known about the potential effect of species introductions in marine environments. Countries at high risk from marine species introductions are those in which most of the trade is done via international shipping-the prime medium for the spread of marine species worldwide. In the USA, zebra mussels (*Dreissena polymorpha*) introduced from Europe via shipping have been spreading through the Great Lakes and clogging waterways. The costs of controlling the spread of the mussel run in the millions. In Tasmania, a Japanese starfish (*Asterias amurensis*) has assumed plague proportions in the Derwent Estuary, threatening mussel and scallop-farming industries. Understanding the risks and controlling the impacts associated with marine invasions is vital for many maritime countries.

The Asian (Japanese) alga *Undaria pinnatifida* has invaded New Zealand, probably through international shipping. It first appeared in 1987 in Wellington Harbour and has gradually spread throughout New Zealand; it is now found along even the southern coast of North Island and in numerous South Island harbours (Miller *et al.*, 1997).

Undaria generally rows in shallow water, from the low tide level out to depths of about 6 metres. The highest densities in Wellington are on artificial substrata such as wharf pylons, jetties, retaining walls and marine farms. It is also found in natural rocky-reef habitats where it conexists with native macroalgae such as *Carpophyllum* and *Macrocystis pyrifera.*

Undaria can spread either as whole plants or by spores. Its long-distance dispersal occurs on the hulls of commercial and recreational boats. Its spores are transported via shipping, for instance in the ballast water of freighters, or attached (as gametophytes) to artificial structures such as aquaculture floats, mooring lines, barges and boat hulls. The spores and gametophytes are hardy and can survive a wide range of conditions including darkness (i.e., in ballast tanks) and desiccation, and still be capable of germination (Miller *et al.*, 1997).

4
Extinction of Species and Loss

Introduction

All human activity is underpinned by biological resources. We exploit plants and animals for food and raw materials at differing levels of sophistication in different societies. We cannot exist in isolation from the rest of the natural world. Biological resources and biological diversity are two quite different things. Globally, we derive most of our food from less than 20 species of domesticated plants and animals, and exploitation as raw materials accounts for a small proportion of total diversity. Then why should we worry about species loss? There are three reasons for concern. The first is that all species deserve respect regardless of their usefulness to mankind. The second reason is that the living systems constitute our life support and that we do not know the precise role that each species plays within this global system. It is unwise to reduce its diversity, lest we also upset its functions. Thirdly, wild nature is highly important to us economically. It provides food, fibre, timber fruits and drugs. The wild relatives of crop plants are essential sources of genetic material. As we face climate change, crop breeders will be going back to the wild more and more to select strains that will arm our cultivated organisms against an uncertain future.

The distribution of living species in the world is not uniform. Species richness increases from the poles to the equator. Freshwater insects, for example, are 3 to 6 times more abundant in tropical areas than in temperate zones. Tropical regions also have the highest richness of mammalian species per unit of area, and the diversity of vascular plant species is much richer at lower latitudes. As many as a hundred species of trees can be found on one hectare of tropical rain forest in Latin American forest. A few hectares of rain forest in Borneo can harbour over 650 species of trees, as many as in the whole of North America. A region in lowland Malaysia near Kuala Lumpur has over 600 plant species per hectare. In comparison, all of Denmark possesses no more than 300 species. Global patterns of species diversity in the marine environment resemble those on land. The number of tunicate (sea squirt) species increases from 103 in the Arctic to some 625 in the tropics. These terrestrial and marine patterns of increasing diversity in the tropics reach their peak in tropical forests and coral reefs.

The study of life has always been an attempt to understand the biosphere and its history, and biodiversity research is undoubtedly one of the most important enterprises in biology today. One major aim of biodiversity research is the calculation of the genetic information content of the biosphere and the identification of species extinction rates accompanying the loss of information.

There are several scientific uncertainties in the measurement of biological diversity. The layman usually equates biodiversity with the flora and fauna of a rain forest and feels that saving the rain forests will prevent the loss of biodiversity. But this gives only an incomplete picture, for biodiversity is by no means confined only to rain forests. It is everywhere, on the land and in the sea. The greater attention given to the species rich tropical forests is not surprising because we believe that perhaps half of all the species on Earth inhabit those forests, and that, as the forests are cleared, this richness of life is at risk. But we must not forget the biological diversity of the seas. Coral reefs have been described as the tropical forests of the oceans. And we must not equate quantity of species with significance; very often even a small number of species that make up the biological diversity of a country can become critical in terms of preserving the fertility and maintaining the biogeochemical cycles of that country. Every area's biological diversity is important to it, and to the world as a whole.

Habitat Fragmentation

Fragmentation which occurs at several different spatial scales can have various short-term and long-term effects. A few species benefit from fragmentation whereas others are exposed to increased risk of extinction. Global biodiversity can be maintained by conserving resources to those species and ecosystems most at risk, and by reversing the processes that place them at risk. In addition, even those species not currently at great risk, should also be saved from the adverse impact of habitat fragmentation. There is a need to develop effective management strategies for landscapes that are already fragmented.

Eight categories of species are likely to be especially vulnerable to the effects of fragmentation : rare species, those with large home ranges, those with limited dispersal and low reproductive potential, species with short life cycle, those that depend on unpredictable resources, ground-nesting birds, species of habitat interiors, and species exploited or persecuted by humans.

Some recommendations for maintaining biodiversity in fragmented landscape are briefly stated below (See Meffe *et al.*, 1997).

1. Conduct a landscape analysis and look for the pattern of habitats and connections. Relate these to the needs of the native species.
2. Find out whether the landscape forms part of a critical linkage of ecosystems at a regional scale.
3. Avoid any further fragmentation or isolation of natural areas. Developments, resource extraction activities, and other land uses should be clustered and minimized so that large blocks of natural habitat remain intact.
4. Minimize edge effects around remanent natural areas by establishing buffer zones with low-intensity land-uses.
5. While the larger unfragmented patches of habitat should be duly conserved, the smaller fragments may also be important as the last refuges for many species in highly fragmented regions and can maintain populations of many species for decades.
6. Identify and protect traditional wildlife migration routes by diverting human activities away from critical wildlife movements areas.
7. Maintain native vegetation along streams, fence rows, roadsides, and other remanent corridors in transects or belts as wide as possible in order to minimize edge effects and human disturbances.

8. Minimize the area and continuity of artificially disturbed habitats that are dominated by weedy or exotic species so as to reduce the potential for biological invasions of natural areas.
9. Small fragments often suffer from disruption of natural processes, such as fire regimes. Active management is necessary to maintain the native flora and fauna of these fragments.

In the case of fragmented habitats, species richness does not always decline on isolated habitat patches, as predicted by island biogeographic theory. In some cases, richness con even increase for some time as species invade from adjacent disturbed areas. In such a case, species composition often shifts towards weedy, opportunistic species while sensitive species of habitat interiors are lost. The matrix in a fragmented landscape is also constantly changing as crops are planted and harvested, as tree plantations go through their rotations, as farming or silvicultural methods change, and as human settlements flourish or decline. Thus, sometimes the external environment of a habitat patch may not be as constant or predictable as per the island biogeographic theory.

Of course, fragmentation does not necessarily always lead to extinction. A species might persist in a highly fragmented landscape in three ways : (i) it might survive, even thrive, in the matrix of human land-use, e.g. some weedy plants, insects, fungi, microbes, and house mice; (ii) it might maintain viable populations within individual habitat fragments; e.g. many plants, microbes, and small-bodied animals with modest area requirements; (iii) it might be highly mobile and able to integrate a number of habitat patches, either into individual home ranges or into an interbreeding population. For example, pileated woodpeckers fly among a number of small woodlots to forage in landscapes that were formerly continuous forest. A species that cannot pursue any of the above three options will eventually become extinct in a fragmented landscape.

As a general rule, reducing the size of a habitat by 90% will reduce the number of species which can be supported in the long run by about 50%. The second reason for the loss of species is overexploitation. Commercial harvesting has been a threat to many marine species. Overexploitation has been the cause of extinction of some large terrestrial animals and well-known species such as the African elephant are under threat today. Pollution is the third reason for the continuing loss of species. Pesticides have affected several species of birds and other organisms. Air pollution and acid rain have been linked to forest diebacks in Europe and North America. Acid rain has resulted in the loss of a number of fish species in European lakes. The fourth reason for the loss of species is the impact of introduced exotic species as they threaten natural flora and fauna by predation, competition or by altering the natural habitat.

Loss of Genes

A species consists of many genes; genetic diversity refers to the variation of genes within species, as expressed, for example, in the thousands of rice varieties in Asia. The genetic variability of many species is diminishing and hence reduces their ability to adapt to pollution, climate change, disease, or other forms of environmental stress. The remaining gene pools in crops such as maize and rice amount to only a fraction of the genetic diversity they harboured only a few decades ago. Many agriculturalists argue that the loss of genetic diversity among domestic plants and animals looms as an even greater threat to human welfare than the loss of wild species, because it is the diversity that will enable crops to adapt to future environmental change.

Wild species and the genetic variation within them make a substantial contribution to the development of agriculture, medicine and industry. Many species help in stabilization of climate, protection of watersheds, soils, nurseries and breeding grounds. It is not possible to estimate the total economic value of the full range of goods and services that biological diversity provides, but a few selected examples are given below:

1. About 4.5% of the GDP in the United States can be attributed to the harvest of wild species.
2. In Asia, by the mid-1970s genetic improvements had increased wheat production by $2 billion and rice production by $1.5 billion a year by incorporating dwarfism into both crops.
3. A 'useless' wild wheat plant from Turkey has been used to confer disease resistance in commercial wheat varieties worth $50 million annually to the United States alone.
4. One gene from a single Ethiopian barley plant now protect California's $160 million annual barley crop from yellow dwarf virus.
5. An ancient wild relative of corn from Mexico can be crossed with modern corn varieties with potential worldwide savings to farmers estimated at $4.4 billion annually.
6. Worldwide, medicines from wild products are world some $40 billion a year.
7. In 1960, a child suffering from leukemia had only one chance in five of survival. Now the child has four chances in five, due to treatment with drugs containing active substances discovered in the rosy periwinkle, a tropical forest pant found in Madagascar.

Biotechnology now offers novel possibilities of increasing the production of food, drugs, energy and speciality chemicals and of improving environmental management. This reinforces the need to maintain the riches possible pool of genes. The loss of biodiversity can cripple the genetic base required for the continued improvement and maintenance of currently utilized species and deprive us of the potential use of developments in biotechnology.

Adequate appreciation of the ecological dimension of the problem of genetic erosion emerged only in the 1980s. Before then, concern for the conservation of genetic resources was mainly in relation to agriculture. The growing environmentalism during the 1980s led to the UN Conference on Environment and Development (UNCED) in Rio de Janeiro (1992), whereafter genetic resources began to be dealt with in two arenas, instead of only one. These arenas were:

1. the agricultural arena, as represented by FAO and IBPGR, and
2. the environmental arena, as represented by the United Nations Environment Programme (UNEP), the International Union for the Conservation of Nature and Natural Resources (IUCN) and the World Wide Fund for Nature (WWF).

Threatened and Endangered Species

It is feared that a quarter of the earth's species could be endangered within the next 30 years. There are various ways of countering the threat to genetic resources and biological variation. Rich natural habitats can be preserved in national parks and reserves and the overexploitation of particular species in nature can be banned. Seeds may be stored for *ex situ* conservation in botanic gardens or gene banks. It is also important to try to preserve an undisturbed environment for the species to live in, for example by reducing pollutant emissions to the biosphere.

Conservation of the earth's living resources is important in maintaining essential ecological processes and life-support systems, preserving genetic diversity and ensuring the sustainable utilization of species and ecosystems (Hanneberg, 1992).

Recently, some integrated conservation and development projects have been started to solve the biodiversity crisis at local level throughout Madagascar. These should stimulate attempts to link developments alternatives with natural resource conservation.

Besides indicating the geographical and taxonomic groups containing most threatened species, lists often endangered species often provide other kinds of information (Tables, 4.1 & 4.2). Data from classifications of threatened species using quantitative criteria provide a new method for estimating extinction rates in diverse vertebrate taxa. This analysis has revealed that during the next century extinction rate might rise to 15-20% in these groups (Mace, 1994). These values are comparable to those based upon extrapolations from species-area curves. However, allocating threatened species categories is only a first step towards developing rational systems for setting conservation priorities.

Table 4.1 : Some parameters used in definitions of some globally threatened species categories (condensed from Mace, 1994).

Taxon	*In danger of extinction*	*Will become endangered*	*Population* size small*	*Recovered from threat*	*Probably threatened but no data*	*Not threatened*
Vertebrates	+	+	+	+		
Swallowtail butterflies	+	+	+	+	+	
Dolphins, porpoises and whales	+	+	N		+	
Amphibians and reptiles	+	+	+	+	+	
Invertebrates	+	+	+		+	
Cycads			N			+

*N=numerical guidelines available.

Table 4.2 : Numbers of plant and animal species to have become extinct since 1600, and presently threatened with extinction (from Smith *et al.*, 1993).

Taxon	*No. extinct since 1600*	*No. presently threatened**
Corals	1	-
Molluscs	191	354
Crustaceans	4	126
Insects	61	873
Fishes	29	452
Amphibians	2	59
Reptiles	23	167
Birds	116	1029
Mammals	59	505
Total (animal)	486	3565
Ferns and fern allies	16	-
Gymnosperms	2	242
Monocotyledons	120	4421
Monocotyledons : palms	4	925
Dicotyledons	462	17,474
Total (plants)	604	23,062

* i.e. Includes IUCN categories of 'vulnerable', 'endangered' and 'probably extinct'.

Recent years have witnessed marked declines among some amphibians populations in many parts of the world, including some species extinctions. These events have generated particular concern because many have occurred in protected areas such as nature reserves and national paks (Halliday, 1998). Berger *et al.,* (1998) identified a pathogen that links declines among frog populations in two geographically distinct parts of the world, Panama and Queensland, Australia. They studied many dead and dying frogs and identified the cause of population declines as they occurred (rather than after the event) and found that the same pathogen, a chytrid fungus, was the cause of mortality at the two localities. The chytrid invades the skin of adult frogs, and kills them by interfering with their cutaneous respiration and water uptake. It particularly infects the pelvic patch, an important site of water absorption in many frogs (Halliday, 1998). The chytrid is a newly discovered genus Chytrids have been known to attack plants and insects, but the new one is the first to be identified as a vertebrate pathogen. Interestingly the chytrid attacks just adult amphibians, probably because it only attacks skin that contains the protein keratin, which does not occur in amphibians until metamorphosis. This feature differentiates the Berger *et al.,* study from previous reports of amphibian population declines, a characteristic of which has been reproductive failure owing to mortality among eggs and larvae (Halliday, 1998).

Extinctions

Species extinctions have been occurring in recent years at a rate exceeding that recorded for the past 65 millions years. This is one of today's most serious global problems and, unlike other global environmental problems, is completely irreversible.

Homo sapiens first appeared about 5,000,00 years ago in the Earth's 4.5 million year history. Our hunter-gatherer ancestors began to exterminate some of the large animals and birds that they killed for food.

This number then began to increase quickly and the extensive land clearing and grazing that characterized early agriculture brought in its trail rapidly increasing extinctions (Raven, 1994). The number of people has now increased to an estimated 5.5 billion.

Homo sapien is just one of an estimated 10 million species on earth, but is estimated to be consuming, wasting or diverting some 40% of the net photosynthetic production on land. Man is using about one-third of the planet's available fresh water. And yet our numbers are continuing to grow and the proportion of young people in developing countries is also rising. As a result of our great impact on forests and other biologically rich communities, we are losing species at up to 10,000 times the natural rate that prevailed before our ancestors first appeared on Earth. The background rate of extinction is about four species a year but we are actually around 50,000 species a year (Raven, 1994). Indeed it is possible that we may lose two-thirds of all living species over the course of the next century.

It is ethically, aesthetically and economically important to minimize or prevent species losses. We use organisms for food, medicines, chemicals, fibre, clothing, structural materials, energy and many other purposes. Only about 100 kinds of plants provide most of the world's food; their genetic diversity needs to be preserved. There are also thousands of other plants, especially in the tropics, that have some edible parts and might be cultivated for food, if we knew them better. But overconcentration on the 20 or so best known food plants makes us neglect the others.

Plants and other organisms are natural biochemical factories. A large number of the world's people depend directly on plants for their medicines : the Chinese use more than 5,000 of the estimated 30,000 species of plants for medicinal purposes. Most of the Western medicines come from the natural products that organisms produce. Very few of the 2,50,000 kinds of plants of the world have been fully examined, so it seems reasonable to suppose that the remaining species may contain hitherto unknown therapeutically important compounds. It is not possible for any chemist to imagine the complex bioactive molecules produced by nature, but he can often proceed with synthetic modifications to improve on what nature has produced.

A novel compound called Michellamine B has been extracted from the African vine *Ancistrocladus korupensis.* It shows some anti-HIV activity. Further studies on its mode of action may facilitate the discovery of other drugs that will be more effective against AIDS.

Yet another reason for being concerned about the loss of biodiversity relates to the variety of essential services provided by natural ecosystems, e.g., the regulation of local climates, the maintenance of atmospheric quality, absorption of pollution, and the restoration and maintenance of soils.

Much of the quality of ecosystem services will be lost if the present episode of species extinction is allowed to run unhindered for much longer.

Over four-fifths of the world's resources are consumed by the people (now less than a quarter of the total) that live in industrialized countries; this overconsumption needs to be checked.

If the problems of population and poverty in the developing world and of overconsumption in the developed world are tackled adequately, several strategies could be used for the management of biodiversity, including the conservation of a reasonable sample of the species living now. One approach is to establish a system of protected areas (biosphere reserves). Another strategy is to preserve the samples of selected organisms outside their natural habitats. For plants, botanical gardens need to be encouraged to form an operational network to conserve plants throughout the world, and a global network of seed banks should also be created. Culture centres should preserve bacteria, algae and fungi.

In the context of extinction, large areas generally hold more species than smaller areas and larger populations persist longer than smaller ones. There is no similar generalization in respect of ecological interactions. If some plant-pollinator mutualisms involving the subtle, complex web of interactions are broken by human actions, this could cause a cascade of extinctions (Janzen, 1987; Gess and Gess, 1991).

Both vertebrate and invertebrate pollinators and dispersers face several threats including poisoning by pesticides, habitat change, invasions of alien animals and plants, and insularization of habitats (Diamond, 1988). These and other forces tend to reduce the diversity of animal mutualists and change the population densities of the survivors. Conceivably, specialists dependent on a few species might be more vulnerable than generalists.

It is possible to assess the risk of plant extinctions by considering the probability of dispersal or pollinator failure, reproductive dependence on mutualism and demographic dependence on seeds. Certain traits can enable us to rank species rapidly according to these three criteria. Plants often compensate for high risk in one of the three categories by low risk in another. For example, self-incompatible plants with rare specialist pollinators often propagate vegetatively. Some systems, e.g. elements of some tropical rain forests, lack compensatory traits and in these the risk of plant extinction from failed mutualism can be fairly high.

Extinction Thresholds

Metapopulations involve a shifting mosaic of presence and absence, with certain patterns of patch occupancy (as opposed to details of within-patch events). A population governed by metapopulation processes is vulnerable to a threshold of habitat availability, below which the metapopulation may be doomed to extinction because colonization is too infrequent to overcome random local extinctions.

Most research relating to the dynamics of a single species in complex landscapes entails simulations that move individuals around in landscapes drawn from actual maps of habitat. However, while it is easy to map vegetation, it is quite difficult to obtain information on dispersal behaviour or animal movements (Kareiva and Wennergren, 1995).

The metapopulation processes of extinction and colonization that are central to the fate of singlespecies also influence communities. Removal of habitat can alter patterns of coexistence because it favours organisms with high colonization ability over less mobile, but more competitive species. This entails declines in biodiversity when too much habitat has been destroyed because some species may fail to colonize vacant patches fast enough.

Kareiva and Wennergren (1995) proposed the following general principles for the guidance of resource managers :

1. Species that act as metapopulations live with a threshold requirement for habitat, below which they face inevitable extinction (long before all of the habitat has been removed).
2. Destruction of habitat can cause dramatic loss in biodiversity that is long delayed, non-linear and conspicuous only after substancial habitat disappearance with surprisingly negligible obvious effects has occurred. Monitoring programmes and trend analysis can mask the risk of continued habitat loss.
3. A good knowledge of how habitats are arranged and how organisms move between them can mitigate the above risks, and determine whether species persist or become extinct, or whether populations fluctuate wildly or remain stable. Management options that emphasize dispersal rates or the deployment of habitat can help to maximize the benefit of conservation efforts.

Population Extinction and Biodiversity

It is difficult to estimate the vulnerability of a species to extinction from the number and size of its populations, but in some cases reduction in number and size of populations dooms a species to extinction long before it becomes scarce in nature, as was true for the passenger pigeon, *Ectopistes migratorius* (Blockstein and Tordoff, 1985).

The two most common definitions of population refer to related kinds of geographic collections of individuals. One is a demographic unit, which is simply an interbreeding group sufficiently isolated from other such interbreeding groups so that changes in its size do not greatly influence the size of near-by groups and vice-versa. The other is a *Mendelian population,* in essence a genetically defined entity that can evolve independent of other such units; i.e., its evolutionary future is not primarily determined by flows of genetic information from other populations. These two kinds of populations often exist as parts of continua, just as many species do. Demographic units may be

Mendelian populations and vice versa, but the two are not necessarily congruent. The important point is that populations are geographic entities within species that may be defined either ecologically or genetically.

In general, while animal species diversity tends to decline as one moves towards the poles, the geographic range of each species appears to increase. The average range of temperate-zone species is between two and three times larger than that of their tropical counterparts. Thus, there are almost certainly more populations per animal species in temperate, subarctic, and arctic regions than in the tropics. Biodiversity as a whole represents billions of populations.

In many parts of the world the extinction of populations, rather than a species, appears to be the most important aspect of the decay of biological diversity. Consideration only of species extinctions can greatly underestimate the rate of loss of organic diversity as a whole. Extinctions of animal populations appear to be more frequent in proportion to species extinctions in temperate and polar regions than they are in the tropics. The average extratropical species might suffer many more population extinctions before becoming threatened globally than could the average tropical species.

It is not wise to downgrade the importance of the extinction of populations on the assumption that other populations of the same or similar species elsewhere can supply the same services. A crucial question in ecosystem ecology currently is the extent to which species diversity may be required to maintain ecosystem services. The quality of local ecosystem services. The quality of local ecosystem services is strongly tied to exactly what populations are in a given area.

The loss of diversity in an area means a loss of 'ecosystem plasticity'. A diverse mix of species which change in proportions and distribution in a forested area has a much better chance of maintaining forest cover than a monoculture (Ehrlich and Ehrlich, 1992).

Causes of Extinction

The chief causes of today's global mass extinction spasm are the activities and overpopulation of *Homo sapiens* (See Flessa, 1986). Species-rich biotas are being destroyed at high rates, especially in tropical lowland forests, but also in other environments including coral reefs and freshwater lakes. Consequently, the biological diversity of all types is declining. Although predictions of future losses vary, all point to extinctions of significant percentages of existing life-forms, most of which are unknown today. One easy and efficient way of halting this loss would be protection of 18 identified 'hot spots; of diversity, small areas of the earth that contain exceptionally high concentrations of species found nowhere else.

Not all species are equally vulnerable to extinction. Rare species and long-lived species are especially vulnerable. Keystone species support many other species that are dependent on them. If these keystones are lost, the vulnerability of their many dependent species increases and cascading extinctions can occur.

Not only species losses but also genetic diversity- the basis for evolutionary change and adaptation-and population diversity-the result of local adaptations-are also being lost rapidly. This problem is particularly serious with respect to losses of crop genetic diversity, which forms the basis for agricultural production. Most of the world's major crops and domestic animals have fairly narrow ranges of genetic variation, and the wild ancestors of some of these species have already become extinct. As a result of this impoverishment, the human sustenance base is becoming more and more vulnerable to pests, diseases and changing environments.

Besides outright habitat destruction, fragmentation and degradation of habitats pose grave threats to biodiversity. Creating ever smaller chunks of forest exposes the remaining habitats to edge effects, invasive species and other troubles that hasten extinctions. Species introduction, overharvesting, pollution and toxification, are some other principal types of threats to biodiversity. Global climate change is certain to add to and complicate all the other problems leading to extinctions.

One problem that has attracted some notice is that conservation and protection are not the sole objectives of national parks and certainly not of many provincial and regional parks, nor a wildlife or biosphere reserve area. Indeed, wildlife refuges are sometimes manipulated to enhance habitat for particular species such as ducks which are hunted to the detriment of non-game birds or habitat diversity generally.

The Evil Quartet

According to Woodruff (1992), genetic factors do not figure among the four major causes of extinction (the Evil Quartet) : overkill, habitat destruction and fragmentation, impact of introduced species and secondary or cascade effects. Thus, although genetic factors are major determinants of a population's long-term viability, conservationists can do more for a threatened population in the short term by managing its ecology. Ecological management remains the best way of conserving genetic diversity.

There is no doubt that genetics has an important role in modern conservation biology. Genetic variation provides the raw material from which adaptation proceeds and is vital for the march of evolution. This variation occurs at three levels : within individuals, within populations and among populations. Loss of variation occurs in small populations through founder effects, genetic drift and inbreeding. Management of genetic variation in nature should proceed with immediate (month, years), moderate (decades, centuries) and long-term (millennia) temporal scales in mind. This biogeographic aspect in an index of natural population genetic structure and probable historical levels of gene flow among populations. Genetics may be best viewed as 'fine-tuning' to be attempted after the 'coarse-tuning' of habitat protection has been accomplished. For the long term, habitat availability must receive the highest priority.

Species Extinction Rates

We are losing at least 27,000 to 50,000 species annually (Raven, 1994; Myers, 1990a, b, 1993). In the future, and in the absence of greatly expanded conservation efforts, independent analyses have proposed (Diamond, 1989; Wilson, 1997) that we face the prospect of losing 20% of all species within 30 years and 50% or more by the end of the next century.

Extinction rates have been estimated in various ways. Usually, estimates of the rate of habitat loss (e.g. deforestation) are combined with those of species numbers to get a general idea of the numbers of species expected to become extinct over a certain time period. These estimates vary greatly because different authors have chosen different baseline estimates of (1) the number of living species in the world (range 3-10 million), (2) the proportion of species found in tropical rain forests (range 25%-75%), (3) the nature of the relationship between habitat loss and extinction, and (4) the rate and extent of habitat loss (range 0.5%-2.0% annual global deforestation). Because of these, extinction estimates suffer from several limitations, e.g. the enormous uncertainty in our estimates of the numbers of species worldwide.

Habitat loss is usually accompanied by habitat degradation, including fragmentation effects. Many studies have shown an increase in local extinctions with fragmentation of tropical forests. Degraded forests may lose species, though perhaps not as rapidly as forests destroyed outright. Usually, losses due to habitat degradation are not included in extinction estimates so there is a tendency to underestimate this component of extinction. Likewise, as cleared areas of forest recover, they will be able to support more species again. Extinction estimates usually ignore recovery processes and thus may overestimate extinction rates.

According to Ehrlich *et al.*, (1995), rate of population and species extinction may be estimated from total consumption of energy (industrial plus traditional) by man. Three assumptions underly this proposition : (1) The rate of extinction is proportional to the rate of habitat destruction because most organisms are adapted to rather limited environments. (2) The rate of habitat destruction is correlated with the scale of human enterprise : the product of the number of people, average consumption and the environmental damage wrought by the technologies used to supply each unit of consumption. (3) Average energy may be used as a surrogate for the above two factors (consumption x technology) (Ehrlich and Ehrlich, 1981). Total energy use is therefore an indicator of trends in extinction rates, and thus could be used to estimate the rates themselves.

Population Extinctions

It is much more difficult to estimate rates of population extinctions than those of species extinctions because populations are more difficult to define than species and no systematic attempt seems to have been made anywhere to keep track of population extinctions in any major taxonomic group.

The impact (*I*) of humanity of Earth's life-support systems may be considered the product of the size of the human population (*P*), average affluence (*A*) (as measured by average consumption), and an index of the environmental damage often technologies (*T*) used to supply each unit of consumption (Ehrlich and Ehrlich, 1990). Average energy use (for which numbers are generally available) may be substituted for $A \times T$ in the equation $I = P \times A \times T$ (Ehrlich and Ehrlich, 1990).

Ecological and climatic systems often show non-linearities, prolonged lags and threshold effects that tend to make projections conservative. Neither species nor population extinctions occur independently. Extinction cascades may well become common place in the next century. Humans now threaten to alter the climate dramatically and already use or destroy roughly 40% of terrestrial net primary productivity, the food supply of virtually all animals and most micro-organisms that live on land (Vitousek *et al.*, 1986). There is no guarantee that technological changes such as those in forest harvesting techniques will not accelerate the expansion of human enterprise and accelerate the exponential rate of diversity loss. It seems logical to infer that projections based on past and current rates of anthropogenic extinction may well be underestimates.

Pivotal Linkages in Ecosystems

Species known as keystone mutualists are rare species which are likely to suffer differentially high rates of extinction (Terborgh, 1986). The Brazil nut tree is a cash crop in Amazonia. It is pollinated solely by a glossine bee, which also pollinates orchids and many other plants. These other plants often supply prime sources of food to sundry other insects, which pollinate further plants, and so on. In this crucial sense, the bees serve as a *Mobile link* species, and their plant host, by virtue of supplying food to extensive associations of mobile links, serves as a *keystone mutualist* (Owen-Smith, 1988; Pimm, 1991).

These pivotal linkages within tropical forest ecosystem are manifested by thousands of plants that through their nectar, pollen and fruits supply critical support for multitudes of insects, mammals and birds. If human disturbance eliminates a keystone mutualist, the loss can often spell a loss of many other species as well. Sometimes these additional losses can in turn trigger a chain of linked extinctions. Eventually, entire series of forest food webs can become lost.

Extinction Epidemic

Biologists often watch the flora and fauna fading away before their very eyes. Coral reefs on which the behavior of fascinating fishes is studied have been destroyed by the sewage from tourist cruise ships. Many habitats of butterflies have been converted to freeways, parking lots or farm fields.

Some biotic diversity is constantly but slowly being generated by natural speciation. But the present-day extinction rates are much higher than the rates at which the natural speciation can compensate for the losses. The extinction outputs greatly exceed the speciation inputs.

To biologists, perhaps the most ominous data pointing to the urgency of dealing with the extinction problem are those relating to the human impact on the planet's total supply of energy produced in photosynthesis-global net primary production.

Extinction Vortices

Smaller populations are particularly vulnerable to demographic and environmental variations and genetic factors that contribute to further population decreases and even doom the population to early extinction. Random environmental variation appears to be usually more important than random demographic variation in increasing the probability of extinction in small populations. This tendency of small populations towards extinction is analogous to a vortex (a whirling mass of gas or liquid which spirals inwards); the closer it gets to the centre, the faster it moves. At the centre of an extinction vortex, the species is lost forever. If the species gets trapped in this vortex, it cannot resist being pulled toward its fate-extinction. This happens when some natural catastrophe, a new disease, or human disturbance reduces a large population to a small size.

Smaller populations tend to lose genetic variability just by chance. This process is known as genetic drift and leads to inbreeding depression and a lack of evolutionary flexibility. Experience with captive animals has suggested that isolated populations should have *at least* 50 to 500 or more breeding individuals to maintain genetic variability.

The central factor in the protection of small populations is determination of effective population size based on the number of individuals that are actually producing offspring. Variations in reproductive and mortality rates and environmental variation often cause small populations to fluctuate randomly in size, leading to extinction.

Protection and judicious management of any rate or endangered species need to be based on a firm grasp of its ecology and natural history. One must know the species environment, distribution, biotic interactions, morphology, physiology, demography, behaviour and genetics. Demographic studies are especially valuable in estimating the long-term stability of populations.

It is feared that some wild species may suffer extinction; these must therefore be saved. This can be done by maintaining them in artificial conditions : this is known as *ex-situ* or *off-site* preservation. These captive colonies can be used later to re-establish species in the wild.

All over the world, major zoos are attempting to house self-maintaining populations of some rare vertebrates, by using modern animal breeding and biotechnology techniques to increase their reproductive rates.

Similarly, botanical gardens and arboreta are engaged in collecting rare and endangered plant species. The seeds of many plant species can be stored for long periods of time under cold conditions in seed banks.

5
Conservation of Biodiversity

Introduction

The conservation of biological diversity constitutes an essential aspect of sustainable development worldwide. It is biodiversity that basically determines the structure and function of all ecosystems (see Hawkes, 1991). It is the foundation on which the future well-being of human society rests. It has a powerful role in building sustainable human societies.

Unfortunately, many countries having the largest concentrations of biological diversity are also characterized by persistent poverty and high rates of population growth; both these factors work against conservation as they increase the pressure for inappropriate and harmful land-use, and limit the ability of individuals and governments to halt the loss of diversity.

In most tropical countries, conservation efforts have been largely influenced by foreign agencies. The traditional conservation model followed by industrial countries and adopted by many developing countries emphasizes custodial management, such as parks and reserves, that ban most human activities within their boundaries. Protected areas have been set aside to preserve the so-called charismatic species (e.g., elephants, tigers, or bears), spectacular vistas or geologic formations, and recreational or historic sites. Some areas have also been reserved for protecting watersheds or reserving timber supplies. Until recently, however, few areas were established expressly for conserving biological diversity (Anonymous, 1994).

Many species hybridize sometimes and reproductive isolation is often incomplete. This is true of both animals and plants, even those with elaborate 'reproductive isolating mechanisms'. Occasional hybridization, even leading to gene flow, does not necessarily vilate a species' integrity.

Plant diversity is certainly the basis of all life, and one of the most troublesome questions today is : what to conserve? Any acceptable answer to this poser is severely limited by the fact that in reality we know nothing about plant diversity in many countries and soil is not possible to select certain areas as the most 'worthy' of conservation. Areas such as the tropical Andes or the Amazon basic are still dark holes of ignorance. For plants, the intensity of collecting effort has perhaps had a greater influence on perceived centres of diversity than does the true nature of that diversity. For example, the heavily collected area around Manaus (on the Amazon near its confluence with the Rio Negro in Brazil) is shown as a centre of endemism in all compendia of plant diversity.

Hot Spots

The term 'hot spot' is generally applied to a geographic area that is particularly high in one or more of the following : species richness, levels of endemism, numbers of rare or threatened species and intensity of threat (Reid, 1998). The term biodiversity hot spot is most commonly used for regions of high species richness. Hot spots are used to identify gaps in the existing network of protected areas, by mapping hot spots of species richness, followed by determining which species are already well conserved in existing protected areas and thereafter mapping the pattern of species richness for the remaining species. By using various selection algorithms, the minimum set of grid cells that encompass the unprotected species is chosen. Hot spots of species rarity or endemism-region rich in species with restricted distribution ranges-can also be used to help fix priorities for bird conservation (Balmford and Long, 1994).

Hot spots can also be defined on the basis of both species richness and endemism. Megadiversity countries (for example, Brazil and Indonesia) are nations that either have extremely high species richness of plants and vertebrates or are relatively less species rich but with very high levels of endemism (e.g. Madagascar and Australia). Finally, hot spots have been defined as regions with the largest number of threatened species, independent of the overall species richness or endemism of the region.

Two key aspects of the analysis of biodiversity hot spots are valuable forsetting conservation priorities. First, because species distribution data for most taxa are either very limited or unavailable, use of the hot spot approach in setting priorities depends on the assumption that patterns of diversity among relatively well-studied 'indicator' groups, such as birds, mammals and plants, are good predictors of patterns of diversity in less-studied groups. Second, research aims to determine the optimal method of analysis for using hot spot information in setting conservation priorities (Reid, 1998).

One important lesson that has emerged from studies of diversity hot spots is that the extent of the biodiversity crisis is often highly localized. A substantial fraction of species diversity can be found in very small regions and most threatened species can be found in smaller regions still. Rarity hot spots covering just 5% of the UK represent 98% of the British species of breeding birds. Richness hot spots in the UK covering a similar area encompass 91% of the butterflies, 92% dragonflies, 95% liverworts, 95% aquatic plants and 87% breeding birds (Prendergast *et al.*, 1993; Reid, 1998).

At the national level, biological diversity can be efficiently protected by including all major ecosystem types in a system of protected areas. These types should include those that are unaffected by human activity as well as those managed or disturbed (such as managed range lands and forests). Another approach is to compare a detailed vegetation map with lands under governmental protection.

In Geographic Information System (GIS), computers are used to integrate the natural environment data with known species distributions (Sample, 1994). This enables identification of those critical areas that need to be included within national parks and areas that should be avoided by development projects. The GIS approach involves storing, displaying and manipulating many types of mapped data such as vegetation types, climate, soils, topography, geology, hydrology and species distributions. Satellite imagery is also useful. Images taken at different times reveal patterns of habitat fragmentation and destruction that need to be promptly considered.

On the basis of criteria for rain-forest plants, Myers (1988) identified 12 tropical hot spots that together include 14% of the world's plant species on only 0.2% of its total land surface. Later on, 8 non-forest habitats were also added, four in the tropics and four in Mediterranean-type climates.

Also 12 megadiversity countries have been identified that together contain 60% - 70% of the world's biological diversity : Mexico, Colombia, Brazil, Peru, Ecuador, Zaire, Madagascar, Indonesia, Malaysia, India, China and Australia (Table 5.1). In these countries, greater conservation attention is warranted. Most workers are agreed on the special need for increased conservation efforts - and the establishment of additional protected areas - in the following regions :

1. *Latin America* : The coastal forests of Ecuador; the Atlantic coast forest of Brazil.
2. *Africa* : The mountain forests of Tanzania and Kenya; the large lakes throughout the continent; the island of Madagascar.
3. *Asia* : South-western Sri Lanka; the eastern Himalayas; Myanmar, Thailand, Kampuchea, Laos, Vietnam and south-eastern China; the Philippines.
4. *Oceania* : New Caledonia.

Wilderness Areas

Large blocks of land that have been minimally affected by human activity, that have a low human population density, and are not likely to be developed in the near future, may well be the only places in the world where large mammals can survive in the wild. These wildness areas serve as controls showing what natural communities can be like with the least human influence.

Table 5.1 : 'Top ten' countries with the largest approx. number of estimated species of selected well-known groups of organisms

Rank	*Mammals*	*Birds*	*Amphibians*	*Reptiles*	*Swallowtail butterflies*	*Flowering plants (x 10^3)*
1	Indonesia 515	Colombia 1721	Brazil 516	Mexico 717	Indonesia 121	Brazil 55
2	Mexico 449	Peru 1701	Colombia 407	Australia 686	China 99-104	Colombia 45
3	Brazil 428	Brazil 1622	Ecuador 358	Indonesia ca. 600	India 77	China 27
4	Zaire 409	Indonesia 1519	Mexico 282	Brazil 467	Brazil 74	Mexico 25
5	China 394	Ecuador 1477	Indonesia 270	India 453	Myanmar 68	Australia 23
6	Peru 361	Venezuela 1275	China 265	Colombia 383	Ecuador 64	S. Africa 21
7	Colombia 359	Bolivia ca. 1250	Peru 251	Ecuador 345	Colombia 59	Indonesia 20
8	India 350	India 1200	Zaire 216	Peru 297	Peru 58	Venezuela 20
9	Uganda 311	Malaysia ca. 1200	USA 205	Malaysia 294	Malaysia 55	Peru 20
10	Tanzania 310	China 1195	Venezuela Australia 197	Thailand Papua New Guinea 282	Mexico 52	USSR (former) 20

Centres of Diversity

There are instances when specific data about whole communities are not available. In these cases, certain organisms can be used as biodiversity indicators. Diversity in birds, for example, can be a good indicator of community diversity in some areas.

Attempts are underway to identify over 200 global centres of plant diversity with large concentrations of species. Also, the International Council for Bird Protection ICBP (1992) has identified over 200 Endemic Bird Areas (EBAs) (Stattersfield *et al.*, 1998). These are localities containing about 245 restricted-range bird species. The localities include islands and isolated mountain ranges that also contain endemic lizards, butterflies and plants, and so represent priorities for conservation : certain EBAs contain no protected areas and require urgent conservation measures.

Protecting habitats is undoubtedly the best method for preserving biological diversity. The human impact on protected habitats vary greatly and compromises need to be made between protecting biological diversity and satisfying human needs. Considerable biological diversity is often found in unprotected multiple-use management areas. About 6% of the Earth's surface today is included in about 8,600 protected areas. And in many countries 10% - 20% of the land is managed for multiple-use resource production. But the future fate of some of these species is in doubt due to small population sizes, the inability of the protected area to supply all of the needed requirements of species, the tendency of many species to migrate, and threats from outside the protected area. Effective preservation can be achieved by means of large and well-managed protected areas that include examples of all biological communities.

Some of the issues being debated by conservation biologists are the following :

1. How large must nature reserves be to protect species?
2. Is it better to have a single large reserve or many smaller reserves?
3. How many individuals of an endangered species need to be protected in a reserve to prevent extinction?
4. When several reserves are created, should they be closed together or far apart, and should they be isolated from one another or interconnected by corridors?

In general, the following principles may be recommended for designing better reserve : protection of complete ecosystem, larger size, unfragmented, larger number (of reserves), maintenance of corridors that interconnect different reserves, diversity of habitats, more or less rounded (spherical) shape (to prevent edge effects), mixing of reserves of various sizes, and regional, planned management of reserves.

Landscape Ecology and Park Design

Landscape ecology is the study of patterns of habitat types on a regional scale and their influence on species distribution and ecosystem processes. A landscape is an area where a cluster of interacting communities is repeated in similar form. Fig. 5.1 illustrates four landscape types in which interacting communities form repetitive patterns.

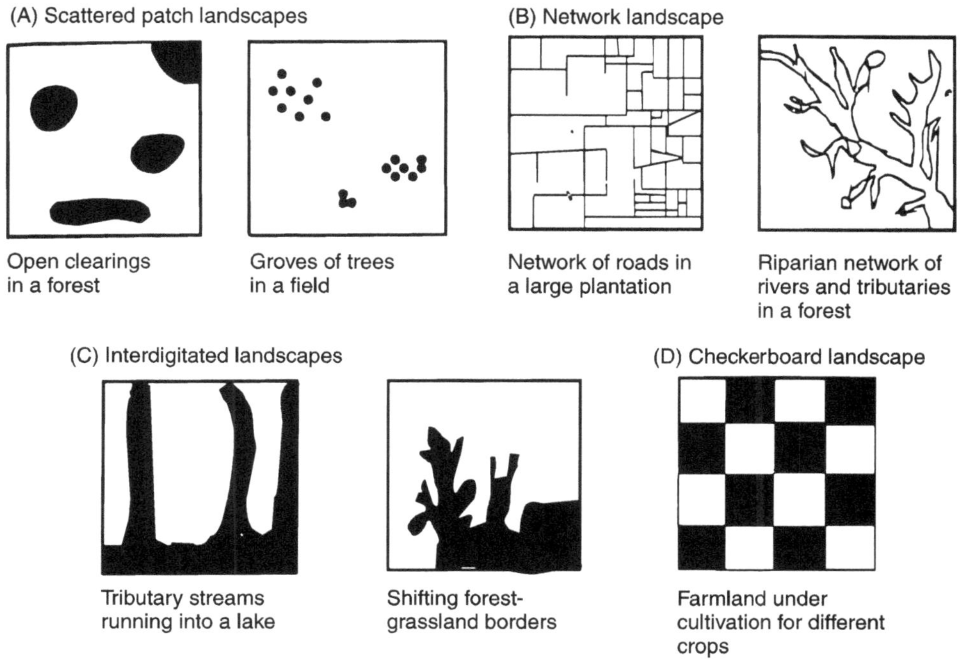

Fig. 5.1 : Four landscape types where interacting communities form repetitive patterns.

For species not confined to a single habitat, or those that live on borders where two habitats meet, the patterns of habitat types on a regional scale are of vital importance. The presence and density of many species is influenced by the size of habitat patches and their degree of linkage. Different landscape patterns also have different effects on the micro climate, pest outbreaks and animal movement patterns. Different land-uses sometimes produce highly contrasting landscape patterns. Forest areas cleared for shifting agriculture, permanent subsistence agriculture, plantation agriculture, or suburban development have completely different appearances from the air, differing distributions and sizes of remanent forest patches, and different species assemblages.

Mapping of Biological Resources for Conservation

The emphasis on the spatial scales of conservation has changed in recent years. Previously, whole forests used to be conserved and taken over under state control. In keeping with this, certain protected areas were set apart. The current emphasis is to identify 'hot spots' at both global and national levels. The shift has been partly due to a lack of knowledge about the distribution pattern of biological resources. Today doubts have arisen concerning the very basis of demarcating these parks and their efficiency in conserving. On the one hand, the strategy of fencing off the areas as conservation sanctuaries is being questioned; on the other, several instances show that the protected areas exclude certain patches with high conservation value (Ganeshaiah and Uma Shanker, 1998). Conservation plans need to be built upon our understanding of the spatial distribution of biological resources, which is generally inadequate. Fortunately, recently developments in the areas of Geographic Information System (GIS) and Remote Sensing (RS) techniques have brought within reach new ways to develop, analyse and monitor the spatial distribution patterns of biological resources. These tools have greatly speeded up the efficiency with which the biological resources can be mapped and managed.

The fact that some countries (including India) do have good data bases on the occurrence and distribution of biological resources (e.g. numerous herbaria), raises the hope that it should be possible to develop initial patterns of spatial distribution of our biological resources. In fact, Mushtak Ali and Ganeshaiah (1998) have already shown the potentiality of these data sets in developing the biodiversity maps of India. Nair and Menon (1998) have mapped the distribution, and estimated the stocks of bamboo resources.

Protected Areas

The global network of international, national and regional protected areas remains an important tool in biodiversity conservation. These areas serve as critical repositories ensuring that at least a minimum of the world's genes, species and ecosystems are conserved (Anonymous, 1994). By 1994, over 8,500 protected areas had been designated worldwide, covering about 8 million km^2 (Table 5.2 & 5.3) (WRI, 1994) which represents only about 6% of the Earth's total land surface. The largest park is in Greenland, covering 9,72,000 km^2, accounting for about 12% of the global area protected. Yosemite National Park in California was established as the world's first protected area. Since then, about 7,000 national parks, reserves, and other protected areas have been established. Additionally, there are over 500 zoos, 1,500 botanical gardens, 60 gene banks, and a small number of scientific aquaria, collectively known as *ex-situ*, or off-site conservation facilities that have helped sustain manyspecies, including some at serious risk of extinction. Between 1950 and 1990, the number and extent of protected areas increased more than five times. Park boundaries follow political lines rather than ecological lines (e.g. the limits of a watershed). Many parks are too small to effectively conserve intact ecosystems or provide adequately for their inhabitants. For example, many park animals have to range beyond the boundaries during certain seasons to find enough food to survive.

All too often, protected areas are imposed on a community with no input from and little regard for the local people, thus creating conflict. Conflict also arises when the benefits of the areas go to society at large, or into government or business coffers, while the costs are borne by local people hose use of the area tends to be restricted.

Some activities outside a park's boundaries may have a negative impact inside the park. When land adjacent to a park is altered, there may be changes in breeding and migration of park species, changes in water availability or quality, or air pollution.

Many zoos, botanical gardens and aquaria started long ago as mere menageries of exotic biological curiosities for public display. Some of them still treat plant and animal species strictly as acquisitions. In the past, seed banks collected a limited number of species and storage methods sometimes failed to maintain the viability of the material.

To respond to today's challenges, many leading zoos, botanical gardens and other facilities have started reorienting their missions through captive breeding and scientific research. Leading zoos and botanical gardens seek to breed and propagate rare species. These facilities can maintain viable population of only a small fraction of the species threatened with extinction. While captive breeding and reintroduction hold so promise, they are expensive and rather limited in scope. These measures cannot control one of the primary threats to endangered species : habitat destruction.

Table 5.2 : Protected and managed areas in the world's geographic regions (from : WRI, 1994)

	Protected areas (IUCN categories I-V)			Managed areas (IUCN category VI)		
Region	Number of areas	Size (km^2)	Per cent of land area	Number of areas	Size (km^2)	Per cent of land area
Africa	704	1,388,930	4.6	1562	7,46,360	2.5
Asia (incl. former USSR)	2399	1,454,910*	5.5	1150	3,10,290	1.1
North and Central America	1752	2,632,500	11.7	243	1,61,470	0.7
South America	667	1,145,960	6.4	679	2,279,350	12.7
Europe	2177	4,55,330	9.3	143	40,350	0.8
Oceania[a]	920	8,45,040	9.9	91	50,000	0.6
World	8619	79,22,660	5.9	3868	3,588,480	2.7

*Only 124 countries currently have protected areas.

[a]Australia, New Zealand and the Pacific Islands.

Table 5.3 : Relative protection of 14 major biomes (after IUCN, 1994)

Biome type	*Total area (km^2)*	*Protected areas*	*Area Protected (km^2)*	*Per cent of area protected*
Tropical rain forests and moist forests	10513,210	506	538,334	5.1
Tropical dry forests /woodlands	17,312,538	799	817,551	4.7
Tropical grasslands /savannas	4,264,833	59	235,128	5.5
Subtropical/temperate rain forests/woodlands	3,930,979	899	366,297	9.3
Temperate broadleaved (deciduous) forests	11,216,660	1507	358,240	3.2
Temperte evergreen coniferous forests	3,757,144 forests	776	177,584	4.7
Temperate grasslands	8,976,591	194	99,982	0.8
Warm deserts/semideserts	24,279,842	300	984,007	4.1
Cold deserts	9,250,252	136	364,720	3.9
Boreal forests and woodlands	15,682,817	429	487,227	3.1
Tundra	22,017,390	78	1,645,043	7.5
Islands	3,25,270	530	322,769	9.9
Lake systems	517,694	17	6,635	1.3
Mountain systems	10,633,145	1277	852,494	8.0

Buffer zones and biosphere reserves have come up. Biosphere reserves have concentric areas zoned for different uses. The 'core zone' is devoted to preserving biodiversity with no human interference. Around the core is a buffer zone' in which some settlement and resource use is allowed, surrounded in turn by an indefinite 'transition area' where sustainable development activities are permitted.

As a general rule, the greater the human-caused disturbance, the greater the loss in biodiversity and the narrower the scope of preserving it. The International Union for the Conservation of Nature (IUCN) developed the following system of classification for protected areas that ranges from minimal to intensive allowed use of the habitat by human (IUCN, 1994).

1. *Strict nature reserves and wilderness areas.* Protect natural organisms and natural processes in an undisturbed state, so as to perpetuate representative samples of biological diversity for scientific study, education, environmental monitoring and maintenance of genetic variation. Included are two subcategories : (la) primarily includes nature reserves, for research and monitoring work, and (lb) primarily includes wilderness areas maintained for recreation, for subsistence economic activities and for protecting natural processes.
2. *National parks.* Large areas of scenic beauty, maintained for scientific, educational and recreational use; usually not used for commercial extraction of resources.
3. *National monuments and landmarks.* Smaller areas designed to preserve unique natural areas of archaeological interest.
4. *Managed wildlife sanctuaries and nature reserves.* Similar to nature reserves, but some human manipulation is encouraged to maintain the community characteristics. Some controlled harvesting is also permitted.
5. *Protected land and seascapes.* These are areas that embody the harmonious interaction of people and the environment through the traditional, non-destructive use of natural resources, while concomitantly providing opportunities for tourism and recreation. Include grazing lands, orchards, or fishing villages.
6. *Managed resource protected areas.* These allow sustainable production of natural resources such as water, wildlife, grazing for livestock, timber and fishing. This is done in such a way that the biological diversity is preserved.

The first 57 biosphere reserves were identified by the Man and Biosphere Programme of UNESCO in 1976. The number has increased since then. Many biosphere reserves were created simply by going the designation to an existing national park or reserve, without adding new land, regulations, or functions nor have agencies been sufficiently funded to make the complicated zoning system really work. Without local involvement in design and management, the only way a buffer zone might work is if the park agency can effectively enforce the regulations (Anonymous, 1994). A novel conservation technique that addresses the root cause of biodiversity loss is bioregional management, a method that goes beyond the buffer zone approach by involving local people, integrating ecological, economic, cultural and managerial considerations at the regional scale.

The integrative approach centres around the idea of managing whole regions biodiversity in mind. Dividing government responsibility into isolated forestry, agriculture, parks, and fisheries sectors fails to reflect ecological, social, or economic reality. A 'bioregional' approach requires cross-sectorial and transboundary co-operation and integration, as well as broad participation by all affected constituencies.

Bioregions are areas with high value for biodiversity conservation in which a management regime is established to co-ordinate land-use planning of both public and private landowners and to define development options that will meet human needs without losing biodiversity.

Dinerstein and Wikramnayake (1993) classified tropical Asian countries into four types on the basis of the per cent of their currently protected rain forest and the per cent of unprotected forest predicted still to be intact in 10 years. Category I countries (Brunei, Indonesia, Bhutan and Malaysia) have protected large amount of forest and will have remaining unprotected forests in the near future. Category II (Thailand, Sri Lanka, India, Nepal and Pakistan) have many parks, but the remaining unprotected forests is disappearing rapidly. New parks should be created immediately. Category III (e.g. Papua New Guinea, Solomon Islands and Laos) have much forest but no protected areas. Category IV countries (Vietnam, Philippines, China and Tonga) have little protected forest and also little forest left to protect; the remaining unprotected areas are degrading rapidly as a result of human activities. Saving the remaining forest patches in these countries is an urgent conservation priority.

Although a large number of heritage sites have been set up throughout the world, many of the sites are facing certain problems. These include development (e.g. logging, mining), tourism, external threats, grazing and cultivations, illegal wildlife harvesting, fire, natural threats or calamities, introduced species, and insufficient management etc. The most serious and common problem in Asian sites is that of insufficient or incompetent management and supervision.

A part has to be managed very efficiently to ensure that original habitat types are maintained. Some species only occur in specialized niches and specific successional stages of that habitat. When land is protected the pattern of disturbance and human usage can change so greatly that many species that previously were found on the site vanish. Natural disturbances, including fires and grazing are key factors in the ecosystem required for the presence of certain rare species. Small parks often lack the full range of successional stages, so many species may disappear for this reason.

Protected areas need to be managed competently so as to maintain biological diversity because the original conditions of the area may have been changed by human activities. The best management sometimes involves doing nothing. Effective management begins with a clearly articulated statement of priorities. Part of protected areas occasionally have to be burned, dug up, or otherwise disturbed by people to create the openings and successional stages needed by some species. This kind of management has proved to be crucial for some endangered butterfly species that need early successional food plants to complete their life cycle.

Keystone resources such as nesting sites and water holes have to be carefully preserved, restored, or even added to protected areas so as to maintain populations of some species.

Local residents and outside visitors are key elements in park management. Quite often a compromise has to be made between banning human use of park resources on the one hand, and allowing unlimited use that allows people to use park resources in a sustainable manner without harming biological diversity, on the other.

Substantial biological diversity exist outside of protected areas, particularly in habitat managed for multiple-use resource extraction. These unprotected habitats are also vital for conservation because protected areas account for only a small percentage of total area. Animal species living in protected areas often forage or migrate on unprotected land where they are exposed to hunting, habitat loss and other threats from humans.

In temperate forests biodiversity can be increased if logging operations minimize fragmentation, and if some alter-successional component, including living trees, standing dead trees and fallen trees are left. These trees are important resources for cavity-nesting birds. In Africa, many large animals live predominantly in range land outside the parks. Local people and landowners often maintain wildlife on their land for a variety of purposes; further incentives are often given to encourage this practice.

Monitoring

Efforts to monitor, assess and conserve biodiversity have been focused largely on documenting species and habitats and on better understanding the biological functions and values of biodiversity (Wilson, 1988). Increasing attention is now being paid to the links between biodiversity and the human or cultural context, but biodiversity remains essentially a biological concept defined in terms of genes, species and ecosystem that underpin human activities (WRI, 1991). Efforts to conserve biodiversity largely in terms of biological inventory, assessment and monitoring may only document the disappearance of species and the destruction of habitats. Moreover, for effective conservation and restoration of biological diversity and productivity in threatened areas, assessment and monitoring must address the different ways in which people value, use, manage and affect the biodiversity of any given area.

Another response, that of Conservation Biology, argues that conserving biodiversity requires improved understanding of the biological variety of plants and animals through more systematic field research on taxonomy and the functioning of biophysical processes.

The conservation biology response has led to the creation of national parks and other types of protected areas, but the view of national parks and related reserves as fortresses is sometimes challenged. A more co-ordinated and integrated set of conservation programmes are now being advocated (Nelson and Serafin, 1992).

Adaptive Management

Monitoring is necessary for ascertaining whether or not our management strategies for forests, range lands or aquatic ecosystems are actually conserving biodiversity. The best way to understanding the impact of our actions is through long-term, systematic monitoring of biodiversity at many levels of organization and spatiotemporal scales. Monitoring should constitute an integral part of management as it can indicate which of our practices are compatible with protecting biodiversity. Conversely, by learning which practices harm biodiversity, such practices can be avoided. The linking of management with monitoring is termed *adaptive management.* It views human activities with some humility and realizes how ignorant we are about biodiversity and how to maintain it. The underlying assumptions of adaptive management as applied to biodiversity have been summarized by Noss and Cooperrider (1994) as follows :

1. Maintaining optimally functioning ecosystems with all their components (i.e., biodiversity) is an overriding goal.
2. Ecosystems are extremely complex and human understanding of them is rudimentary.
3. Human activities may have severe and largely unpredictable effects on ecosystems, and these effects can be irreversible or require centuries for restoration.

4. Management should therefore be conservative, erring on the side of minimal risk to ecosystems.
5. Careful, systematic monitoring of ecosystems and how we affect them can help us learn how to avoid causing further harm to them.

Monitoring is the foundation stone of adaptive management because without it one cannot learn to adapt. Adaptive management may be easily applied to a problem of any scale. It is a cyclic process.

Range management : It is unfortunate that forestry has changed into a service profession for the timber industry. But foresters are not the only natural resources professionals to have facilitated this change. Range management may be considered a counterpart of forestry for lands without trees. Like forestry, , range management has developed in response to the idea that range lands were deteriorating-from overgrazing. Livestock production has been the primary focus in range land management in many advanced countries. Range management involves optimizing the returns from range lands in those combinations most desired by and suitable to society through the manipulation of range ecosystems. As with forestry, the emphasis is on products and production (returns), on improvement over nature (optimizing) and on technology (Table 5.4, 5.5).

Wildlife management : This also has a remarkable history parallel to that of forestry and range management. *Wildlife management* includes both fisheries and wildlife management (Noss and Cooperrider, 1994). In some parts of the oceans, humans have virtually annihilated salmon fisheries through dams, logging and other habitat destruction, and then tried to undo this by huge expenditures (but limited success) to raise and release salmon in hatcheries. Like foresters and range manager, wildlife and fisheries manages have also focused on raising a single species or only a few species such as deer, ducks and pheasants for hunting, and bass and trout for fishing.

According to Noss and Cooperrider, because biodiversity conservation is largely an issue of how we treat the land, any national strategy for biodiversity must be land-based; the real threats are logging, grazing, road building, mining, dam building, agriculture, housing development and off-road vehicles. It is necessary to interact with nature gently and more sustainably. Any rational strategy must duly recognize the linkages between proximal threats to biodiversity and the ultimate causes : human population growth, poverty, misperception anthropocentrism, cultural transitions and economics. It must give incentives for controlling population and resource consumption.

Land management can have good, bad or neutral effects on biodiversity. Good management serves to protect biodiversity from harm or restores an ecosystem previously damaged. It is neutral if it mimics or substitutes for natural disturbance-recovery processes. Negative management is one that leads to biotic impoverishment.

Biodiversity Assessment for Conservation

Identification of prospective conservation areas is usually based on such surrogate information as richness of indicator taxa, taxa restricted to a given area, or higher texon richness (that is, genus or family richness). It is generally agreed that conservation areas should strive to sample regional features, using sets of grids (Florey *et al.*, 1994; ICBP, 1992) that contain all species in a taxon at least once; the complementarity principle ensures that conservation areas represent all species efficiently and that rare species are included (Jaarsveld *et al.*, 1998). The outcome of such a complementarity analysis is not only a good basis for the efficient conservation of the focal taxon, but may also be more widely applicable to other taxa.

Table 5.4 : Some forestry practices and their effects on biodiversity (after Noss and Cooperrider, 1994)

Practice	*Purpose*	*Effect on biodiversity*
Planting of exotics or genetically 'improved' tree species	Improved yield of commercial trees	Replacement of native species; loss of genetic purity or genotypes of native stock
Pesticide spraying	Protection of commercial trees	Reduction of secondary effects on-non-target insect species and vertebrates; ecosystem disruption
Clear-cutting and reforestation	Maximum utilization of existing tree biomass and speeding up new forest growth	Artificial disturbance cycle truncated succession; loss of species richness; loss of structural and functional diversity
Clear-cutting and even-aged management	Efficient forest regulation maximum profit from growing trees	Truncated succession; loss of structural diversity
Swidden	Site preparation for new forest	Loss of structure, biomass and Nutrients from forest ecosystem
Tree thining	Increased growth of commercial trees	Reduced structural diversity in forest
Maximum production of commercial timber and pulpwood	Maximum forest output; highest profit	Reduced structural and species diversity of forest; deterioration of forest ecosystem

Table 5.5 : Some common range management practices and their impact on biodiversity (after Noss and Cooperrider, 1994)

Practice	*Purpose*	*Effect on biodiversity*
Reseeding with exotic species	Increased forage production for livestock	Displacement of native species
Brush or tree removal	same	Replacement of woody species with grasses or other herbaceous species; loss of structural and species diversity
Water development	Increased water for livestock; better livestock distribution	Usurpation of water required by wildlife species; extension of livestock impacts into weakly grazed landscape
Fencing	Control of livestock movement	Impairment of movement of some wildlife species
Predator control	Livestock protection from native predatory animals	Extirpation of predators; reduction or eliminationof non-target species
Salting (leaving salt on range)	Better livestock distribution	Spreading of livestock impacts into previously ungrazed or lightly grazed landscape
Pesticide/herbicide spraying	Control of harmful insects or noxious weeds	Reduction in non-target organisms and plant species diversity; disruption of ecosystem
Maximizing livestock utilization of forage	Maximum livestock production; high profit	Decreased forage availability for wildlife species; reduction of species diversity; ecosystem degradation

It has been commonly observed that richness 'hot spots' (highly species-rich areas) and 'cold spots' (areas poor in species) rarely coincide; nor do hot spots and rare (restricted range) taxa generally coincide.

Five basic principles of reserve design that have influenced conservation efforts are listed below (see Wilcove *el al.*, 1993).

1. Well-distributed species are less prone to extinction than those confined to small portions of their ranges.
2. Large habitat areas containing many individuals of a given species are more likely to sustain that species than are smaller habitats with only a few individuals.
3. Blocks of habitat in close proximity are preferable to those that are widely dispersed.
4. Contiguous, unfragmented blocks of habitat are better than fragmented blocks.
5. Habitat between protected areas is more easily traversed by dispersing individuals the more closely it resembles suitable habitat for the concerned species.

Conservation Targets

About one-half of the world's species may already have been lost (Soule and Sanjayan, 1998). This has prompted the near-term protection of at least 10% of the total land area in each country or in each ecosystem, so as to double or triple the land area now designated as national parks or biosphere reserves. However, protection of only 10% of the Earth's ecosystems could still make at least one-half of all terrestrial species vulnerable to anthropogenic extinction sooner or later. The situation is particularly serious in the tropics, which contain some two-thirds of the world's terrestrial plant and animal species. The relatively undisturbed tropical forests (area already reduced by about half during the past five decades of this century) is currently shrinking at a rate of about 0.8% annually. Only 5% of the tropical rainforest biome is protected. The increasing global demand for tropical goods is accelerating rates of habitat conversion in developing nations.

It appears that lands outside strictly protected reserves everywhere will be greatly diminished in their capacity to sustain native species and ecosystem by 2050, by which time human population may have more than double (Sole and Sanjayan, 1998). Therefore to deal with this problem a much larger area than 10% (perhaps as much as over 40%), may need to be set aside-aformidable task indeed in light of the fact that even achieving the 10% target in many countries today would be a herculean venture.

Some general desirable attributes of conservation measures are outlined below :

1. They must be firmly grounded in the natural sciences, but should also involved the social sciences, humanities, law, education and economics.
2. They should be implemented at several scales of time and space.
3. They should aim to harmonize species-oriented and ecosystem-oriented approaches.
4. They should take into account biological attributes and processes at all levels of the biological hierarchy.
5. They should work across national, cultural and ecological boundaries.
6. They should strive to address local community development and conservation needs in an integrated manner.

Biotechnology has spurred today's new-found urgency. Advances in biotechnology raise the monetary stakes because any species can potentially be 'raw material' for genetic engineering. Pharmaceutical companies are once again interested in natural products research, biotechnology has made screening of natural products a low-cost, potentially lucrative activity. Moreover, when a promising chemical is found, large quantities of it can be produced by transferring the proper genes to bacterial 'hosts'. Pharmaceutical companies now realize that even their most creative chemists may be no match for nature. Two of the most promising drugs in clinical trials have come from plants : the anticancer drug taxol derived from the yew and an antimalarial drug derived from a Chinese herb known as Qunghaosu.

Biotechnology has increased biodiversity's value to agriculture. Genes for pesticides resistance may be transferred from bacteria to plants. The distant relatives of modern crops-once considered useless for breeding purposes-contain unique traits that can now be bred into elite varieties. Biotechnology makes it possible to quickly tailor crop varieties to local soil, climate, or rainfall conditions so they can meet the needs of marginal farmers. Customized crops could also be critical to protecting the world's food supply.

Genetic resources have traditionally been treated as though they were a common heritage of mankind-free to all who could use them. The growing importance of biotechnology has forced a reassessment of the ownership issue. Nations should assert national sovereignty over genetic resources regulate collection, and decide whether to adopt intellectual property rights protection for knowledge about these resources (Reid, 1992). Two decades ago, biodiversity conservation would have been limited to saving genes, species and habitats. But now a new conservation framework has emerged, based on saving biodiversity, studying it, and using it sustainably and equitably (Reid, 1992). The steps needed go beyond maintaining protected areas and seed banks to include reforming agricultural, forestry and technological policies (See Fig. 5.2)

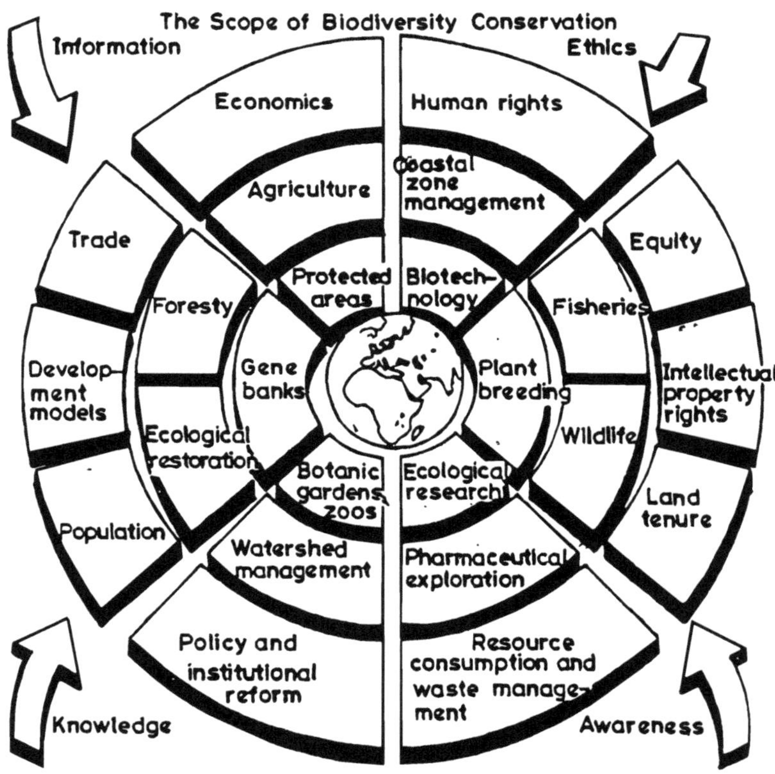

Fig. 5.2 : The scope of biodiversity conservation (after Reid, 1992).

Species-focused conservation efforts area essential

but they must be complemented by efforts to conserve habitats and ecosystems in natural or seminatural states.

Conserving biodiversity over the long run requires an understanding of past and present human land-use and its effects in a holistic way. Biodiversity in an area is the product of the history of interaction between human use and the environment. This legacy includes not only changes in climate, hydrology, geology and other biophysical factors, but also changes in human activities, culture, perceptions, attitudes and values, technology and institutions. Broader inventory, assessment and monitoring systems are needed to collect the range of biophysical and human information that is necessary for planning and managing biodiversity are related matters (Nelson and Serafin, 1992).

Conservation Strategies

The following actions can be taken to promote the conservation and sustainable use of biological diversity : (a) measures to protect particular habitats as National Parks, Biosphere Reserves or other protected areas; (b) measures to protect particular species or groups of species from overexploitation; (c) measures to promote *ex-situ* conservation of species in botanical gardens or in gene banks; and (d) measures to curb the contamination of the biosphere with pollutants.

There are three global objectives of living resources conservation : (a) to maintain essential ecological processes and life-support systems; (b) to preserve genetic diversity; and (c) to ensure the sustainable utilization species and ecosystem. Biodiversity needs to be conserved as a matter of principle, as a matter of survival and as a matter of economic benefit. The objectives can be achieved by taking action at several different levels. (In view of the growing threats to biological diversity, the time has come to appreciate the Earth's biological resources as assets to be conserved and managed for the benefit of all humanity.

Identifying the Specifics

According to Scott *et al.* (1995), biodiversity management may start by (1) compiling a list of plant and animal species, their numbers, distribution patterns and habitats; (2) identifying and mapping for each biological community, existing and potential successional stages; and (3) judging the extent and severity of human disturbances. By comparing these parameters with historical conditions, managers may be able to quantitatively assess biological diversity. There is no doubt that biodiversity concerns are intimately linked to the effect of human disturbances and associated species losses. Lautenschlager (1997) has devised the following system to identify and prioritize major specific concerns (resources, management practices, ecosystems) from local (district or smaller) through larger (region, province, country) scales.

1. Resource (Natural asset, usually biotic) and specific concern. The specific concern can include some ecological process, or a more general concern.
2. The human activity associated with the specific resource of concern, e.g. fire suppression, a silvicultural system, or environmental changes caused by development activities.
3. The ecosystem(s) in which this human activity is commonly affects this resource; may range from small ecosystems (a wetland type in a certain forest type or region) to extensive and widely distributed ecosystems.

When the above components are combined, a concern matrix (Table) results.

When agreement among experts about prioritizing concerns, at all scales, cannot be reached, there may be a consensus. Local and regional biological priority lists constitute the first task in identifying larger scale priorities.

In the face of expanding human population there is an urgent need for increasing both the area and intensity of natural-resource management worldwide. Growing concern about the intensity of natural-resource management in several developed contries has led to increasing natural-resource extraction from tropical or subtropical countries (Kimmins 1992). However, biotas adapted to these more stable ecosystems are commonly less adapted to disturbance than those in northern ecosystems are commonly less adapted to disturbance than those in northern ecosystem where natural fluctuations are widespread and severe. Both species diversity and threats to diversity decrease with increasing distance from the equator (Pimm *et. al.,* 1995). Forest harvesting and slash-and-burn agriculture in tropical and subtropical moist forests are usually rich in endemics and contribute significantly to reductions in global species.

Species have long been known to originate by reproductive isolation. Polyploidy involves the multiplication of entire chromosome numbers within individual species or within hybrids of species. It isolates the new breed from its ancestor in one step. This instantaneous mode has generated some 40% of the present-day plant species and a much smaller number of animal species. Equally important is geographic (or allopatric) speciation, the origin of intrinsic isolating mechanisms in two or more daughter populations while they become isolated by a geographic barrier, such as desert basin or mountain range. Yet another mode of speciation is non-polyploid sympatric speciation, in which new species emerge from the midst of parental species even when individuals of both populations are close enough to intermingle during part of their life cycles (Enrlich and Wilson, 1991). Members of the parental species feed upon and mate in the vicinity of one kind of plant; they give rise to an alternate host race that shifts to a second species of host plant growing nearby; the two races, which have become isolated by their microhabitat differences, diverge further in other traits that reinforce reproductive isolation. Sympatric speciation has probably had a key role in the origin of large numbers of insects and other invertebrates.

Certain forms of speciation can occur rapidly, within one to a few generations. And when species meet, they can displace one another genetically within a few generations, reducing competition and the likelihood of hybridization.

The Precautionary Principle

We are all painfully aware of a woeful paucity of knowledge. But the problem of continuing loss in biodiversity is so urgent that we must act according to the precautionary principle - i.e., we must not delay a decision to act simply because science cannot at present answer all the questions. Enough is known about how and where biodiversity is being lost to start setting priorities for its protection and to develop practical mechanisms right now.

There is a strong need for a List of Recorded Species, computerized and readily available. Such a list should indicate not only formal aspects of taxonomy, but also ecological information about the species (estimated geographical range; life history and habitat; physical size of the organism, abundance/rarity; etc.). Such a list could serve as an indispensable tool for making judicious decisions or choices.

Fortunately, India is a land of diversities, having very varied topography and soils. Climate both West to East and South to North varies significantly. We have many harsh or extreme habitats, some of which harbour species at the limit of their viability. Onto this pattern of variation has been imposed the consequent regional differences in rural land-use and the pressures of a fairly long history of natural resource extraction. Such a high degree of variability warrants a much more scientific and systematic approach to conservation and its management than adopted hitherto.

It needs to be emphasized that, unlike other scientific endeavours, the task of understanding the magnitude, causes and consequences of biological diversity has a time limit : half the remaining tropical rain forests are doomed to vanish within the next half a century and, as they vanish, no one knows how many species will be lost forever.

There is a serious lack of reliable survey data in many developing countries. Remote sensing is partially filling the gap but the data are inappropriate for species-level monitoring, and further there is also the serious problem of 'ground truthing'. Faced with the dearth of comparable data on biological resources in the developing world, a World Conservation Monitoring Centre (WCMC) was established in Britain in 1988. WCMC's role is to support programmes for the conservation and sustainable utilization of biological resources by providing quantitative data in a format that is directly applicable for improved land-use management. As the development pressures upon the world's biological resources inexorably grow, so the need for reliable, up-to-date information correspondingly increases. Using the latest advances in information on threatened plant and animal species; habitats of conservation concern, particularly tropical forests, coral reefs, and wetlands; critical sites within these habitats for the conservation of biodiversity; threats to the global network of national parks and protected areas; and the international trade in endangered species. Another project at the global level in which WCMC is currently engaged is the monitoring of the world's tropical forests (Anonymous, 1991).

Loss of forest does not just mean loss of area, but also some loss of condition of the forest through unsustainable use. WCMC is developing the methodology to classify and map forest conditions, which can then be linked to forest auditing against preset management targets. By overlaying such forest maps with species distribution maps covering a wide cross-section of taxa, such as forest primates, endemic birds, snakes and lizards, and swallowtail butterflies, it should be possible to identify 'hot spot' areas of high species diversity. Such information could then be relayed to various agencies to ensure that such sites are adequately surveyed and their ecological integrity maintained through appropriate forest management programmes.

Restoration Ecology

Conservation biologists can effectively help restore original species and communities in damaged and degraded ecosystems. Restoration is the process of intentionally alternating a site to establish a defined, indigenous, historic ecosystem with the objective of emulating the structure, function, diversity and dynamics of the specified ecosystem (Society of Ecological Restoration, 1990). In many cases, business houses are legally required to restore habitats degraded by such activities as strip mining or water pollution. Dumping of sewage into rivers and estuaries by municipalities and chemical pollution on military bases are two examples of governmental activity that damage biological communities. Restoration efforts sometimes constitute a part of the mitigation process in which a new site is created or rehabilitated as a substitute for a site damaged by development (Zedler, 1993).

Sometimes entirely novel environments such as reservoirs, canals, landfills and industrial sites are created by human activity. Neglect of these sites often results in their becoming dominated by exotic and weedy species, which are of no value from a conservation perspective. But if in these sites the native species are reintroduced, native communities can sometimes be restored.

Restoration ecology provides theory and techniques to restore these various types of degraded ecosystems. The following four approaches can be adopted for restoring biological communities and ecosystems (Cairns and Heckman, 1996).

1. *No action,* if restoration is too expensive, if the ecosystem can recover on its own, or if previous attempts at restoration were not successful.
2. *Replacement* of a degraded ecosystem with another productive type. This is sometimes termed 'habitat creation'.
3. *Rehabilitation* of at least some of the ecosystem functions and some of the original species (e.g. replacing a degraded forest with a tree plantation or replanting a degraded grassland with a few species that can survive). The focus is on dominant species critical to ecosystem function.
4. *Restoration* of the area to its original species composition and structure by means of site modification and reintroduction of the original species.

Ecological restoration simply means restoring an ecological system to its original state. Restoration can be viewed as a distinct form of conservation management, differing from 'preservation', 'conservation' or even 'management' itself. These various forms of manipulation all involve attempts to compensate in a specific, ecologically effective way for alterations typically caused by human activities.

Rehabilitation is a broad term that refers to any attempt to restore structure or function of an ecological system, without necessarily attempting complete restoration to any specified prior condition; for example, replanting of sites to prevent erosion.

The term reclamation denotes rehabilitative work carried out on the most severely degraded sites, such as sites disturbed by open-cast mining or large-scale construction. Though reclamation work often falls short of restoration in the fullest sense (a copy of a native ecosystem is not achieved), it is a necessary first step in the process of restoration under such conditions.

Recreation : Recreation attempts to reconstruct an ecosystem, wholesale, on a site so badly disturbed that virtually nothing is left to restore. The new system may be modelled on a system located outside the range of the historically system, or may be established under conditions different from those under which it occurred naturally. These efforts do not amount to restoration *sensu strictu* but can yield insights into the systems involved and the conditions that support them that can be invaluable in restoration efforts.

Ecological recovery : This involves letting the system alone, hoping that it will regain desirable attributes through natural succession. Such a zero order approach to restoration may or may not work and is best regarded as a key component of restoration-the contribution of the system itself. In such cases the restorationist seeks to complement and reinforce natural processes.

Perspective in Community and Ecosystem Restoration Ecology

It is not easy to measure restoration at the community level because of the variability found in most natural communities. This needs special attention to restoration of community function (e.g., trophic structure) rather than merely restoration of particular species. Much debate is underway concerning the possible benefits and limitations of using species composition or biodiversity measures as end-points in restoration ecology (Palmer *et al.,* 1997). Since re-establishment of all native species is usually impossible, there is need for research on the relationship between species richness and community stability of restored sites and functional redundancy among species. Efforts targeted at restoring system function must take into account the role of individual (keystone) species, particularly if some species play a disproportionate role in processing material or are strong interactors. Some attempts are being made to understand the relationship between physical habitat structure and restoration ecology. In those cases wherein community development is highly predictable, it may be feasible to manipulate a speedup in natural succession processes.

Species diversity is the central pillar of community ecology, in particular the creation and maintenance of local and regional biodiversity (e.g. Huston, 1994). It is generally agreed that restoring biodiversity is desirable for ecological, applied and aesthetic reasons. It is desirable to know the minimum numbers and types of species necessary for proper community functioning. Is it possible to increase community or ecosystem stability by addition adding more species (enhancing diversity) or particular species in the restoration process? If the answer to this is affirmative, it implies that restoration success may depend on careful consideration of community level attributes, not just a focus on single species or clusters of 'desirable' (e.g., endangered) species (Palmer *et al.,* 1997). According to May (1973), diverse systems are less stable than simpler ones : the more diverse a community, the more complex the web of species interactions and thus the larger the effect disturbances would have on the system. Diversity can make individual species more vulnerable to extinction, but total community or ecosystem properties (e.g. energy transformation, biomass) usually tend to stabilize, since some species compensate functionally for others (Naeem *et al.,* 1994; Tilman, 1996).

Restoration efforts also need to take it account a proper ecosystem perspective which provides a framework within which other aspects of the ecology of restoration may be incorporated. The restorationist must consider the placement of the project in the landscape-its boundaries, its connections or lack thereof to adjoining ecosystems, and its receipts and losses of materials and energy from its physical surroundings (Ehrenfeld and Toth, 1997). These characteristics delimit types of biological communities that can be established on the site. The ecosystem perspective also providers restorationists with conceptual tools for structuring and evaluating restorations. These include the mass balance approach to nutrient, pollutant, and energy budgets; subsidy/stress effects of inputs; food web architecture; feedback among ecosystem components; nutrient efficienty transfers, primary productivity and decomposition; and disturbance regimes. The large spatial scale of mot restorations and ability to manipulate species, oil, water, and even the landscape could allow ecosystem-level experiment to be conducted that could not be performed otherwise (Ehrenfeld and Toth, 1997).

Dynamic Conservation by Fisherfolk

In situ conservation of aquatic biodiversity is central to the strategies of *nurture fisheries* adopted by artisanal fishing communities and differs markedly from the modern industrial approach to fishing, using *capture fishery strategies. Nurture fishery* technologies are necesarily selective, passive, low-energy and ecologically efficient. They do not bring in the biggest catch in the short term, but do ensure the sustainability of the fishery through their harmonious interaction with the dynamics of the ecosystem, on which depends the aquatic biodiversity (Mulvany, 1996).

Some agricultural production comes from traditional methods which conserve fish stocks *in situ.* Aquaculture contributes to genetic diversity by the development of domestic breeds, and shares with agriculture the need to conserve this man-made diversity along with wild aquatic diversity. Combination of diverse fish species of different food habits and ecological niches allows optimum use of available food in the fish pond. Multispecies aquaculture (polyculture) is based on the harmonious stocking of different varieties of fish species at different levels of population. Planktonic feeders (at the pond surface and in mid-water) feed directly off the 'phytoplankton' and 'zooplankton' produced by the natural productivity of the pond. Their faeces enhance the productivity of the pond, as does pond manuring. Fish species which feed on larger organisms in mid-water (small fish, insects etc.) and the pond bottom (snails, worms etc.) are also stocked. Consequently, energy flow and transformation are extremely efficient. Sustainability of this method of conservation *in situ* can be promoted by decentralized production, using diverse, environmentally sensitive techniques. Protection from introduced species and diseases is also required.

Safeguarding of Aquatic Biodiversity by Fisherfolk

The management systems of *capture fisheries* are technically based whereas those of *nurture fisheries* are socially and spiritually inspired (Baines, 1995). Community-based managementis vital to sustaining the fishery as it is geared towards sustaining the aquatic ecosystem and the fishery resource base. It advocates harvesting rather than hunting approaches in the exploitation of aquatic resources.

Over the last two decades several attempts have been made to create management systems that recognize traditional rights. Both the Territorial User Rights in Fisheries (TURFs), which confer rights to fish in particular waters, and Customary Marine Tenure (CMT) systems, which define the ownership of a particular fishery, have promise if developed in a manner that ensures that rights are conferred on local communities rather than individuals (Fairlie *et al.*, 1995). Any strategy that entrusts responsibility to individuals is destined to fall in achieving long-term security. It is the community as a whole that is crucial to ensuring sustainability and equity (Mulvany, 1996).

6

General Aspects of Biodiversity

The Value of Biodiversity

The well being and survival of human beings are dependent on millions of species of plants, and microbes. These include species and varieties of crops, livestock and wildlife. Biological diversity is part of our daily lives and livelihood and constitutes resources upon which families, communities, nations and future generations depend. Biodiversity has numerous values in agriculture, medicine, food and industry. Biodiversity maintains ecological balance and evolutionary processes and has spiritual, cultural, aesthetic and recreational values.

Millions of species which constitute biodiversity have their own intrinsic value as recognised by religions and faiths in India. Traditionally Indians have celebrated this diversity of life. Conservation and sustainable use of biodiversity is ingrained in the Indian ethics and culture. Even today millions of people derive their daily sustenance from forests, rivers, grasslands and seas. The practices and beliefs of forest dwelling tribes, nomadic communities, fisherfolk and farmers remain closely interwined with the biological diversity they live amidst.

Rapid depletion in biodiversity is largely due to disempowerment of local communities, commercial exploitation, fragmentation and degradation of habitats. This loss of biodiversity is being felt through greater water shortage, lock of fodder, fuel, and other livelihood resources. And with this there has been serious erosion in the knowledge base of people, especially the men and women of local communities and of cultural values linking people with nature.

General Introduction

The term biodiversity is formed by a contraction of the term biological diversity. The term biodiversity was coined by Walter and Rosen in 1985. There are many definitions available for biodiversity. Biological diversity means the variability among living organisms from all sources.

Biodiversity refers to the variety among living organisms and the ecosystem complexes in which they live. It includes diversity of forms right from the molecular unit to the individual organism and biosperic levels. In the simplest sense, biodiversity, may be defined as the sum total of species richness, i.e. the number of species of plants, animals and micro-organisms inhabiting in a given habitat.

Types of Biodiversity

Broadly speaking, the term biodiversity includes three different but closely related aspects. They are genetic, species and ecosystem level diversities.

Genetic Diversity

It refers to the variation of genes within species. This constitutes distinct population of the same species or genetic variation within population or varieties within a species. Genetic diversity resides in the variation in the sequence of the four base pairs, such as components of nucleic acids constitute the genetic code. Other kinds of genetic diversity can be identified at all levels of organizations, including the amount of DNA per cell, chromosome structure and number. New genetic variations arise in individuals by gene and chromosomal mutations, and in organisms with sexual reproductions, these variations are spread through the population by recombination.

Only a small fraction of the genetic material (less than 1%) of the higher organisms is outwardly expressed in form, function and variation. The purpose of Junk DNA which not expressed is still unclear. It is estimated that there are 10^{-9} different genes distributed across the worlds biota. These genes do not make an identical contribution to overall genetic diversity. These genes which are responsible for fundamental biochemical reactions, undergo very little change through evolution in different taxa.

Species Diversity

It refers to the variety of species within a region. Such diversity could be measured on the basis of number of species in a region. Perhaps it is customary to refer the living world in terms of species, hence the term biodiversity is commonly used as a synonym of species diversity. It refers to species richness, in terms of number of species in a site or habitat. Discussion of global diversity is typically represented in terms of global number of species in different taxonomic groups. An estimated 1.5 to 1.7 million species have been identified and described to date. The total existing organisms on the earth at present vary from five to thirty million. A conservative working estimate suggests there might be 12.5 million species.

Ecosystem Diversity

In a ecosystem, there may exist different landforms, each of which supports different and specific vegetation. Ecosystem diversity in contrast to genetic and species diversity is difficult to measure since the boundaries of the communities which constitute the various sub-ecosystems are elusive. Ecosystem diversity could best be understood if one studies the communities in various ecological niches within the given ecosystem, each community is associated with definite complexes. These complexes are related to composition and structure of the biodiversity.

Global Biodiversity

We believed that there may be 5 to 30 million species of organisms exist on our earth. These include 3,00,000 species of green plants, 8,00,000 species of fungi, 40,00,000 species of insects, 3,60,000 species of microorganisms and many invertebrates and vertebrates. According to some recent estimates, the number of insects alone may be as high as 10 million, but many believe that it is more likely to be around 5 million. Global diversity reserves and country-wise distribution of animal species is shown in Table 6.1 respectively.

The tropical forests are regarded as the richest in biodiversity. Scientists are of the opinion that whatever be the absolute numbers, more than half of the species on the earth live in moist tropical forests which is only 7% of the total land surface.

The following reasons have been put forward with regard to the high species diversity in tropics:

- In tropics conditions for evolution are optimum and for extinction are meager.
- In tropics species diversity was conserved over geological time. This is because of low rates of extinction prevailing there.
- Biological diversity is the result of interaction between climate, organisms, topography, parent soil materials and the heredity. The tropical region is the ideal place for such an interaction.

Table 6.1: Countrywise distribution of total number of animal species (upto 10th rank)

Country	Mammals	Country	Birds
Indonesia	515	Columbia	1921
Mexico	449	Peru	1703
Brazil	428	Indonesia	1519
Zaira	409	Ecquador	1447
China	394	Venezuela	1275
India	372	Bolivia	1250
Peru	361	Malaysia	1200
Columbia	359	India	1200
Uganda	311	China	1195
Tanzania	310	Brazil	672
Country	**Reptiles**	**Country**	**Amphibians**
Mexico	719	Brazil	516
Australia	686	Columbia	407
Indonesia	600	Ecquador	358
Brazil	467	Mexico	282
Columbia	383	Indonesia	280
India	353	China	265
Ecquador	345	Peru	251
Peru	297	Zaira	216
Malaysia	294	USA	216
Thailand	242	Venezuala	197
Country	**Butterfiles**	**Country**	**Angiosperms**
Indonesia	121	Brazil	55000
China	104	Columbia	45000
India	77	China	27000
Brazil	74	Mexico	25000
Myannamar	68	Australia	23000
Ecquador	64	S. Africa	21000
Columbia	59	Indonesia	20000
Peru	59	Peru	20000
Malaysia	56	India	15000
Mexico	52	All others	2000

Biodiversity in India

The Indian subcontinent (8° - 30' N and 60° - 97.5' E) having a geographical area of 329 million hectares is quite rich in biodiversity with a sizable percentage of endemic flora and fauna. This richness in biodiversity is due to immense variety of climate and altitudinal conditions coupled with varied ecological habitats. These vary from the humid tropical Western Ghats to the hot desert of Rajasthan, from the cold desert of Ladakh and the Icy mountain of Himalya to the warm coasts of peninsular India.

The country has over 1, 15,000 species of plants and animals already identified and described. In addition, the country is the very important centre of biodiversity and the place of origin of over 167 important cultivated plant species, and some domesticated animals. To name a few the following crops arose in the country and spread throughout the world, namely rice, sugarcane, Asiatic vignas, jute, mango, citrus, banana, several species of millets, cucurbits, some ornamental orchids, several medicinals and aromatic plants. In fact, our country has been recognised as one of world's top 12 megadiversity nations. This region is also a secondary centre of diversity for grain amaranthus, maize, red pepper, soyabean, potatoes and rubber plant.

There are several examples of plant germplasm from India contributed to plant improvement. Two cases are well known:

1. Nearly 20 varieties of rice contain useful genes from wild rices of Kerala, which are responsible for resistance in cultivated rices.
2. The contribution to the cantaloup (musk melon) industry (14,000 ha) of California by powdery mildew-resistant genes is from the wild musk melon of our country.

Among the flora, the country can boast of 45, 000 species which accounts for 15 per cent of the known world plants. Of the 15,000 species of flowering plants, 35 per cent are endemic and located in 26 endemic centers. Among the monocotyledons, out of 588 genera occurring in the country, 22 are strictly endemic. The family Poaceae has the highest endemism both by genera and species. The north eastern region could boast of being unique treasure house of orchids in the country, the abode of about 675 species out of 1,000 available in the Indian peninsula and against 17, 000 species the world over. The important Indian orchids include *Paphiopedilum faireyanum* (Lind.) pfitz., *Cymbidium aloiflium* sw. and *Aerides crispum* Lind., etc.

Our country is also rich in faunal wealth. The country has nearly 75,000 animal species, about 80 per cent are insects. In animals, the rate of endemism in reptiles is 33% and in amphibians it is 62 per cent. Further there is wide diversity in domestic animals, such as buffaloes, goats, sheeps, pigs, poultry, horses, camels and yaks. Domesticated animals too have come from the same cradles of civilization as the major crops.

There are no clear estimates about the marine biota though the coastline is 7,000 km long with a shelf zone of 4,52,460 sq. km and extended economic zone of 20,13,410 sq. km. There is an abundance of seaweeds, fish, crustaceans, molluscs, corals, reptiles and mammals. Information regarding other flora and fauna are fragmentary. Hundreds of new species may be present in out country awaiting discovery. The Western Ghats in penisular India, which extend in the southern states, are a treasure house of species diversity. Out of the described 15,000 species of the flowering plants in India about 5,000 species occur on the Western Ghats of Kerala, 235 are exclusive to this region. It is estimated that almost one-third of the plant varieties found in India have taken refuge in Western Ghats of Kerala alone.

Species Diversity and Ecosystem Stability

Species diversity or diversity of genes within species increases its ability to adapt to adverse environmental conditions. When these varieties of populations of these species are destroyed, the genetic diversity within the species will diminish to a minimum.

Ecosystem stability has been considered to be related to the cycling and recycling of nurtients, which in turn, increases the efficiency of the resource use in the ecosystem. The high species diversity thus may be instrumental in cycling and recycling of nutrients and thereby achieving the stability of the ecosystem. The survival and well being of the present day human population depends on several substances obtained from plants and animals. The nutritional needs of mankind are also met by wild and domesticated animal and plant species, Indeed, the biodiversity in wild and domesticated form, is the source for most of humanity's food, medicine, clothing and housing, much of the cultural diversity and most of the intellectual and spiritual inspiration. It is believed that ¼th of the known global diversity, which might be useful to man in one way or other, is in serious risk of extinction. This calls for an integrated approach for conserving global biodiversity.

Establishment of nature reserves and biospheres with lot of biophysical variability, maintenance of corridors with different nature reserve for the possible migration of the species in response to climate change, etc., are the immediate steps to be taken for conserving the biological diversity. An international consensus on establishing global net-work for gene banks, microbiological resource centers, and marine parks is also important. At the same time conservation must be coupled with socio-economic development, especially in countries where population pressure threatens the national biotic resources.

Measures of Biodiversity

Alpha Biodiversity (within the habitat)

No matter now the data is finally represented, biodiversity eventually boils down to the censusing of individuals. An ornithologist typically walks along a transect and records the number of different birds found in that area. This method of recording is essential in biodiversity assessment. But having recorded the number of species and their abundance, the ornithologist has to use certain methods to be able to compare across different areas. The most commonly used techniques are Diversity indices. There are different approaches to calculate these indices, each with its own assumption and biases, hence care should be taken to determine which index is best suited for the purpose.

Site A and Site B both have the same number of species and the total abundances are also equal. However, Site B is more diverse because the species are more evenly distributed. The diversity index incorporates both the number of species (i.e. Species richness) as well as their abundance. Such an index is called a Heterogeneity Index. The most commonly used index, the Shannon-Wiener index, is derived from information theory.

Some of the diversity indices are explained in this chapter.

1. *Shannon-Wiener Index*

(a) Formula

$H' = -\Sigma_{P/i} \, In_{P/i}$

Where P_i proportion of individuals of the species.

(b) Interpretation

This index assumes that individuals are randomly sampled from an 'Indefinitely large' population. The index also assumes that all species are represented in the sample. It is moderately sensitive to sample size and places more weightage on richness, i.e. if there are rare species recorded, the index tends to give a higher value. The value obtained from a sample is in itself of no significance. The index becomes useful only while comparing two or more sites.

2. Simpson's Index

(a) Formula

Simpson's Index = 1 – D

Where D = Epi^2

and P_i is the proportion of individuals of the ith species.

(b) Interpretation

Simpson's index is heavily weighted towards the most abundant species in the sample and is less sensitive to species richness. It is moderately sensitive to sample size.

3. Fisher's Alpha

(a) Formula

$S = \alpha (1 + N/\alpha)$

Where S is the number of species in a sample

N is the number of individuals.

α, the index of diversity, is computed by an iterative procedure.

(b) Interpretation

The procedure for fitting the model is to calculate the number of species expected in each abundance class and compare that with the number of species actually observed using a goodness of fitness (χ^2 or G test). This index's sensitivity to sample size is low and gives more importance to richness.

4. Brillouin Index:

(a) Formula

$HB=(In\ N!-\Sigma in\ n_il)\ N$

Where N is the total number of individuals and M is the number of individuals of the species.

(b) Interpretation

This index describes a known population, there is no room for uncertainty while using this index. It places more emphasis on species richness and is moderately sensitive to sample size. However, calculating the Brillouin Index can be time consuming and complex and can give misleading answers due to its dependence on sample size.

5. Avalanche Index

This index was recently introduced in an attempt to capture the taxonomic and ecological diversity that is ignored by indices commonly used by ecologists worldwide.

(a) Formula

$$A_i = \sum_{i=1}^{n}\sum_{j=1}^{n} (P_i d_{ij} P_j)$$

Where P_1 and P_j are frequencies of ith and j^{th} species and

$$d_{ij} = \left[\sum_{k=1}^{c} (X_{ik} - X_{jk})^2\right]^{1/2}$$

Where X_{ik} and X_{jk} are the values of the ith and i^{th} species for k^{th} character and c is the number of traits for which information is available.

(b) Interpretation

In addition to using species richness and abundances, this index also uses the biological and ecological differences among species comprising a community. This index requires that the exact taxonomic position of the species in question need be known as its eventual value depends on this feature (Ganeshalah *et al.*, 1997).

Beta Diversity (*between habitat diversity*)

Beta diversity refers to the comparison of diversity between habitats. Beta diversity can be defined as a measure how different (or similar) a range of habitats or samples are in terms of the variety (and sometimes the abundances) of species found in them (Magurran, 1988). One common approach to Beta diversity is to look at how species diversity change along a gradient.

As in the case of Alpha diversity, there are many measures of Beta diversity. These are commonly termed as Similarity indices. Some of these indices are:

1. *Morislta-Horn Index of Similarity*

(a) Formula

$$C_{mh} = \{2\sum a_{nl}.b_{nl})\}/\{(da+db)aN.bN\}$$

Where aN is the total number of individuals in Site A and a_{ni} is the number of individuals in the it i^{th} species in A.

and $da= (\Sigma an_i^2)/aN^2$

(b) Interpretation

This index is not strongly influenced by either species richness or sample size. However, it is highly sensitive to the abundance of the most abundant species. The value obtained from the index ranges from 0 to 1: the value '0' implying that the two samples are totally dissimilar and '1' implying that they are totally similar.

2. *Jaccard's index of Similarity:*

(a) Formula

$$CJ = J/(a + b - J)$$

Where is the number of species round in both sites and a is the number of species in Site A and b is the number of species found in Site B.

(b) Interpretation

This index assumes a value of '1' between sites of complete similarity and '0' for sites of complete dissimilarity. This index is easy to calculate, but the disadvantage is that the coefficients take no account of the abundances of species.

3. *Sorensen's quantitative Index of Similarity*

(a) Formula

$$C_N = 2_{JN}/(aN+bN)$$

Where aN is the total number of individuals in Site A, bN is the total number of individuals in Site B, and jN is the sum of the lower of the two abundances recorded for species found in both sites.

(b) Interpretation

This version of Sorenson's index was modified to take into account the abundances of the species as well. This index also returns a value of '1' for sites of complete similarity and a value of '0' for sites of complete dissimilarity.

More information on the above indices and their measurements can be found in Magurran (1988).

Gamma Diversity

The sum of alpha and beta diversities of the ecosystems is an expression of the biodiversity of a landscape, which is considered as Gamma Diversity (Daniels *et al.* 1996).

The Species-Area Relationship and the Theory of Island Biogeography

There is a good correspondence between the number of species and the extent of area they are found in. Roughly speaking, larger the area, more the number of species found in that area. Mac Arthur and Wilson (1967) took this idea a step forward in their *equilibrium theory of island biogeoraphy*. The main point of this theory is that the number of species on an island is determined by a balance between immigration and extinction, and that this balance is dynamic, with species continuously going to become extinct and being replaced by the same or a different species (Townsend *et al.*, 2000).

While the original theory specified islands, similar work has been carried out in mainland areas where a certain habitat type is enmeshed in a matrix of the other habitat type. An example for a virtual island is the grassland-shola matrix found in South Indian forest ecosystems. (Ganeshaiah *et al.*, 1997).

Hotspots of Biological Diversity

The simplest method of suggesting target areas for conservation action is to identify countries with the highest numbers of species (greatest species richness) and those countries with highest species richness are called as magadiversity countries (Myers, 1988).

Though life teems in every corner of the world, this is not spread uniformly. Certain regions, such as the tropics harbour a multitude of species. Areas where high levels of species richness, threat and endemism coincide are termed as hotspots (Myers, 1988, 1990; Gaston and Williams, 1995, Myers *et al.*, 2000). There are 25 recognised hotspots in the world. These hotspots harbour many rare and endangered species. Two criteria help in defining hotspots namely rich endemism and the degree if threat. To qualify as a hotspots an area must contain at least 0.5 per cent or 1500 of the worlds 3,00,000 plants species as endemics (Myers *et al.*, 2000).

Loss of Biodiversity

The loss of biological diversity is a global crisis. Biological extinction has been a natural phenomenon in geological history. But the rate of extinction was perhaps one species every 1000 years. But man's intervention has hastened the extinction rates. Between 1600 and 1950, the rate of extinction went up to one species every 10 years. Currently it is perhaps one species every year.

The destruction of the world's tropical forests, which are disappearing at an alarming rate, is one of today's most urgent global environmental issues. A rich species diversity is slowly being lost for ever. About 154 mha of tropical forests equivalent to almost three times the size of France have been destroyed. If the current rate of deforestation continues scientists estimate that roughly 5 to 10 per cent of tropical forest species may face extinction, within next one or two decades. The extent and loss of tropical forests in different ecosystems is shown in Table 6.2.

Table 6.2 : Loss of Tropical Forests in Different Ecosystems

Types of forests	During 1990	Annual decline (1981-90)
Rain forests	713790	0.6%
Moist deciduous	519779	0.9%
Hills & Mountain	201417	1.1%
Dry deciduous	178579	0.9%
Very Dry	59742	0.5%
Deserts	8086	0.9%

Rainforests, the home to half of the world's life forms, continue to be destroyed at the rate of over 1,00,000 sq. km every year. This loss of biodiversity has immediate and long term effects on human survival. The majority of the world's population still depends on wild plant and animals for their daily food, medicine, housing and household material, agriculture, fodder, fuelwood and intellectual stimulation.

The loss is even more direct in the case of domesticated biodiversity. Traditional farmers of the world have developed an incredible variety of crops and livestock. This too has been eroded over the last few decades, as lakhs of traditional crop strains and hundreds of domesticated livestock breeds being replaced by a handful of laboratory-generated hybrids or by dominant cash crops.

Estimating the precise rate of loss of biodiversity is a herculean task because no monitoring system, whatsoever, systematic is likely to match the huge diversity in its existing form. Thus much of the basic information is lacking especially in the places like tropics and the hotspots. At least 10% of India's wild flora and fauna are threatened with many on the verge of extinction.

India's biodiversity is one of the most significant in the world. As many as 45,000 species of wild plants and over 77,000 of wild animals have been recorded, which comprise about 6.5 per cent of the world's known wildlife. An assessment of wildlife habitat loss in tropical Asia in 1986 showed that the country had only 6,15,095 sq. km out of its original wildlife habitat of 30,17,009 sq. km i.e. loss of about 20 per cent. In the last few decades India has lost at least half of its forest, polluted over 70 per cent of its water bodies, built on or cultivated much of its grassland, and degraded most of its coasts. Under such circumstances, none can say how many species have already lost.

The country has several problems such as over population, large number of cattle heads, growing demand for land, energy and water supply. Unplanned developmental works and over exploitation of resources have made its living resources most vulnerable. Of the worlds 12 top priority biodiversity hot spots, India has two within its boundaries. Over exploitation has not only resulted in shortage of various materials, but also left our biodiversity exposed to various ecological threats. Over emphasis on timber logging has affected many animal species. Faunal losses have been mainly attributed to over-exploitation of certain species for trading purposes, habitat alteration and destruction, and pollution of streams, lakes and coastal zones.

Wildlife Status Categories

***Vulnerable*:** These are the species whose population numbers are decreasing and are likely to become more severely threatened with time and in near future, they may enter into the category of endangered species, if unfavorable conditions in the environment continue to operate.

***Endangered*:** The species with few individuals because of unfavorable environmental or human factors and that its natural regeneration is not able to keep pace with exploitation or destruction by natural and unnatural means. If the same factors continue to operate unabatedly then soon the species would become extinct, e.g. Indian Rhinoceros, Asiatic lion, Great Indian Bustard and white winged duck of northeast India.

***Rare*:** The species (or taxa) with small world population that are not at present endangered or vulnerable, but are at risk. Such species are usually localized within restricted geographical pockets or habitats or are thinly scattered over a more extensive range. Rare species have a population of less than 20,000 individuals. Some species are naturally rare and have never occurred in greater numbers, yet they are able to maintain these numbers. Other species become rare through man's action or other unnatural forces.

***Extinct*:** Species that are no longer known to exist in the wild but survive under captivity or cultivation. Generally, the term Extinct is used for the species that are no longer known to exist in the wild. e.g. the Cheetah and Lesser Indian Rhinoceros.

Threatened: It is a broader term that is used for species that fall into any of the vulnerable, rare and endangered categories.

Threatened Animals

Though the IUCN in 1988 listed 23 species of mammals as endangered or vulnerable to extinction, 75 species are totally protected as listed under schedule I of the Wildlife (Protection) Act 1972. Forests of the Western Ghat is famous for their endemic fauna and flora. The lion-tailed macaques, the Nilgiri Langur, the Malabar large-spotted civet, the Nilgiri tahr, etc. are some important endangered species. The Hoolock gibbon is the only ape found in the hilly forests of north-eastern India. Among the 19 primate species, 12 are endangered. The Manipur brown-antlered deer (Thamin), once distributed in some parts of the northeast, is perhaps the most threatened species. There are some 50 to 100 of them in the Keibul Lamjao on the bank of Logtak Lake sanctuary of Manipur.

About 427 endangered plant species have been listed by the BSI for the red data book. This contributes about 20 per cent of the flouristic composition of higher plants of India. Some examples are given below.

Eastern Himalaya

Acer laevigatum, A. molle, Anglics anubigena, Bunium nothum, Carum villosum. Pimpinella wallichi, Panax pseudogineseng, Calamus inermis, Phoenix rupicola, Lactuca cooperi, Berberis affinis, Engelhardita wallichiana, Coptis teeta, Aquillaria agallocha. Boehmeria terapensis.

North-eastern Himalaya

Heracleum burmanicum, Peucedanum sikkimensis, Pimpinella evoluta, P. flaccida, Trachelospermum auritum, Ilex embeloides, I. khasiana, I. venulosa, Amorphophallus bulbifer, A. sylvaticus, Dioscoria laurifolia, D. arbiculata, Hopea shingkeng, Musa velutins, Hedychium aurantiacum, H. gracilimum, H. dekianum, H. gratum, H. greenii, H. hookeri, H. marginatum, H. rubrum.

According to Dr. G. M. Oza, President of the Indian Society of Naturalists, there are atleast 30 mammals in India classified under the category of threatened species and our country tops in the list of countries with endangered mammalian species. After India, the US and China jointly occupy the second place with 29 mammals in both the countries on the threatened species list. Brazil, with 25 endangered species ranks third in the list. Further in the oriental region, the country stands first in the threatened mammalian species. Indonesia with 20 threatened species ranks second and Nepal and Thailand, with 14 threatened mammals jointly occupy the third place. Both Burma and Pakistan have 13 threatened species of mammals and Sri Lanka with only four threatened species ranks fifth in the list.

There are 1175 species of the birds, including 42 endemics and 14 of these are considered threatened. N. J. Collar and P. Andrew in their book, Birds to Watch', have identified 70 species of threatened birds, 63 on the mainland, four on the Andaman Islands and thereon the Nicobar Islands. A number of threatened endemic birds are from northeast. The wood-pigeon of the Western Ghats, the white-winged wood duck of northeast India and the forest owlet of the Satpura hills are among the threatened birds. IUCN *Red Data Book* includes Bengal florican and cheer pheasant and Great Indian Bustard as most endangered bird species.

The reptile fauna comprises 435 species including 238 species of snakes, about 30 kinds of turtles and many crocodiles and lizards. The tropical forests of Western Ghats alone support some 20 species of snakes. In addition to their commercial exploitation for skin, the dwindling forest ecosystem also exposes these creatures to danger of extinction. The four monitor lizards-the common Bengal monitor, the water monitor, the desert monitor and the yellow monitor-have been declared as protected. Marine turtles, freshwater tortoises and crocodiles have also been exploited. The green turtle, the loggerhead and the Olive Ridley are the most widely consumed species. All the four species of sea turtles and five of mud turtles have been declared protected. The gharial, the saltwater crocodile and the marsh corcodile (mugger) are exploited for their valuable skins. The three are included in IUCN list of threatened animals. But the population of crocodiles has been increased through breeding centers in Uttar Pradesh, Andhra Pradesh, Tamil Nadu, Orissa and West Bengal.

The amphibian fauna comprises 182 species, which includes salamanders, caecilians, frogs and toads. There is a high degree of endemism and out of 112 endemics, 84 occur in the Western Ghats and 20 in the northeast. Three of the Indian species, namely the Himalayan newt, *Tylatotrition verrucosus*, the Malabar tree-toad, *Pedostibes tuberculosus* and the Garo Hills tree-toad, Pedostibes kempi have been identified as rare endangered species by B. K. Tikader, in his *Threatened Animals of India*. The first mentioned is the only species of salamander known in the country and are distributed in Sikkim, north West Bengal, Arunachal Pradesh and Manipur. The Malabar tree toad was recently rediscovered in the Silent Valley.

The endangered species of mammals, amphibians, reptiles and birds are listed below.

Mammals (Number : 85)

- Andaman wild pig (*Sug Scrofa* and *S. amanensis*)
- Bharal (*Ovis nahura*)
- Binturong (*Arctictis binturong*)
- Bison or gaur or mithun (*Bos gaurus*)
- Black buck (*Antelope cervicapra*)
- Blue whale (*Balaenoptera musculus*)
- Brown antlered deer or Thamin (*Cervus eldi*)
- Capped langur (*Presbytis pileatus*)
- Caracal (*Felis caracal*)
- Cetecen species (all)
- Cheetah (*Acinonyx jubatus*)
- Chinese Pangolin (*Manis pentadactyla*)
- Chinkara or Indian Gazelle (*Gazella gazella bennetti*)
- Chital or spotted deer (*Axis axis*)
- Clouded leaopard (*Neopfelis nebulosa*)
- Crab eating macaque (*Macaca irus umbrosa*)

- Desert cat (*Felis libyca*)
- Dugong (*Dugong dugong*)
- Fishing cat (*Felis viverrina*)
- Flying squirrels – all species of genera : *Petaurista*, Eupetaurus, *Belomys* and *Hylopetes*
- Four horned antelope (*Tetraceros quadricornis*)
- Fresh water or gangaitic dolphin (*Patanista gangaitica*)
- Gaint squirrels (*Ratufa macroura, R. indica. R. bicolor*)
- Golden cat (*Felis temmincki*)
- Golden langur (*Presbytis geei*)
- Gorals (*Nemorhaedus goral, N. hodgsoni*)
- Himalayan black bear (*Selenarctos thibetanus*)
- Himalayan brown bear (*Ursus arctos*)
- Himalayan crestless procupine (*Hystrix hodgsoni*)
- Himalayan ibex (*Capra ibex*)
- Himalayan tahr (*Hemitragus jemlahicus*)
- Hispid hare (*Caprolagus hispidus*)
- Hog badger (*Arctonyx collaris*)
- Hoolock gibbon (*Hylobates hoolock*)
- Humpbacked whale (*Megaptera neavangliae*)
- Hyaena (*Hyaena hyaena*)
- Indian elephant (*Elephas maximums*)
- Indian lion (*Panthera leo persica*)
- Indian wild ass (*Equus hemionus* khur)
- Indian wolf (*Canis lupus pallipes*)
- Kashmir stag or hangul (*Cervus elephus hangul*)
- Leaf Monkey (*Presbytis melalophos*)
- Leopard cat (*Felis bengalensis*)
- Leopard or panther (*Panthera pardus*)
- Lesser or red panda (*Wilurus fulgens*)
- Lion tailed macaca (*Macaca silenus*)
- Loris (*Loris tardigradus*)
- Lynx (*Felis lynx isabellinus*)
- Malabar civet (*Viverra megaspila*)
- Malay or sun bear (*Helarctos malanus*)
- Marbled cat (*Felis marmorate*)

- Markhor (*Capra falconeri*)
- Mouse deer (*Tragulus meminna*)
- Musk deer (*Moschus moschiferus*)
- Nilgai (*Boselaphus trgaocamelus*)
- Nilgiri langur (*Presbytis johni*)
- Niligitha (*Hemitragus hylocriux*)
- Otters (*Lutra lutra, L. perspicilata, Aonyx cynerea*)
- Ovis ammon or nyan (*Ovis ammon hodqsoni*)
- Pallas cat (*Felis manus*)
- Pangolin (*Manis crassicaudata*)
- Pig tailed macaque (*Macaca memestrina*)
- Pygmy hog (*Sus sylvanius*)
- Ratel (*Mellivora carpensis*)
- Red fox (*Vulpes vulpes*)
- Rhinoceros (*Rhinoceros unicornis*)
- Rusty spotted cat (*Felis rubiginosa*)
- Sambar or barking deer (*Cervus unicolor*)
- Serow (*Carpricornis sumatraensis*)
- Sloth bear (*Melursus ursinus*)
- Slow loris (*Nycticebus coucang*)
- Small travencore flying squirrel (*Pteromys fuscocapillus*)
- Snow leopard (*Panthera uncia*)
- Spotted linsang (*Prionodan pardicolor*)
- Swamp deer or gond (*Cervus duvauceli* and other species)
- Takin or mishmitakin (*Budorcas taxicolor*)
- Tibetan antelope or Chiru (*Panthelopus hodgesoni*)
- Tibetan fox (*Vulpus ferrilatus*)
- Tibetan gazelle (*Procapra picticaudata*)
- Tibetan wild ass (*Epuus hemionus* kiang)
- Tiger (*Panthera tigris*)
- Urial or Shapru (*Ovis vignei*)
- Wild buffalo (*Bubalus bubalis*)
- Wild dog or dhole (*Cuon alpinus* and other species)
- Wild yalk (*Bos grunnies*)

Amphibians

- Himalayan newt or Salamander (*Tryletotriton verrucosus*)
- Malabar tree toad (*Pedostibes tuberculosus*)
- Garo hills tree toad (*Pedostibes kempii*)
- Viviparous toad (*Nyctophyrne tuberculosa*)

Reptiles (Number : 20)

- Agra monitor lizard (*Varanus qriseus*)
- Atlantic ridley turtle or Kemps ridley turtle, (*Lepidochelys kempii*)
- Barred oval yellow monitor lizard (*Varanus flavescens*)
- Estuarine crocodile or salt water (*Crocodylus porosus*)
- Fresh water crocodile (*Crocodylus palustris*)
- Gharial (*Gavialis gangeticus*)
- Ganges soft shelled turtle (*Trionyx gangeticus*)
- Green sea turtle (*Chelonia mydas*)
- Hawksbill turtle (*Eretmochelys imbricata inslacata*)
- Indian egg eating snake (*Elachistodon westermanni*)
- Indian soft shelled turtle (*Lissemys punctata punctata*)
- Indian tent turtle (*Kachuga tecta tecta*)
- Large Bengal monitor lizard (*Varanus Bengalensis*)
- Leathery turtle (*Dermochelys coracea*)
- Loggerhead turtle (*Caretta caretta*)
- Olive back loggerhead turtle (*Lepidochelys olivacea*)
- Pythons (All species of genus *python*)
- Three keeled turtle (*Geoemyda tricarinata*)
- Tortoise (All member of the families : *Testudinidae* and *trionovechidae*)
- Water lizard (*Varanus salvator*)

Bird (Number : 38)

- Andaman teal (*Anas gibberifrons albogularis*)
- Assam bamboo Patridge (*Bambusicola fytechii*)
- Bazas (*Aviceda geordoni* and *A. leuphotes*)
- Bengal florican (*Eupodotis bengalensis*)
- Black necked crane (*Grus nigricollis*)
- Blood pheasants (*Ithaqinis cruentus tibetanus, I . cruentus kukseri*)
- Brown headed gull (*Larus brunnicehalus*)

- Cheer pheasant (*Catreus wallichii*)
- Comb duck (*Sarkidiornis melanotos*)
- Forest spotted owlet (*Athene blewiti*)
- Great Indian bustard (*Choriotis nigriceps*)
- Great Indian hornbill (*Buceros bicornis*)
- Hooded crane (*Grus monacha*)
- Hornbills (*Ptilolaemus tickelli, Aceros nipalensis, Rhyticeros undulatus ticehursti*)
- Houbra bustard (*Chlamydotis undulata*)
- Humes bar backed pheasant (*Syrmaticus humiae*)
- Indian Pied hornbill (*Anthracoceros malabaricus*)
- Jerdons courser (*Cursorius bitorquatus*)
- Lammergeier (*Gypaetus barbetus*)
- Large falcons (*Falco peregrinus, F. biarmicus, F. chicquera*)
- Large whistling teal (*Dendrocygna bicolor*)
- Monal pheasants (*Lophophorus impejanus*)
- Mountain quail (*Ophrysia superciliosa*)
- Narcondom hornbill (*Rhyticeros undulatus narcondami*)
- Nicobar megapode (*Megapodius freycinet*)
- Nicobar pigeon (*Caloenas nicobarica pelewensis*)
- Peacock pheasant (*Polyplectron bicalcaratum*)
- Peafowl (*Pavo cristatus*)
- Pink headed duck (*Rhodonessa caryophyllacea*)
- Sclaters monal (*Lophophourus sclateri*)
- Siberian crane (*Grus leucogeranus*)
- Spur fowl (*Gallaperdix* spp.)
- Tibetan snow cock (*Tetraogallus tibetanus*)
- Tragopan pheasants (*Tragopan melanocephalus, T. Blythii, T. satyra, T. temminckii*)
- White bellied sea eagle (*Haliaeetus leucogaster*)
- White eared pheasant (*Croosoptilan crossoptilan*)
- White spoon bill (*Platalea leucorodia*)
- White winged wood duck (*Cairina scutulata*)

The status of many invertebrate species is not known. Lack of data is perhaps why Schedule I of the Wildlife (Protection) Act 1972 includes only a few of the species the Cocount crag of Andaman and Nicobar Islands, the Himalayan dragonfly and 130 species of butterflies. A number of species, beetles and butterflies are also becoming rare or vulnerable to extinction, either due to excessive exploitation or habitat destruction. Two important species of butterflies which are involved in the trade are, Kaiser-I-Hind, the swallow-tail butterfly found along the Eastern Himalayas, and the Bhutan Glory, distributed in Assam, Sikkim, Manipur and Nagaland. Besides the well-known scred chank of India, Trochus and Turbo are the shell fish that deserve special mention and trade controls.

Threatened Plants

Plants of Ornamental Value

- *Aerides crispum* Lindl (Orchidaceae) is an orchid with showy flowers; grows in certain regions of peninsular India.
- *Cymbidium aloiflium* Sw. (Orchidaceae) has not only attractive flowers but is reported to be medicinally useful and hence in great demand.
- *Diplomeris hirsuta* (Orchidaceae). The species is seriously threatened as it grows in parts of Darjeeling and Meghalaya only and is vulnerable to landslides and forest clearance.
- *Pabhiopedilum fairieyanum* (Lindl.). Pfitz. (Orchidaceae) or Lady's Slipper Orchid. A ground orchid which occurs in Arunachal Pradesh has been badly disturbed due to constant clearance and burning of forests for "Jhum cultivation". It is so rare that it was once labelled as the long lost orchid.
- *Qaphiopedilum drurui* (Orchidaceae) an orchid of medicinal value found in Kerala, has become very rare.
- *Rhododendron edgewortythii* Hk, f, (Ericaceae) is a showy epiphyte, reported to have been collected only on three occasions during the last one hundred years.
- *Symplocos chenqapae* Raiz. & Sahni (Symplocaceae) is an ornamental tree with fragrant flowers, confined to Nicobar Islands.

Plants of Medicinal Value

- *Acorus calamus* (Araceae) or 'Safed Bach' . The plant species which grew abundantly in Kashmir , Manipur and Nagaland, only a few years back, is now being imported to meet the medicinal demand of the country.
- *Atropa acuminata* Royle ex Lindl. (solaanaceae) or 'Sag-angur' . This species is a source of some active principles. The plant found to grow in Western Himalayas is rare now. Several species of this genus have also become rare.
- *Dioscorea deltotdea* (Dioscoreaceae) . The tubes of the plant species are used to cure rheumatic illness and for opthalmic problems. The species found to grow in the western Himalayas, Punjab and Kashmir is now facing extinction due to over exploitation .
- *Drosera* species (Droseraceae) . A small insectivorous plant which is not only a botanical curiosity but also famous for its medicinal value . The population of the plant has become rare now due to over-exploitation.
- *Rauwolfia serpentina* Benth . Ex. Kurz (Apocyanaceae) or 'Sarpagandha'. The roots of the plants species are highly valuable in medicine. The alkaloid, reserpine prepared from the roots has its value in treating high blood pressure. The species is nearly at the verge of extinction due to over-exploitation for commercial production of reserpine.

- *Saussurea lappa* Clarke (Asteraccae) or 'Kuth'. Its roots are highly valuable in medicine. The species found to grow in higher altitudes of the Himalayas has become rare.

Plants of Scientific Value

These plants exhibit some morphological interesting features and hence, are having academic interest also.

- *Balanophora involurata* Hk.f . (balanophoraceae) is a fiesty tuberous parasite on roots of trees, occurring in the Himalaya. It has been collected only on a few occasions so far.
- *Dischidia benagalensis* Coleb (Asclepiadaceae) is climber found in eastern Nepal and Sikkim. Leaves develop into pitchers which coollect humus and debris and into these, special branched roots develop from the petiole for absorption.
- *Nepenthes khasiana* Hk. f. (Nepenthaceae) is an insectivorous pitcher plant. The leaf tips of the plant develop into curious pitcher like structures arhced oval by lids. The pitchers are up to about 25 cm long. This plants is a special botanical curiosity. The population of this species, restricted to the khasi hills in Meghalaya, is not enough even to supply educational institutions of our country. The native habitat of the plant is now a gene sanctuary.
- *Sapria himalayana* Griff. (Rafflesiaceae) occurs as root parasite in Mishmi Hills of Arunachal Pradesh. Its flowers resemble the well known Reflesia flower and are up to about 35 cm in diameter.

Plants of Phytogeographic Importance

Most of the Indian endemic plant species show restricted distribution and are rare which need protection:

- *Ceropegia jainii* Ansari & Kulkarni (Asclepiadaceae) is an erect tuberous herb endemic to Ratnagiri in Maharashtra.
- *Frerea indica* Dalz. (Asclepiadaceae) grows in hilly regions of Maharashtra and has showy flowers.
- *Glyphochola mysorensis* (Jain & Hem) Clayton (Poaceae/Graminae) is a grass endemic to Karnataka.
- *Helicanthes danser* (Lorantahceae) is a genus of parasitic plants, confined to the mountain of peninsular India.
- *Jainia nicobarica* Balak (Rubiaceae) is a plant endemic to Andaman & Nicobar Islands.
- *Manisuris divergens* (Hack.) Ktze (Poaceae\Graminae) is a grass probably first collected in peninsular India in 1032. Its second collection was in 1963, after 13 decades. It is confined to Western Ghats.
- *Willisia warm* (Podostemaceae) is a genus of small herbs which grows on rocks in swift-flowing system of Western Ghats.

Trees of Forestry Importance

While most trees of silvicultural importance are carefully looked after by the Forest Departments, some rare species do need protection:

- *Dysoxylum malabaricum* Hedd. (Meliaceae) is a large tree that grows in Western Ghats. Its timber is ornamental.
- *Pterocarpus santalinus* Linn. F. (Fabaceae/Papiolionaceae) or 'Rakta chandan', commonly called Red Sandal, the valuable wavygrained timber. It yields insoluble blood-red dye, santalin, used for dyeing cloth, leather and wood. It grows in certain localised regions in Andhra Pradesh and its protection and expansion is very necessary.
- *Decussocarpus wallichianus* (Presa) de Lauben (Podocarpaceae) is the only conifer occurring wild in peninsular India.
- *Phyllostachys bambusoides* Sieb. & Zucc. (Poacea/Graminae) is common in China and Japan, but reported in India only recently. It is economically important.
- *Pinus gerardiana* wall (Pinaceae) or 'Chilgoza'. The fruits of the tree are used as a delicious dry fruit. Its distribution is restricted to a small region in Western Himalaya.
- *Santalum album* Linn. (Santalaceae) or 'Chandan'. Occurs commonly but its uncontrolled illicit exploitation and attack by spike disease warrant care and protection to the species.

Causes for the Loss of Biodiversity

a. *Destruction of habitat*: The natural habitat may be destroyed by man for his settlement, grazing grounds, agriculture, mining, industries, highway construction, drainage, dam building, etc. As a consequence of this, the species must either adapt to the changes, move elsewhere or may succumb to predation, starvation or disease and eventually die.

b. *Hunting*: Commercially, wild animals are hunted for their products such as hides and skin, tusk, antlers, fur, meat, pharmaceuticals, perfumes, cosmetics and decoration purposes. For example, in Africa, in recent years 95% of the black Rhino population have been exterminated by poachers for their horn. Today, rhino horn fetches more than $ 15,000 in the pharmaceutical market. In the last seven years alone, over one third Africa's elephants have been wiped out for some 3,000 tones of ivory. In the international market the cost of ivory is about $ 150 per kg, whereas in the Indian market it varies between Rs. 2,000 to 2,500 per kg. Carving and sale of African ivory has been banned in India from April 2, 1992. In January, 1987 the Central Government also banned the trade in Indian ivory (Agrawal, 1993).

The scarlet macaw, once common throughout South America, has been eliminated from most of its range in Central America. Several species of spotted cats such as the ocelot and Jaguar have been jeopardized by the demand for their fur. In 1962, nearly 70,000 whales were slaughtered. However, international trade in whale products is banned now.

In our country, rhino is hunted for its horns, tiger for bones and skin, musk deer for musk (medicinal value), elephant for ivory, Gharial and crocodile for their skin, and jackal for thrivin fur trade in Kashmir. According to WWF-Report, the spiny tailed lizard of the Indian desert, locally known as *Sanda*, is being hunted to extract the oil used for making '*Sanda ka tel*' for its so-called aphrodisiac properties.

Officials of TRAFFIC-India say poaching of the Indian tiger has risen because of increasing demand from south-east Asian countries and China, where pharmaceutical factories consume the bones of 100 tigers each year. Smuggling of tiger bones and skine is a lucrative business. One kg of tiger bones fetches $ 90 in India and $ 300 in the international market. Hunting for sport is also a factor for loss of animal biodiversity.

c. ***Over exploitation*** : This is one of the main cause of the loss of not only economic species, but also biological curiosities like the insectivorous and primitive species and other taxa needed for teaching or laboratory work (like Nepenthes, Gnetum, Psilotum, etc). Commercial exploitation of biodiversity has invariably meant its overuse and eventual destruction. This was true in the case of whales, into plywood as of the whales, that were hunted for tallow, in the oceans.

Plants of medicinal value like *Podophyllum hexandrurm*, *Coptis teeta*, *Aconitum*, *Discorea deltoidea*, *Rauwolifia serpentina*, *Qaphiopedilum druri*, etc., and horticultural plants like orchids and rhododendrons come under the over-exploitation. For instance, excessively harvesting of marine organisms such as fish, molluscs, sea-cows and sea-turtles has resulted in reduction of these animals.

d. ***Collection for Zoo and Research***: Animals and plants are collected throughout the world for zoos and biological laboratories for study and research in science and medicine. For example, primates such as monkeys and chimpanzees are sacrificed for research as they have anatomical, genetic and physiological similarities to human beings.

e. ***Introduction of Alian Species***: Native species are subjected to competition for food and space due to the introduction of exotic species. For example, introduction of goats and rabbits in the Pacific and Indian regions has resulted in destruction of habitats of several plants, birds and reptiles.

f. ***Control of Pesta and Predators***: Predator and pest control measures, generally kill predators that are a component of balanced ecosystem and may also indiscriminately poison non target species.

g. ***Pollution***: Pollution alters the natural habitat. Water pollution especially injurious to the biotic components of estuary and coastal ecosystem. Toxic wastes entering the water bodies disturb the food chain, and so to the aquatic ecosystems. Insecticides, pesticides, sulphur and nitrogen oxides, acid rain, ozone depletion and global warming too, affect adversely the plant and animal species. The impact of coastal pollution is also very important. It is seen that coral reefs are being threatened by pollution from industrialization oil transport and offshore mining along the coastal areas.

Noise pollution is also the cause of wildlife extinction. This has been evidenced by the Canadian Wildlife Protection Fund. According to a study Arctic Whales are seen on the verge of extinction as a result of increasing noise of ships, particularly ice breakers and tankers.

h. ***Deforestation***: One of the main causes for the loss of biodiversity is population explosion and resultant deforestation. Deforestation mainly results from population settlement, shifting cultivation, development projects, demand for fuelwood, demand of wood for industry and other commercial purpose.

In India, the rate of deforestation is 13,000 sq. km. annually. If this rate of deforestation continues, one can imagine the ultimate fate of our forest and biological richness. It is presumed that in coming years, the global loss of biodiversity from deforestation alone would be 100 species everyday.

i. ***Other Factors***: Other ecological factors that may also contribute to the extinction of plant and animal diversity are as follows:

1. Distribution range. The smaller the range of distribution, the greater the threat of extinction.
2. Degree of specialization. The more specialised an organism is, the more vulnerable it is to extinction.
3. Position of the organism in the food chain. The higher the organism in food chain, the more susceptible it becomes.
4. Reproductive rate. Large organisms tend to produce fewer off springs at widely intervals.

Socio-economic and Political Causes

The socio-economic and political causes of biodiversity loss vary from region to region. In recent times, they can be linked to governmental and international support for industry, forestry, agriculture and energy programmes. The promotion of subsidy for big mono-cultures of soyabeans did not make place to feed hungry Brazilians, but to feed fat cows here in the common market, cows that are producing what even the technocrats of the European market call seas of milk and mountains of butter. It was this policy of promoting cash crops in the south that destroyed one of the most well adapted, healthy and ecologically safe peasant cultures that it was in the south of Brazil.

In south-East Asia, it is the tropical timber industry that is cutting down the last primary rainforests, against the will of local people (Nectoux and Kuroda, 1989; Colchester, 1989; Hong, 1986). The role of western culture, i. e. use and throw-away culture that causes the liquidation of natural resources for short term profit, is also one of the important causes of deforestation, and thereby the loss of biodiversity in industrialized countries.

The introduction of exotic species of plants and animals causes major disturbances to ecosystems because the checks and balances that hold the equilibrium in the native ecosystem do not apply to the introduced species. In developing countries like India, the development policies and projects have rarely been sensitive to the need for biodiversity conservation, and that of the local communities.

Biotechnology, Biodiversity and IPR

In the last twenty years, all over the world, there has been a revolution in the field of biotechnology. New discoveries and inventions in the area of isolation and manipulation of genes, better understanding of biological molecules and the advent of recombinant DNA technique have increased our knowledge. Biotechnologists all over the world have made efforts to create transgenic crops which will withstand the pests as also have enough resistance to withstand environmental stress. In fact, biotechnology is inherently knowledge-intensive and having strong infrastructure would led to value area of agriculture, animal husbandry, fisheries, forestry and medicine.

Biotechnology is not a miracle solution to the problem of biodiversity crisis. Rather, the use of biotechnology in the production of uniformity in plants and animals has threatened not only the life forms but also rendered entire community or ecosystem unstable. Further, indiscriminate and unregulated uses of genetically modified organisms pose a threat to mankind. In fact, uniformity (or homogeneity) in life forms accelerates the loss of biodiversity. The institutional structure that controls the biotechnology, therefore, should not over shadow those institutions that deal with conservation of Biodiversity.

Productivity and Diversity

Productivity goes against diversity as it creates imperative for uniformity and homogenization. This has generated the paradoxical situation in which modern plant improvement is based of the logic of uniformity and homogenization. Green revolution, for instance is based on high productivity and low biodiversity. There is no need to combine high productivity and high genetic diversity to enhance yield as well as to provide insulation against environmental stress and pollutants. Over the last few decades lakhs of traditional crop strains and hundreds of domesticated livestock breeds have been replaced by a handful of laboratory-generated hybrids or dominant cash crops. Similarly, forestry schemes introduced monocultures of commercial species like teak, eucalyptus and bamboo, and pushed into extinction the diversity of local species. Agriculture modernization fisheries, commercial forestry and animal husbandry thus produce uniform crops and domesticated livestock's and destroy the diversity of local species which fulfil local needs. Such a strategy of productivity increase based on the logic of destruction of biodiversity is no longer desirable as it will ultimately lead to loss of biodiversity.

Politics of Patents and Intellectual Property Rights

Biotechnological processes use life forms of derivatives therefore to make or modify products or processes for specific use. Under Intellectual Property Rights (IPRs), transformed microorganisms, plants and animals can be patented and become exclusive private property. The north has always used third world germplasm as a freely available resource and modified it. The issue of patent protection for modified life forms raises a number of unresolved political question about the ownership and control of the genetic resource. By simply manipulating the life forms one does not acquires the patent or property right, because the modified life forms do not arise from nothing but from existing life-forms which belonging to others. Also biotechnology does not create new genes, but merely relocates genes already existing in the organism. The advanced capitalist nations wish to retain free access to the developing world's storehouse of genetic diversity, while the south like to have the proprietary varieties of the worth's industry declared a similarly public good. The north, however, resists this democracy. US has freely taken the biological diversity of third world to earn millions of dollars, none of which have been shared with third world countries, the original owners of the biological resource. For instance, an American Industry earned $8 million a year in 1962 simply by increasing the soluble solid contents of a wild tomato variety, *Lycopersicum chomrelewskii* taken from Peru. None of these profits was shared with Peru, the original contributor of the genetic material. The Convention on Biological Diversity is also not clear on this. Industrialized countries, particularly the US interpreted key clauses of the treaty in a manner that would protect the interest of its own biotechnology industries. This is a clear set-back to the developing countries, who stand to lose the benefits due to them.

In absence of a proper biotech base, a developing country cannot match an industrial country although the former may be far richer in Biodiversity. However, the Convention on Biodiversity helped to place the subject matter of technology transfer and IPRs on the top of the agenda of policy and decision makers. Furthermore, access to genetic resources and transfer of technology is treated on the same plan. On the issue of IPRs, the basic requirement of the Dunkel proposals is that inventions in all branches of technology shall be patentable, whether products or processes, if they meet the three tests of being new, involving innovative steps and being capable of industrial applications. It has also been provided that microorganisms will be patentable. In respect of plant varieties there is a separate obligation to provide them protection by patents or by an effective sui-generis system, or by a combination of the two. Sui-generis protection implies a system different from other categories of intellectual property protection (such as patents) and is a class by itself. Dunkel text, thus does not compel to patent seeds (i.e. plant varieties). So far we are concerned, seeds are also not patentable in India today, and we do not have any intention of changing this system. However we will adopt our own system for the protection of plant varieties under which we may provide certificates for plant breeders right. The farmers rights include their using the seeds for their own needs or for exchange in the village community according to their traditional custom. Since farmers right will be fully safeguarded under system of protecting the plant breeders right, there is no truth in the allegation that the farmers will not be able to retain the seeds for their own use and that they will have to buy seeds every year from multinational companies. Furthermore, India is not in favour of the patenting naturally occurring life forms/germplasm.

The extension of IPRs to plant varieties either in the form of patents or in the form of plant breeders rights is bound to result in increased in prices of seeds, greater domination of agriculture by multinational companies and slower diffusion of new varieties. These would be in sharp contrast to the experience of the green revolution where the new varieties of seeds evolved by the government institutions percolated down to the fields in a short span of time with very little cost to the individual farmer. Our farmers will have to face great hardship due to the new regime. India calls for the removal of distortions in the IPRs regimes in areas related to prior existence of knowledge in nutrition. Mr. Kamal Nath, late Environment and Forest Minister, referred to the harsh fact that IPRs regimes in these areas resulted in a virtual denial of benefit flows of financial return and markets to those very communities who had by their sustainable life styles preserved these systems and the natural resources on which they were based. It is said that the market value of allopathic medicines derived from plants used in traditional remedies was calculated to be over $ 43 billion annually. Less than 0.01 percent of profits had gone to the indigenous people who had led the researchers to them.

The south had not merely preserved much of its precious biogenetic resources, but also the knowledge and practices about their optimum and sustainable utilisation. Access to these resources have to be regulated and careful exercise, in keeping with the objectives of the convention and with due compensation to such people who have preserved their resources.

How to recognize and measure the value of indigenous knowledge, is one of the basic problem in deciding the compensation and for protection of farmers IPRs. As a result of the persistent North-South split, the CSD could able to move forward on this contentious issue. However, the decision of knowledge, innovations and practices of indigenous and local communities is an important step in the direction of the protection of traditional knowledge and practices of the indigenous and local communities relevant to conservation and sustainable use of biological diversity.

Biosafety Protocol

India has demanded a clear comprehensive and equally binding international protocol on bio-safety under the convention on biodiversity. Addressing the first meeting of parties to the Convention on Biodiversity, at Nassau, Bahamas, on December 8, 1994 Mr. Kamal Nath demanded immediate and adequate safeguards against hasty experimentation and use of genetically modified organisms, since these have unimaginable repercussions.

7

Action Plan for National Biodiversity Strategy

The Need

India harbours have enormous diversity of plants and animals, both domesticated and wild, and an impressive variety of habitats and ecosystems. This diversity also ensures the essential ecological functions on which life depends, including a steady supply of clean water, nutrient cycling, and soil maintenance. This diversity is the treasure-house from which future food needs, cures for deadly diseases, and elements for knowledge and technology will be researched.

The diversity is seriously threatened by human activities leading to severe shortages of water, fodder, fuel and natural disasters such as floods, landslides and famines. There is a need to prepare detailed action plans at state, regional, and sub-state levels to conserve and sustainably use biodiversity to protect the cultural, scientific, spiritual traditions and innovations related to it, and to equitably share the benefits arising from sustainable use. The participation of women and men from all walks of life, both within and outside the government, rural and urban, is essential for preparing an implementable action plan.

In the year 1999, the Ministry of Environment and Forests prepared a National Policy and Macro level action strategy on biodiversity through a consultative process. That document was a macro level statement of policies, gaps and strategies needed for conservation and sustainable use of biological diversity. There is a need to prepare detailed action plans at sub state, state, regional and national levels based on this frame work of the document. Towards this end, the Ministry has accessed funding from the global environment facility (GEF) for preparing the National Biodiversity Strategy and Action Plan (NBSAP).

The NBSAP project envisages the assessment and stock taking of biodiversity-related information at various levels, including distribution of endemic and endangered species and site-specific threats and pressures. Key features of this project include an emphasis on gender sensitive decentralized planning, and the use of interdisciplinary working groups to involve all sectors concerned with biodiversity conservation.

Government agencies, non-governmental groups and village communities are already contributing towards conservation of biodiversity. Such a process of consolidation and new initiatives through a series of plans and strategies at local, state and national levels, cannot be successful without public participation. It is proposed to prepare the NBSAP through a process of widespread consultation and participation across India.

Introduction

The national biodiversity strategy and action plan is proposed to be developed through a participatory planning process involving all major stakeholders, as it is stressed in the original project document signed by the Government of India and UNDP. To enable this, it is necessary to develop a clear outline of how this process is to be carried out. This document is presented in following ten sections:

1. Describing the approach underlying the process (Section 1);
2. Stating what is strategy and action plan (SAP) (Section 2);
3. Stating the scope of the process (Section 3);
4. Identifying the specific outputs expected from the NBSAP process (Section 4);
5. Clarifying the overall institutional structure of the project (Section 5);
6. Delineating the specific steps and methods by which the process of preparing a NBSAP will be carried out (Section 6);
7. Laying out the activities and responsibilities of the participating institutions (Section 7);
8. Identifying the broad themes and topics that will form the content of NBSAP, as also some indicative sub themes within these broad themes (Section 8);
9. Identifying some major concerns for co ordinating teams to keep in mind (Section 9);
10. Listing the immediate, medium and long term activities of the technical and policy core group (Section 10).

Approach Statement

Biodiversity forms the basis for the continuous evolution of species. Diversity also forms the back bone of human societies and cultures, in terms of ecological functions it provides, and the myriad survival and livelihood needs it meets. A national biodiversity action plan should therefore be able to protect these critical characteristics and functions. It is proposed that the National Biodiversity Strategy and Action Plan (NBSAP) will take the following as its basic tenets.

a. Scope

The term "biodiversity" is being taken in its holistic sense, to encompass all levels of biodiversity, ecological and evolutionary processes.

1. Natural ecosystems: e.g. forests, grasslands, wetlands, deserts, mountains, coastal and marine areas, including the historical taking place in such ecosystems.
2. Wild species and varieties: species of plants, animals, and microorganisms existing in their natural state, and the genetic variation within each of these species.
3. Agricultural ecosystems: e.g. farmlands, pastures, capture fisheries, aquaculture, including historical changes in land use patterns.
4. Domesticated species and varieties: species of crops, livestock (including poultry), captive-bred fish, pets, and micro-organisms in ex-situ collections, and the genetic variation within each of these species.

b. Aspects of Coverage

It embodies the following aspects:

1. Conservation of biodiversity at ecosystemic, species and genetic levels (through in-situ, ex-situ, in-vitro and other measures), and of the associated ecological functions, with a special focus on endemic, threatened, and otherwise critical elements.
2. Sustainable use of biological resources, implying their use in such a manner as will not imperil their long term existence, or will not in other ways threaten biodiversity.
3. Social, economic, political and inter generational equity, including decision making with regard to conservation and use, in sharing of benefits, in access to and protection of knowledge and information relevant to biodiversity and so on.
4. Ethical, cultural, scientific and economic dimensions, including the right of species and ecosystems to survive; the primacy of survival and livelihoods based on biodiversity and of creating a stake for people in conservation and sustainable use and the implications of research and use of human genetic diversity. It also includes various cultural ways of relating to nature and the role of biodiversity in economics and technology.

c. Process of Preparing the NBSAP

National level policy process in the past have often been limited to a small number of experts and consultants and as a result often end up in remaining on paper. It is proposed that the NBSAP will be highly participatory in nature, reaching out to a large number of village level organisations and movements, NGOs, academics, scientists, government officers from various line agencies, the private sector, the armed forces, politicians, and others who have a stake in biodiversity. The NBSAP should be seen as much as a process as a final product, and if a truly participatory approach is achieved, it will evoke much wider ownership of the process and product, and go a long way in ensuring the implementation of what is recommended in the final NBSAP. It is also important that this should be seen as a truly national process that is going on at sub state, and national levels.

In particular, the following actions will be most critical:

1. Workshops and public hearings at the level of some selected districts and talukas, seeking input from farmers, pastoralists, fisherfolk, and other user communities who are directly dependent on or traditionally relating to biodiversity.
2. Coordination and consultative meetings amongst a variety of agencies of government at central and state levels.
3. Consultations with political decision makers at various levels.
4. Seeking inputs from a wide range of expert and experienced individuals and organisations working on various aspects of biodiversity, including scientists, social activists, academicians, students, industrialists, cultural leaders, politicians, etc.
5. Maximizing the use of local languages in all of the above.

d. What is a strategy and action plan?

Strategy and Action Plan (SAP) should consist of the following:

1. Identification of ongoing initiatives regarding this issue.
2. Identification of key factors involved.
3. Identification of major gaps in coverage.
4. Delineation of strategies needed to plug gaps and enhance the effectivity of ongoing initiatives.
5. List of measures and actions (short and long terms) needed to implement these strategies, with prioritisation in terms of importance and immediacy.
6. Identification of key elements needed for implementation of institutional structures, funds, expertise/human resources, policy/legal measures, monitoring etc.
7. Time frame for implementation.

e. Scope of NBSAP Process

Though the ultimate aim of the NBSAP process is the production of a comprehensive, ready-to-act-on national action plan, this will be the end result of a group-on planning process. An attempt will be made to stimulate and facilitate the preparation of strategy and action plans (SAP) for the following:

1. Some representative or unique districts or sub state regions within each state.
2. Each state and union territory of India (to be prepared by State Steering Committees).
3. Eco-regions that cut across states (to be prepared by Eco-regional Working Groups).
4. Some broad cross cutting themes.

This type of exercise will ensure that the rich details that would arise in the process, for geographical units and themes, will not get lost in the national SAP, which will necessarily be a broad level document putting together the common and unique elements of the above. In addition, prioritisation of species, areas, and measures for conservation and sustainable/equitable use, can be done at various levels, so that national and local priorities are harmonized. The stress will be on laying out both what is desirable in terms of strategies and the actions to achieve this.

f. Outputs of NBSAP Process

The NBSAP process should aim at producing not just a national level action plan, but a series of other action plans at local, state, inter state and thematic levels. This is because the national level plan cannot possibly incorporate all the rich details that will be produced in the process and each theme and region requires a detailed plan in itself. These plans will also contain, integrated or attached, project proposals for further action (including for further funding and other support). The NBSAP process will, therefore, over the next two year (2000-2002), result in the production of the following outcome:

1. Local/regional (sub state) strategy and action plans.
2. State level strategy and action plans.
3. Inter state, eco-regional strategy and action plans.
4. National strategy and action plans.
5. Separate, detailed thematic strategy and action plans.

Several brief review papers on specific topics, to be integrated into relevant action plans and/or brought out as independent papers. Detailed project proposals for action (including funding support), integrated into the above or attached with them.

g. Step by Step Process of NBSAP Preparation

January-June 2000

1. Draft outline of the process.
2. First meeting of the TPCG, NPD, etc.
3. Finalisation of the outline and budged re-allocation.
4. Presentation of outline to the SC.
5. Revision of outline, if necessary.
6. Finalisation of Approach Statement and Announcement Brochure.
7. Selection of nodal agencies and coordinators for sub-state, state, thematic, and eco-regional processes.
8. Formal launch.
9. Intimation to state government and other key actors/partners.
10. Issuing of advertisements, letters, announcements and using a media campaign, seeking widespread public participation.
11. Assessment of available data/information, expertise, processes, etc., at national level, to priorities further actions and fund allocations.
12. First national workshop, to present outline, with key participants.
13. Formation of thematic and inter-state eco-regional working groups.
14. First meeting at state level, and formation of state steering committees.
15. First meetings at sub-state level, and formation of local advisory committees.
16. Assessment of available data/information, expertise, processes, etc., at sub-state, state and national levels.
17. First meetings of thematic and inter-state eco-regional working groups (May/June).
18. Round of local/sub-state level workshops, public hearings, etc.
19. Training/capacity-building exercises with SSCs, LACs and WGs on methodology, themes etc.

July 2000-March 2001

20. Formulation of draft SAPs by sub-state, state, eco-regional, and thematic groups (including necessary workshops, literature reviews, fresh assessments, capacity-building exercises).
21. Dissemination of all SAPs to relevant people/agencies, including in local languages.
22. Issuing of public notices about the availability of draft SAPs.
23. National Biodiversity Mela.

April-June 2001

24. Evaluation of sub-state and state level SAPs through workshops.
25. Revision of sub-state, state, eco-regional, and thematic SAPs.
26. Incorporation of above, along with other reviews/assessments, into national SAP.

g. Broad Activities for each Executing Agency and Working Group

Each working group, state-level nodal agency, etc. will have to undertake the following in relation to their stated scope (theme or geographical unit):

1. Assessment of existing reports, action plans.
2. Identification of available information and data.
3. Prioritisation of further steps based on above, to focus on major gaps, build on existing knowledge and available plans/reports, and avoid duplication.
4. Identification of available expertise and experiences, both in formal and informal sectors.
5. Soliciting of inputs from a wide range of individuals/agencies, through:
   a. letters,
   b. public meetings and workshops,
   c. advertisements,
   d. print and electronic media,
   e. folk media.
6. Capacity-building exercise where relevant (e.g. for the local/sub-state/state level participants, on authentication of data, monitoring, etc.)
7. Assessment of all relevant sectorial plans and policies.

Note: It is critical that in all these activities, there be maximum participation of all sectors (governmental agencies, NGOs, local communities, independent experts, private sector, armed forces, politicians, etc.) especially through the following:

1. Making the process of working fully transparent.
2. Inviting public inputs at every step.
3. Making all relevant information available to the public.
4. Using local langauges in all key documents and events.
5. Respecting the output of lower level (e.g. sub-state) SAPs and information, and integrating them into higher level (e.g. state and national) SAPs.
6. Allowing for a diversity of opinions and approaches to be reflected in the process and in the final SAPs.

h. Broad Themes for NBSAP Working Groups

The NBSAP process will focus on some broad thematic areas that cut across national and state level process, as also commission some reviews on specific sub-themes within these broad themes. Across all the thematic areas, as also in the work of all executing agencies of the NBSAP, the following approaches will be centrally in targeted:

1. People's participation and empowerment to take part in decision-making at all levels, from planning to implementation to monitoring, including the use of indigenous knowledge and practices.
2. Gender and equity issues.
3. Participation of the corporate sector, private and public.
4. International issues, including trans-boundary cooperation for conservation and benefit sharing.

i. Major Concerns

1. How to meaningfully involve other ministries/departments at the centre and in the states. [Centrally involve the Planning Commission, and State planning bodies; suggest possibility of future funds for NBSAP follow-up; link EIA/clearance procedures to integration of biodiversity into sectorial plans];
2. How to ensure participatory process at state level? [Careful choice of steering committee and nodal agencies, insistence on involving all sectors];
3. How to get maximum involvement of local people, given difficulties of language, reach, logistics? [Translate all key documents, ask grassroot groups to mediate, carry out a widespread process of identifying key local actors; encourage/facilitate/input into, level workshop and public hearings];
4. How to judge reliability and authenticity of data/information? [Ask all working Groups and SCs to focus on this];
5. How to reconcile fundamental differences in goals, approaches and strategies? [Maximize possibility of dialogue, clearly state genuine differences];
6. How to ensure that an effective follow-up mechanism is put into place, and that the NBSAP and other SAPs do not gather dust in godowns? [Discuss follow-up with all participants from the start, integrate some solid action proposals into the SAPs attempt acceptance of SAPs into state and national plans and budgets; strengthen people's mobilization processes through NBSAP; build up large-scale support for plan and follow-up processes].

Acronyms/Abbreviations used in this chapter

BCIL	:	Biotech Consortium India Limited
D/LAC	:	District/Local Advisory Committee
EWG	:	Interstate Eco-regional Working Group
IESAP	:	Interstate Eco-regional Strategy and Action Plan
LSAP	:	Local/Substate Strategy and Action Plan
NBSAP/NSAP	:	National Biodiversity Strategy and Action Plan

NGO	:	Non Governmental Organization
NPD	:	National Project Director
PP	:	Project Proposal
SAP	:	Strategy and Action Plan
SC	:	Steering Committee
SSAP	:	State Strategy and Action Plan
SSC	:	State Steering Committee
TPCG	:	Technical and Policy Core Group
TSAP	:	Thematic Strategy and Action Plan
TWG	:	Thematic Working Group

8
Gene Bank Conservation

Introduction

Development in biotechnology, notably in recombinant DNA, is limited by the availability of genetic material. Further, the emergence of biotechnology as a powerful tool of change in the world makes the preservation of programmes. Gene banking provides a method of conservation of genetic diversity, in addition to in-situ and ex-situ conservation methods. In fact, it is not a new concept, but simply long term storage of seeds at sub-freezing temperatures as gene banks, which are the means to ensure and secure rare varieties. The National Bureau of Plant Genetic Resources, New Delhi, has already developed cold gene storage models, where seeds can be used for maintaining genetic diversity. To preserve the diverse of genetic resources, particularly of threatened species and those the seeds of which are not viable for longer periods, banking is one the suitable methods for conservation. This calls for an international consensus on establishing global network of gene banks.

Soil is a natural gene bank to which seeds are added year after year. When a wild plant produces seeds, not all the seeds will grow the following season. Much of the seeds drop to the ground and wait there patiently for right conditions to permit germination. In fact man made gene bank is a kind of freezer that preserves seeds, pollen and spores. Chest freezers are the best choice for small gene banks. Upright freezers, on the other hand are not suitable as the cold air rushes out when the door is opened. Chest freezers with a temperature of -15°C are adequate for this purpose. Operating and maintenance of the equipment does not required heavy investment. This simplicity of a gene bank makes it a viable project for plant enthusiasts, ecologists and NGOs who are interested in biodiversity conservation.

Theory of Gene Bank Preservation

The idea behind gene banks is really quite simple i.e. seeds of many plant species remain viable longer when their moisture is reduced and stored at low temperature. Cryogenic preservation is no more than suspended animation at sub-freezing temperatures and low moisture content. The process requires little more than a household chest freezer, storage containers made of glass, metal or plastic, a simple desiccant or drying agent to dehydrate the seeds.

Seeds are tiny organisms that stay alive so long as there is a food reserve inside the seed to change vital chemical reactions that keep the seeds alive. These chemical reaction rates are dependent on the ambient temperature and the water content of the seed. The higher the temperature, the faster the reactions will take place and the less time the food reserves will last. The converse is also true, i.e. for every 5°C drop in temperature, the life of the seed will double. For example, onion seeds with 10 per cent moisture are viable for 16 weeks at 35°C but will live 78 years at 0°C. Dropping the temperature down to 15°C would increase longevity to 624 years, provided the ice crystals formed at 10 per cent moisture level do not harm the seeds when frozen. Clearly, if seeds could be stored at sub-freezing temperatures, they might be viable for centuries. Cryogenic theory, thus, suggests that extremely low temperatures allow seeds to remain viable for many centuries, but such tremendous life-spans practically may not necessary to be achieved. Too much water in a seed will form ice crystals as the temperature drops below the freezing level. Fleshy or oily seeds cannot easily be frozen and must be preserved by some other methods.

Normally, the life-span of a seed doubles with each 1 per cent decrease in water content. A minimum of 4 per cent water content appears to be required by seeds to stay alive. International standards for long term seed storage suggest that an average seed should contain five per cent water and be stored at 18°C. When both temperature and water content are reduced, the two factors multiply. For instance, when temperature drops 5°C water decreases by 1 %, the seed lives four times long.

Gene Banking

It is accomplished in the following steps :

Collection and Drying of Seed Sample

Before the collection of seeds, first step is to ascertain the viability of seeds as well as the size of seed sample. A variety of stains and tests can be used. The simplest method is to determine the percentage of seed germination. Botanists usually strive for about 95% germination. Dye, 2, 3, 5-triphenyltetrazolium chloride is the best stain for viability test of seeds. Viable seeds respire and thus are capable of changing colourless tetrazolium dye into highly coloured (red) compound by chemical reduction.

The species which have low rate of germination are the endangered species. The size of the seed sample depends upon its use. When sample is required to conserve the whole germplasm, that is full range of variation within the population, then the sample should contain roughly 10,000 seeds. On the contrary, a small sample of 20 to 100 seeds might be sufficient, if it is required to retrieve a few plants for illustrative specimens, breeding experiments or genetic research. The Gene banker is advised to have the seed samples in duplicate or even in triplicate to facilitate the easy removal of desired amount of seeds from the freezer whenever is needed.

Store clean seeds (without chaff and stalks) only and not be capsules to save the limited space available in the freezer and make it easy to inspect and observe. For collection of seeds harvest the pods as soon as they start to ripen or split. Put the seeds in a container, and expose them to air for few weeks. The seeds held in the gene bank at the Royal Botanic Garden Kew, are dried in a conditioning room with a temperature of 16°C and a relative humidity of 14 per cent.

Seeds can be dried simply by storing them for several days in a closed container with a desiccant/drying agent. The most familiar drying agent is silica gel. Another product which readily absorbs moisture from the air is Drierite, a calcium salt. Drierite comes with a colour indicator, when it turns blue, it can still absorb moisture. It is important to test the drying agent before using it to dry seed. A handy test for assessing its freshness is to use a strip of filter paper saturated in a solution of cobalt chloride. The paper strip is blue when dry, and pink as it becomes moist. When you find that the drying agent is stale, rejuvenate it by baking it on a Teflon-coated baking sheet for two or three hours at 60°C or above.

When the drying agent is ready to use, find any container that can be sealed. About 2.5 cm of drying should be placed on the bottom of the container and covered with either cheese cloth or wire gauge, permitting air to reach the drying agent. Open containers holding seeds should be put into the desiccator and the desiccator sealed. Open it at least once a day to stir the seeds. During the first two days seeds will lose most of their water content and a little more is lost during the next five or six days. This is the most practical method of drying seeds for gene bank.

Higher temperatures 35°C or above accelerate the metabolic reactions and so affect the vitality of seeds. It is therefore, desirable to dry the seeds at gentle room temperature. Freeze drying under vacuum has been considered the most suitable technique for drying seed sample. However, it requires extra care as a result of formation of ice crystals at freezing temperature and strong vacuum used to dry the material.

Storing

Containers made up of metal, glass and plastic can be used for storing purposes. Large samples of big seeds, such as beans and peas, store very well in metal containers. Glass containers, the second choice, are more fragile but this method is relatively cheap. In-expensive disposable glass test tubes are easy to get from scientific or medical supply houses. End of the tube can be sealed with a propane torch, without significantly heating up the seed sample. With glass container it is easy to view the frozen sample. The third type of containers are envelopes made of foil-laminated plastic that can be heat-sealed. These envelopes have an inner lining of polyethylene and outer lining of aluminium or some other metal. While sealing the vacuum device can be applied.

Labeling

Care should be taken to use labels that must resist sub-freezing temperatures, for long periods of time. It is advised to use ink that will not fade soon. Further dual system of labels is preferred. One label should be placed on the outside of the sample and another should be placed inside. As an extra precaution, a record of the sample's self position should be maintained in a book or computer. The name of the sample should be printed on a metal foil label. There is also an idea to number the samples and to have the permanent record of identification. Records should be precise and document the quality and viability of the seeds stored.

Pollen and Spore Banks

Preservation of pollen and spores is of significant value for conservation of biodiversity of important flowering and spore bearing plants. The procedure for institution of pollen and spore banks is almost similar to that of gene banks. Cryogenic technique is useful in preserving pollen from flowering or cone bearing plants, and spores from non-flowering plants, such as ferns and mosses. Pollen are so tiny that thousands and even millions can be stored in a very small vial. Pollen preservation is thus advantageous over seed preservation as it gives opportunity to preserve the full range of variation within the population in a very simple manner. However, plant scientists are able to make use of pollen from certain plant species only to grow them into whole plants. We expect that development in science and technology, in near future, will solve this problem. Pollen can also be used in hybridization programme whenever needed. When seeds are used to develop hybrids they must be retrieved from the bank and then planted. It takes several years for the plant to be useful in hybridization. On the other hand, pollen bank can be an extremely powerful tool in plant breeding since it frees breeders from the tyranny of time. Also, it is useful in self-sterilised plant species. Pollen should be collected from fresh flowers only. Fern spores, which are similar in size to pollen, can be treated just like pollen. Even the spores of tropical ferns like staghorns, *Platycerium* species, can be germinated after drying and freezing. Pollen are stored in gelatin capsules, the type used to hold medicine. The stamen, which contains pollen, is removed from the flower, inserted into the capsule and shaken several times to deposit the dust like pollen on the capsule wall. The stamen is then withdrawn. The name of species and date of collection may be written on the capsule with a water proof marker. Dry the pollen by exposing the capsules to the air inside a frost-free refrigerator for 24 hours. Store the capsules so exposed inside individual plastic containers that have a little Drierite in them to prevent moisture from moving through the gelatin and then place these containers in the bank.

Gene Banks at Global Level

The National Seed Storage Laboratory

The United State's National Seed Storage Laboratory (NSSL) is located at the University of Colorado in Fort Collins. It does have some areas for sub-freezing storage, but most of the accessions are stored at 5°C.

The Vavilov Institute

Vavilov Institute in Leningrad is the primary Russian gene bank. It has 25 experimental stations around the country. The facility houses an enormous collection of seeds and living plants, ranging from wild species and primary crops to breeding lines of modern crops, though most of its work is in plant introductions.

The Izmir Centre

The lzmir Centre, one of the most important centers of biodiversity, was established in Turkey in the mid-1960s. The seeds of the area were sampled extensively and the centre proved to be very important for cereals particularly, wheat and for forage legumes such as alfalfa. Within ten years more than 13,000 seed collections from Turkey itself and at least 25000 collections from other countries had been assembled. The collections at the Institute are vital. Political instability in the region may affect the long-term usefulness of the centre.

The International Rice Research Institute

The International Rice Research Institute (IRRI) at Los Banos in the Philppines was started in 1960 to collect rice varieties for breeding. It is one of the most successful examples of a gene bank. It has about 60,000 rice accessions. Some of the rice species have been developed into high yielding dwarf rice varieties that now feed much of the world. To day scientists at IRRI are attempting to breed varieties resistant to insects and disease, as well as have the ability to grow in drier climates or brackish water. Despite the huge number of accessions already located at the IRRI, some experts fear that valuable wild rice strains that would have contributed much to the new breeding stock already have been lost.

The Royal Botanic Gardens at Kew

The Royal Botanic Gardens at Kew is the first genuine gene bank for wild species, and is the model for other banks. Seed units were set up at the Jodrell Laboratory at Kew and at the satellite garden at Wakehurst Place, Sussex. Seeds for the bank were obtained from cultivation at Kew as well as collection from other botanic gardens.

The Bank in Irvine

The seed bank at the University of California, Irvine (UCI) was primarily aimed at to collect South African bulbous and corms plants. Families Iridaceae (the Iris family) and Liliaceae (the lilies) were selected for this particular work. The gene bank at UCI has several components: a living, growing, plant collection, and cryogenic seed and pollen banks.

Other Gene Banks

Gene banks to preserve endangered and regional plant species have been initiated by the University of Dublin, Ireland the Polish Academy of Sciences and the Polytech University in Madrid. This was a co-operative effort among countries in the mediterranean region aimed at saving wild species.

9

Information on Hot Spot

A The Silent Valley : A Success Story

A hydal project of Kerala Government was dropped due to the concern about its impact of environment and endangered species. The story of Silent Valley reveals that it was dropped only to provide due protection to the Lion tailed macacaque and some other endangered and endemic flowering plants of Western Ghats.

The Kerala Forest Department documented a commemorative volume of the natural wonders of the silent valley and the campaigns that helped to preserve them.

Silent valley represents India's most precious environmental heritage and the visitors to this place, usually go back with a fund of fond memories. Naturalists *vs.* Vijayan and M. Balakrishnan who carried out the first systematic scientific survey of the Silent valley in 1977 said recently that they had less to fear from the forests fourlegged denizens the wild boar, the leopard, the panther and so on than from the employees of the Kerala State Electricity Board, who are posted to deal with intruders when there was considerable conflict between the board and environmentalists, who opposed the proposal to build a hydroelectric project in the Silent Valley. Suresh Elamon the wildlife photographer trekked for weeks on end on the fringes of the valley trying to catch a glimpse of the rarest of a rare butterflies, the travancore evening brown, which was last sighed in 1930 in the Periyar Tiger Reserve. The saga of Silent Valley is made up of such individual quests, which have helped to preserve this unique crucible of nature virtually intact.

Prompted by the widespread interest shown by conservationists during a seminar organised to make the first decennial of the silent valley National Park in 1995, the Kerala Forest Department brought out in the second Week of October an anthology of 38 articles which together record the history of the silent valley the struggle by non governmental organization in India and abroad to save from being overwhelmed by hydal project, its regeneration in a period of a decade as a protected National Park together with a hard core authoritative documentation by naturalist and conservationists on the valleys unique flora and fauna. The silent valley was a dense tableland of tropical evergreen forest, approximately 90 sq. km. in area and hemmed in on all sides by steep hills of the Niligiri ranges (Fig 9.1). The Kunthipuza traversed north to south descending 1500 metrs in the process. It rained here almost throughout the year recording over 3000 mm of annual rainfall, with nearly 900 mm in July alone when the southwest monsoon is at its peak. Temperatures were always around 20°C approximately.

The British rulers of the Malabar notified it as a reserve in 1914. The suitability of the Valley for a hydal project was one of the kept secrets of British India, but survey parties penetrated the forest only in the 1950s. Finally in February 1973, the Planning Commission approved a 240 MW

Silent Valley Hydro-electric Project, with an initial outlay of Rs 25 Crores. Incidentally, this was the time when Mrs. Indira Gandhi was addressing the Stalkholm Conference about the concern for environment.

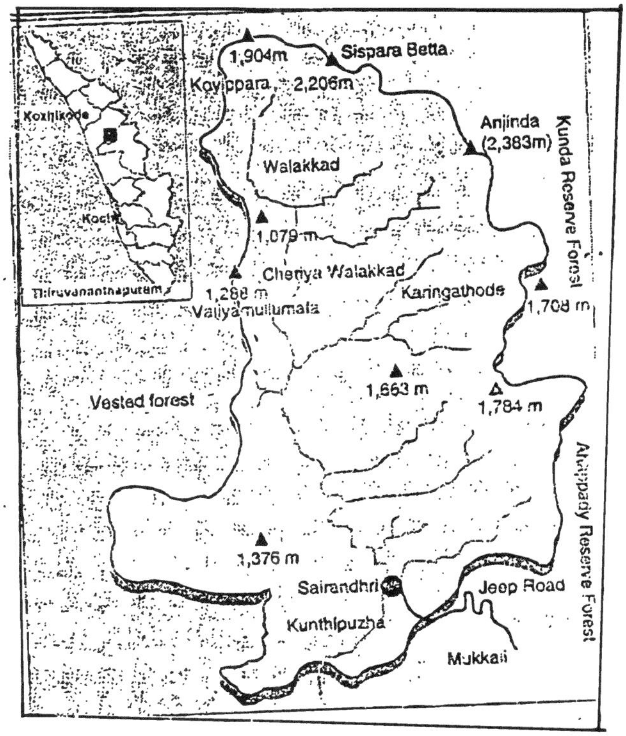

Fig. 9.1 : Silent Valley National Park

But the government agencies were ignorant of the what Silent Valley held, individual naturalists, especially botanists and zoologists from colleges in northern Kerala, who had been trekking the valley for years in search of rare specimens were appalled at the developments. Around them grew the spontaneous moves to save the silent valley, M.K. Prasad a Botany lecturer in Kozikode as the convenor of the Environmental brigade of the Kerala Sastra Sahitya Perished (KSSP) brought the organization into campaign. Dr. S. Sathish Chandran Nair and Prof M. P. Parameswaran carried the message across the state border to scientific societies in Chennai. Malayalam poet Sughathkumari inspired by some of the KSSPs pamphlets, led a parallel movement among Keralas literati, which soon attracted well-known personalities of the state. As both Prasad and Sughatkumari recall in their contributions the Silent Valley protection campaign spread rapidly largely owing to the involvement of an academic literary combine.

Meanwhile nature lovers outside the state were networked. Dr. Salim Ali, Dr. Zafar Futehally from Mumbai, Dr. Madhav Gadgil and Dr. Parthasarathy from World Wildlife Fund for Nature, India, from Bangalore, Dr. Romulus Whitekar from Chennai joined this movement. They helped to add a national and international dimension to the issue. Flooded with telegrams from nature lovers and international organizations such as IUCN and WWF buoyed by a helpful on the spot study by the Principal Secretary in the Ministry of Agriculture, Dr. M. S. Swaminathan. Mrs. Indira Gandhi, the then Prime Minister, took the political route-out of this potential development versus environmental crisis. She appointed Prof. M.G.K. Menon, Secretary, DST to carry out an appraisal of the issues involved.

Meanwhile, the save Silent Valley campaigners took their struggle to the courts arguing the landmark suit filed by Joseph John on behalf of the Kochi-based, an NGO, Friends of Trees. Mrs. Indira Gandhi chose to read in the M.G.K. Menon report, which was presented in 1983, a clear message against the hydal scheme. In a swift order the state government was asked to call off the project and declare the valley as National Park. Mrs. Indira Gandhi, who took the crucial administrative decision, did not live to see the latter part of the order implemented. The National Park was formally inaugurated by Rajiv Gandhi in September 1985.

After the storm over, the Silent Valley had blown over a decade of intensive stock-taking was launched. Botanist K.S. Manilal says that the silent valley harbors at least 108 varieties, many of them spectacularly large. In his article on orchids of the Tropical Botanical Garden and Research Institute (TBGRI) near Thiruvananthapuram lists every one of these species. Some, like the bicolor *Acanthephippium,* are found only in south India and Sri Lanka.

The Silent Valley is a repository of medicinal plants is brought out in the review by Sathish Kumar along with P. Pushpangadan, Director of the National Botanical Research Institute (NBRI) in Lucknow. Nearly 80 per cent of the drugs listed in standard pharmacopoeias and 66 per cent of the spcies and aromatic plants used the world over are found in India. The Silent Valley is a valuable source of some important genetic variants. At least 21 flowering plants discovered in the valley are new to science. The TBGRI is currently using tissue culture techniques to multiply some hybrid orchid varieties.

Regular surveys by the Forest Department and some students sponsored by funding agencies have established that the valley's animal population has thrived in the years since it was notified as a National Park. The Valley's most famous resident the lion tailed macaque according to KFRI estimates, now have around 275 members spread out in 14 troops (1996 figures) and an abundance of their favourite food, the fruit of the *Gullenia exarillata* tree has ensured that they do not migrate.

Large mammals - the elephant, the tiger, the sloth bear and the wild boar are also reported on the fringes and in the adjacent forests. The dense vegetation doesn't support many of these animals in the core of the valley, where lesser carnivores such as the civet, the jungle cat and the mongoose thrive. The Valley harbours 211 bird species. Goerge Mathew's study of the butterflies and moths, admits that his records detailed as they are only scratch the surface.

b. Lion Tailed Macaque (*Macaca Silenus*) : A Most Eendangered Monkey of India

Introduction

The Lion-tailed macaque is an endangered primate species, endemic to the tropical rain forest of the Western Ghats. The total wild population is estimated less than 4,000 individuals distributed in the three southern Indian states of Kerala, Karnataka and Tamil Nadu. However, Silent Valley and Ashambu hills are the only two viable habitats in their entire range of distribution (Green and Minkowski, 1977).

Various stochastic events and catastrophes along with human interferences have affecting seriously the Lion tailed macaque population through the ages. Factors such as habitat fragmentation, reduced habitat size, isolation of populations leading to inbreeding depression and vulnerability to random events make the Lion-tailed Macaque a highly endangered species despite its relatively large scattered population.

Methods

Status and ecology of Lion-tailed macaque was studied in Silent Valley National Park for a period of three years from 1993 to 1996 by Joseph and Ramachandran from KFRI,Peechi. This chapter is mainly based on their work published in *Indian Forester*, October 1998. Repeated census walks were conducted on foot through the existing trails and elephant paths randomly in the evergreen forests. Frequent steps were made to get the characteristic contact call of the Lion-tailed macaque in the vicinity of the transact. The located troops were followed and continuous observations were made until each troop was cited twice in a year.

Animals were classified as four major categories such as adult males, adult females, sub-adult males and immatures. Adult males were identified with stouter body and long canines. The tail tufts were having slightly lifted carriage. Adult females were identified by the presence of elongated nipples and baggy breasts. Although, sub-adult males were having fully developed musculature and comparatively shorter canines. Rest of the individuals including sub-adult females, juveniles and infants were grouped into a single category as immatures.

Results

Fourteen distinct troops were identified from silent valley and adjacent areas when the field study ended in 1996. Among the fourteen different troops eight of them ranged east of Kunthipuzha river and the rest towards the western side

Two troops each in Punnamala (PUN I; PUN II), Aruvampara (AVI; AVII), and Panthanthod (PANII;PANII) regions and one each in Sairandri (SAI) and Parathod regions (PAR) formed the resident troops East of Kunthipuzha river. Three troops in Chembotty areas (CHEIA;CHE IB; CHE II) two troops in Nilikkal areas (NKL I; NKL II) and one troop in Puchapara (PUC) constituted the population west of river Kunthipuzha. All the troops were distributed at an elevation ranging between 700 to 1500 m.

Table 9.1 shows the status and population structure of Lion-tailed macaque troops in Silent Valley forests. A total of 275 individuals were observed in the final census. Among these 8 per cent were adult males and 45 per cent were adult females. The sub-adult males constituted 6 per cent and 41 per cent of the total population formed immatures. The adult sex ratio estimated was 5.64 females per male.

Table 9.1: Population Density of Lion-tailed Macaque

Troop name (Location)	AM	SAM	AF	IMM	Total
SAI (Sairandri)	2	3	16	13	34
PUN I (Punnamala)	2	2	10	6	20
PUN II (Punnamala top)	1	1	7	12	21
PAN I (Panthanthod)	2	1	9	8	20
PAN II (Panthanthod)	1	1	6	8	16
AV I (Aruvampara)	1	2	7	4	14
AV II (Aruvampara)	1	-	5	7	13
PAR (Parathod)	1	-	4	4	9
PUC (Puchapara)	1	1	6	8	16
CHE IA (Chembotty)	2	1	7	6	16
CHE IB (Chembotty)	2	1	10	7	20
CHE II (Chembotty)	2	1	11	7	21
NKL I (Nilikkal)	2	3	17	14	36
NKL II (Nilikkal)	2		9	8	19
Total	22	17	124	112	275

AM = Adult male, SAM = Sub-adult male, AF = Adult female, IMM = Immature.

Out of the nine troops monitored in 1993 five of them, increased their total in 1994, while remaining three unchanged. There was a reduction in the troop size near Attapady from 21 to 13 in 1994, it is because the animals were killed by the Muduga tribals. A total of 11 troops monitored were increased their population size in 1995. Out of 12 troops monitored in 1996 two troops fissioned before 1996 census. Fifty per cent of the troops increased their number and remaining not changed.

Discussion

The present study is connected to the Western Ghats area of Kerala state, where more than 50 per cent of the population is located. In the silent valley areas monkey are distributed between the altitudes 700 to 1500 m elevation. The occurrence of associated food plants viz., *Gullenia palaquium, Meusa palaquium, Meusa calophyllum,* may have influence on the distribution of lion tailed monkeys.

The announcement of the area as a National Park in 1984 is a creative step and subsequent protective measures helped the survival of these monkeys. The low lying areas of Silent Valley have undergone drastic changes in the past and 1/5 of the areas became degraded grasslands due to forest fires. Trapping of nine individuals by the Muduga tribals in 1993 clearly indicates the vulnerability of these animals in the borders areas of the Park. Such stray incidents are restricted to the peripheral areas, not affecting the core forests of the park. The slow and gradual expansion of the macaque habitat during the recent years has positive implications for the population growth of the animals. This also envisages for the immediate steps for extending management strategies for long survival of the species.

Suggestions

1. The National park and adjacent areas are important habitats for the lion-tailed macaque. There is need to conserve the entire stretch of the adjoining rain forests for the long term survival of these monkeys.
2. Fire control steps should be taken and monitored continuously in the summer. Fire affected areas should be regenerated by planting the food plant seedlings for the monkeys and other primates.
3. The tourism zone be marked out side national park and anthopogenic pressure should be minimized in the core areas.
4. The monitoring of the endangered primates should be continued so that effective management actions will be taken regularly.

C. Legal Conservation of River Dolphins

The river dolphins are in great danger of extinction all over the world. The riverine habitat is highly degraded and is under heavy pressure everywhere as the human population is increased by leaps and bounds. Approximately 10 per cent of the human population of the world lives along the Yangtze river, the native domain of the Chinese river dolphin *(Lipotes vexilifer)* and industrial traffic on this river is geometrically increasing as China is building her future. Similarly River Ganga is the home of Indian Gangetic dolphin *(Platinista gangetica),* is being greatly affected by the industrial and agricultural development in India, Bangladesh and Nepal. The Amazon and Orinoco River systems in South America are the habitats of American river dolphin are increasingly victim to the rapid loss of rain forests. If the present trend continues there may be a little time to do anything to save and conserve several of river dolphins.

To cope up with these growing problems those countries have already passed the laws to protect river dolphins, but protection may not be possible because of the difficulty of enforcing these laws in remote areas and with limited resources. The present chapter describes the legal provisions to protect the river dolphins in various countries where river dolphins occur.

National Legislation Concerning Protection of Habitat

Various types of legislations may be relevant to the protection of habitats of the dolphins. Among these are laws that establish protected areas such as parks and sanctuaries and laws that regulate the environmental implications of various developmental projects.

INDIA

The Wildlife Protection Act, 1972 was enacted for the protection of habitat, which also provides designation for sanctuaries. The Environmental Protection Act notification 1986 was also passed for the regulation and protection of environmental factors and handling the polluting and hazardous materials. This law also aims to protect our environment from the further degradations. There is also provisions to punish those who endanger the environmental safety and human well-being.

The new act (The Environmental Protection Bill 57-C/1986) appears to be re-written version with some modifications and deletions and is superior in its approach to the earlier ones.

But this law does not specifically provide for a procedure to assess the environmental impacts of various industrial projects or other sorts of development. However, it is possible that this legislation could be amended to establish such a procedure. Environmental impact reviews could be useful for protecting river dolphins, as industrial pollution, water diversion and dam construction are considered to be major factors influencing the decline of Gangetic dolphins (*Platinista gangetica)* in India (Kumar *et al.*, 1995 and Kumar and Singh, 1996) since environmental pollution is considered to be a threat to *P. gangetica,* laws such as this may be relevant to river to dolphin conservation.

PAKISTAN

A river dolphin reserve was established in 1974 by the government of Sindh covering 170 km between the Guddu and Sukkur barrages on the Indus river. The subpopulation in this stretch of the river is believed to contain over 80 per cent of all known Indus dolphins (Khan and Niazi, 1989).

BANGLADESH

The Third Schedule of the Bangladesh Wildlife (Preservation) Order, 1973 provides for the establishment of sanctuaries but till date none has been designated.

BHUTAN

Although no information on habitat protection laws was available, there exists a large sanctuary adjacent to the Indian boarder that might protect the river dolphin habitat.

CHINA

The People's Republic of China Environmental Protection Law was enacted to ensure rational use of the natural environment, to prevent the environmental pollution and ecological destruction and to promote economic development. The environment as defined in the law includes water and aquatic organisms. Chapter 1 defines the principles of the law. Whenever projects are to be undertaken, including new project or those involving the rebuilding or expansion of existing ones, an environmental impact report shall be submitted to the Environmental Protection Department and other relevant departments for examination and approval before any planning is commenced (Para-6).

This law and other statutes and regulations relating to Environmental Protection apply to foreigners (Para-9).

Chapter-2 describes measures to be taken to protect the environment. Among those applicable to the river dolphins are : Comprehensive scientific surveys must be carried out before the new construction of small, medium or large-scale water works, etc. Measures for the protection and improvement of the environment shall be conscientiously adopted to prevent the destruction of the ecological system (Para-10).

Aquatic or land organisms must be protected, developed and rationally used. Over harvesting and destruction causing extinction must be prohibited (Para-II).

However, it is unclear whether the river dolphins are considered to be aquatic organisms or not.

Chapters-4 and 5 describe the duties of governmental agencies, including international co-operation, environment monitoring, scientific research and environmental education.

ARGENTINA

Argentina passed a law in 1980 to protect certain area. In addition to National Parks, these include Natural Monuments, defined as regions, object or fauna and flora of aesthetic, historic or scientific interest should receive complete protection. National Reserves may be established to conserve ecological systems.

VENEZUELA

This country has created protected areas by Executive Decree, under the Forest, Soils and Water Law of 1965. The Organic Environmental Law of 1976 establishes the importance of conservation. Over 7.6 million hectares area are protected and 280,000 hectares are totally protected as faunal refuges. If destruction of aquatic and forest habitat is a threat to river dolphins, existing legislation may provide some measures of protection.

COLOMBIA

Colombian National Park Legislation contained in Law 2a of 1959, Decree 2420 of 1968, Decree Law 2811 of 1971, Decree of 16 March, 1977. Among the types of protected areas established under the present legislation are National Parks, Natural Reserves and National sanctuaries. Law 30 of 1980 approves the agreement for the conservation of flora and fauna of the Amazonic territories of Colombia and Peru, which was signed in Lima on 30 March, 1970.

PERU

Decree to Ley No. 21147 authorizes the establishment of national parks and other conservation areas. Total protection is afforded *Inia and Sotalia* in the Pacaya National Park in the Upper Amazon basin.

BRAZIL

Lei No 4771, 1965 is Brazils major forestry law, which applies to the preservation, exploitation and management of forests. This law provides scope for the establishment of National Parks, and Biological Reserves where modification of any type is prohibited except for authorised scientific purposes. Biological reserves are created for safeguarding, exceptional natural qualities, flora and fauna for educational and scientific purposes.

NEPAL

According to Shrestha (1993) the river dolphins *(P. gangetica)* in Nepal have been caught traditionally for meat and oil, which is used for burning and medicinal purposes and to attract fish. Hunting of river dolphins is prohibited by the National Park and Wildlife Conservation Act and punishable by a fine and or imprisonment.

BOLIVIA

Bolivias principal wildlife law is Ley de Vida Silvestre, No. 12301. An implementing decree, Decreto Supremo No. 16605 June, 1979 declares a total ban on the hunting and trade of listed native wildlife species considered to be in danger of extinction. Among the species listed is *Inia geoffrensis.* According to Fuller and Swift (1985), the validity of the decree has been questioned because of some scientific errors in the listings. *Inia geoffrensis* is erroneously classified under pisces.

ECUADOR

The 1981 Ley Foresataly de Conservation de Areas Naturales Vida Silestre (Ley No. 74) governs conservation and use of living resources except fish and marine species. In the decree, these restrictions only apply to protected species. It is unclear whether the river dolphins are specifically protected. Presently however, Ecuador prohibits the commercial export of indigenous species.

GUYANA

Under Guyana Fisheries Regulation (Section 30) Act the hunting of a few aquatic species is controlled. The regulations require that the hunting or capture of listed species be done under a license from the fishery department of the ministry of agriculture. *Inia* could presumably be listed on the schedule of regulated species by administrative action.

AMERICA

The U.N. National Marine Fisheries Resource is formally considering petitions to list *Lipotes vexilifer* as an endangered species under the U.S. Endangered Species Act (UEESA). The provisions of the UEESA apply both to trade of river dolphins and to protection of their habitat. The U.S. Marine Mammal Protection Act (USMMPA) also inhibits the import of marine mammal parts in the U.S.A.

International Agreements and Programmes

During the past 3-4 decades the earth is facing environmental degradation problems in terms of increased green house gases, global warming, depletion of tropical forests pollution of air, water, soil and extinction of plants and animals. These have created unpleasant changes in climate. sea level, monsoon, winds and ozone layer. Some important programmes undertaken to protect the wildlife are mentioned herein.

Western Hemisphere Convention, Washington, 1940

Argentina, Brazil, Chile, Ecuador, Peru, Uruguay and Venezuela formed a party and arranged a convention. The parties agreed to adopt laws to protect and preserve such species and to regulate their trade. Species needed special protection are listed. The only Cetaceans listed are four species of great whales. This convention might become useful for the conservation of *Inia* also.

International Importance of Wetlands

This was conducted in Ramsar in 1971 to focus the recognition of the ecological as well as economic, scientific, cultural and recreational importance of wetlands. The Ramsar convention aims to curtail the loss of wetlands. The convention has proved to be useful tool for preventing deleterious changes occurring in wetlands. Although the emphasis is on the protection of wetland birds, specially water foils and their habitats, many of these areas are also important for river dolphins particularly *Inia.*

World Heritage Convention

This convention was organised by the countries namely Argentina, Bangladesh, Bolivia, Brazil, Colombia, Ecuador, Guyana, India, Nepal and Pakistan in Paris in 1978. This convention was concerned to the protection of world cultural and natural heritage. The main purpose of this society was to protect precisely delineated natural areas of outstanding universal values from the point of view of science conservation. The parties agreed to develop scientific studies and take necessary legal scientific steps to protect their natural and cultural heritage.

Convention on International Trade in Endangered Species

This convention was held in Washington in 1973 to prevent the over exploitation of flora and fauna due to international trade. This summit developed a system of trade control by passing many resolutions. Species needed special protection are listed in the CITES appendices and the *Lipotes vexilifer* and *Platinista* are included in this list.

Convention of Flora and Fauna of Amazon Territory

This agreement took place between the two countries, Peru and Brazil in 1975. It aims to preserve flora and fauna of Amazon territories.

Bonn Convention

This was held in Germany for the conservation of migratory species of wild animals. The objective was to protect the species that migrate across and outside national boundaries. This treaty is potentially useful for addressing river dolphins.

Brasilia Treaty

This summit was attended by the representative from Bolivia, Brazil, Colombia, Ecuador, Peru, Venezuela, Guyana and Surinam in 1978 for Amezonian co-operation. Although the treaty is concerned primarily with socio-economic concern, it also recognizes the need for rational use of Amezonian flora and fauna.

Lima Agreement

This meeting was organised by Peru and Colombia in 1979 to conserve the diversity of Amezonian region. Under this agreement certain organizations were assigned the function of promoting and putting into practice the scientific inter-change and other measures to ensure proper management of natural resources and protected areas.

Man and Biosphere Programme

Man and Biosphere (MAB) aims to establish protected area network for endangered species. MAB uses existing national and international legal and administrative mechanism. MAB is admitted by the UNESCO. Several of the countries where dolphins still occur have already established biosphere reserves and those countries are Bolivia, Colombia, and Peru.

SAARC Summit

South Asian Association for Regional Cooperation (SAARC) is an organisation of seven south Asian countries viz., India, Pakistan, Nepal, Bangladesh, Bhutan, Sri Lanka and Maldives Islands. Its objective is to strengthen regional cooperation among the partner countries for the welfare of their people and to maintain eco-balance of the environment.

Earth Summit

Though the first earth summit was arranged in Stalkholm in 1972, but the officially announced and well known U.N. conference on Environment and Development (UNCED) was held at Rio de Generio (Brazil) from June 3 to 14, 1992. Heads and representative from 178 countries participated in the discussions to adopt global plan to save the earth from degradation. This summit also resolved the protection of threatened and endangered flora and fauna of the world. The main aim of the summit was to act quickly and wisely on the problems of ecodegradation of the biosphere.

Status of Indian Dolphins

The gangetic dolphins are under potential threat of extinction because of industrial pollution and other stress from construction dams, diversion of river flow, navigation, fishing and poaching activities. If the present trend of ecodegradation of the Ganga continues, this species will gradually disappear and may become extinct. The government of India has also planned to fund the population research on the dolphins in the Ganga waters. A number of workers (Peterle, 1932; Mohan, 1989; Jones, 1992; Kumar, 1995) have studied the problem and made various suggestions.

d. Gir Forest : A Last Home for Asiatic Lion (*Panthera leo persica*)

Gir is a single largest tract of forest in the Saurashtra region of Gujarat state and is synonymous with the Asiatic lion, *Panthera leo persica*. The lion might entered India from the west and was found in large numbers in the states of Punjab, Haryana, Rajasthan, Uttar Pradesh, Madhya Pradesh, Gujarat and Jarkhand states. The last lion surviving outside the Saurashtra was killed in 1884. The Gir forest in Gujarat has shrunk from 3070 sq.km to 1884 sq km in 1880, due to expansion of agriculture and destruction of habitat. In the early part of this century the Gir forest was connected to Girnar, Mitiyala, Barda, Alech hills, Dhank and Chorwad by corridors of rough semi-wooded forests, grasslands and sparsely populated villages. This made the lions to move freely in the area.

Lions disappeared from Barda and Alech hills towards the mid of 19th century and disappeared from Girnar and Mitiyala by 1963 and 1955 respectively. At the time of notification of Gir forests as a lion sanctuary in 1965, lions were found only in the compact forest of the Gir. Gradually the size of wildlife protected area was increased to 1412 sq. km with the national park surrounded by the sanctuary. The rest of the 403 sq.km peripheral forests constitute the buffer zone of the wildlife reserve. The most important aspect of the Gir forest is that it has become a very stable ecosystem with tremendous regenerating, self supporting and sustaining capacity due to its rich and diverse flora and fauna. The Gir is an unique ecosystem with over 4000 plants, 32 mammals, 24 reptiles, 310 birds and more than 2000 insect species and other invertebrates.

Conservation Through Single Species Project

A study by Paul Joslin (1972) revealed that Gir displayed a sadistic evidence of accelerated degradation of the ecosystem, and he has warned that if nothing was done to arrest the rate of decline in the number of surviving Asiatic lions, the species would become extinct within two decades. Recognizing the serious danger to the lion and pitiable conditions of Maldharis, the Gir lion sanctuary project for initially five years was prepared in 1972. It was implemented at a total cost of 58 lakhs. The work carried out under the Project Lion on the following lines:

1. Fencing of the sanctuary by constructing a rubble a wall of 400 km along its periphery.
2. Constructing barricades along the water course traversing the periphery of the sanctuary.
3. Establishing checkposts across all public roads, passing through the sanctuary.
4. Shifting of all maldharis families from the national park and majority of the families have been shifted outside the Park, who are given cultivable lands, grazing lands, and housing sites. Out of the 845 maldharis families about 2/3 have been moved and re-settled out side the park at different places.

Serious problems like grazing, poaching, lopping, and cutting of trees etc., were tackled by the Department of Forests and development programmes like, soil and moisture conservation, and plantation etc. were implemented. The habitat conditions improved and in many places of the Gir become thick after the launching of the project. Constructing of water harvesting structures and artificial water holes has increased the availability of water to the wildlife.

Positive Changes in Wildlife Density

The implementation of Gir lion sanctuary and park has improved and increased the availability of food and water. Six consecutive census have been carried out at an interval of five years after launching the Project Lion. The lion population along with other wildlife has increased consistently (Table 9.2)

Table 9.2 : Growth of Wildlife Numbers in the Gir Park

Year	Lion	Leopard	Hyena	Ungulates (Herbivores)	No of ungulates/ lion
1968	177	NA	NA	NA	NA
1974	180	142	63	9635	53.5
1979	205	161	84	14964	73.0
1985	239	201	192	16905	70.7
1990	284	212	97	31489	110.9
1995	304	268	137	38221	125.9

NA = Not available.

The herbivores include six major species namely Chital, Sambar, Blue bull, Chinkara, four horned antelope, and wildboar. There has been a consistent increase in the carnivores and herbivores. Availability of ungulates per lion has increased from 53.6 in 1974 to 125.7 in 1995 which makes wild kills easily available to carnivores. The study of Ravi Challam indicated that the food pattern of the lion has changed in favour of wild prey as 65 per cent of the lion diet was recorded as wild animals whereas 35 per cent was made up of livestock kills.

Migration of Lions

Lions disappeared during middle of this century from the forests outside the Gir. The habitat of the Gir improved and the lion population gradually grew with the increased level of protection and conservation. Before the census in 1990 lions were occasional visitors to Girnar, Mitiyala and coastal forests, but they started visiting the neighbouring forests again. The situation has changed then and this carnivore start moving into the former territories. At present there are four satellite populations of lions. The second generation of migrated lions have made Girnar and coastal forests as their home range.

It has been observed that the majority of the migrated lions were sub-adults probably they were compelled by the dominant males to leave the pride and the territory. A long strip of coastal plantation supports a small population of blue bull and wild boar. One pride of 16 lions in Kodinar Dhamrej Sutrapada and another 10 lions in Rajula and Junabad depends on cattle and blue bull. Lions were brought back to the Gir in the past but they repeatedly migrated to coastal forests in search of food and territory and settled in these areas permanently after 1990. The fourth satellite population of lion is in the Mitiyala forests of Bhavanagar forest division, and is in the process of settling.

Carrying Capacity

The carrying capacity of an area depends upon the availability of food, water and conduciveness to reproduction and space. Food is not a major limiting factor in the Gir as the population of important ungulates increased at the rate of 14.2 per cent per year during the last three decades, but shortage of space may resulted in territorial fighting and migration. The population has remained almost the same level. It was 267 in 1990 and 262 in 1995 in the National Park and Sanctuary during the past 10 years.

Conservation Strategy

Greater Gir ecosystem or Extended Gir in the new territories of lions has become the need of the hour in the interest of the increasing population of lions. Proper strategies to expand the Gir present area of 1814 sq. km. to 2370 sq. km. in the near future by covering Girnar (179.5 sq. km), Mitiyala (19.4 sq. km) Coastal forests (110.1 sq. km), Barda (187.4 sq. km) and Vidis (60 sq. km) areas. In addition to the forest lands, wastelands and panchyath lands connecting Gir, the above forests should be developed and managed as corridors for lions. The following measures are required for the management of the habitat :

1. The habitat of the Gir should be improved and expanded by arresting the degradation of peripheral forests and wastelands.
2. The Maldharis still living inside the sanctuary and park should be attracted to re-settle outside.
3. The statellite population at the four sites should be managed by increasing the prey base.
4. The Girnar and coastal forest should be developed and the herbivore population should be managed in those areas.
5. Barda forest which was considered as the alternative home for the lions should be rehabilitated and prey base for the lions should be increased.
6. For herbivores *in-situ* breeding programme should be developed to increase the ungulate population in Girnar, Mitiyala and the coastal forests.
7. Necessary management practices should be followed to facilitate natural migration of lions in new areas. The problematic animals should be utilised for the breeding purposes in captivity.
8. Management of Girnar, Barda, Mitiyala, coastal forests and other neighbouring forests with the management of the Gir should be integrated. The concept of greater Gir ecosystem management should be evolved, not only for the conservation of asiatic lion, but also to provide ecological security to the region. Management of wasteland, panchyath land, grassland and forests connecting Gir to new habitats should be covered under a management plan as Greater Gir ecosystem.

E. Musk Deer-Moschus chrysogaster Musk Live Extraction

The Himalayan musk deer-*Moschus chrysogaster* is primitive ruminant commonly found in the Himalayan region between 2400 to 4300 m in Nepal. The main threat to musk deer is its musk extraction. The musk sac is located between the reproductive organs and the umblicus. It opens to the exterior through an orifice which lies anterior to that of urethra. Traditionally the musk pod is incised out after killing the deer. The musk can also be extracted by simple operation. But this method is time consuming and causes much stress in the deer.

Mukti N. Shrestha (1988) has explained a simple method of musk extraction from live deer and it is as follows. An 18 months old male was physically restrained and secured to expose the umblical region. A silver scoop with one large and one small groove at each end was sterilised and umbricated with antibiotic cream. Holding the musk sac with the left hand, the scoop was inserted gently with a rotating movement. The edge of the scoop is smooth and rounded to facilitate easy insertion and to prevent injury to the musk gland. This method was repeated at different ages and of the buck for the musk and it was extracted and analysed for its quality.

The musk can be extracted annually during the rutting season by this method from adult males in the wild after catching them with nets and they may be released after the harvest of musk. This particular method will avoid killing the buck, which create imbalance in the sex ratio and endangering the animal to extinction. In this method the local village folk may be involved and they may be given the benefits, so that the villagers themselves will become watchdogs against poaching and ensuring the survival of the endangered musk deer even in India and elsewhere.

Project on Great Indian Bustard *Ardiotis nigriceps*

Great Indian bustard is one of the rarest birds of the world. It was existing in the short grass plains of Uttar Pradesh, Tamil Nadu, Sindh area of Pakistan, and in the west of Orissa. Unabated shooting and destruction of its habitats exterminated the bustard from most of its range. By 1980 it survived only in six states of India in small populations. These states include Madhya Pradesh, Rajasthan, Gujarat, Maharastra, Andhra Pradesh and Karnataka. The Tourism and Wildlife Society of India has organized an international conference on Bustards in 1980 at Jaipur, which initiated special conservation project on Great Indian bustard. It has also alerted all the states to undertake conservation measures and eight protected areas for the bustard have been established (Rahamani, 1987, 1989).

In the year 1980 the bustard population was estimated to be around 500-1000 with half of them surviving in the Rajasthan (Rahamani, 1993). Besides the Great Indian bustard, three more species of bustards also occur in India. They are the Houbara (*Chlamydotes undulata*), the Lesser Florican (*Sypheotides indica*) and the Bengal Florican (*Houbaropsis bengalensis*).

Madhya Pradesh

The bustard population declined all over its range in Madhya Pradesh. There were four known bustard areas; Karera, Gatigaon, Pohri and Panna. In the mid 1980s, the total bustard population in these four areas was estimated to be around 50, with Karera and Ghatigoan having 30-35 birds. In the late 1980s there has been a progressive decrease of bustards in these sanctuaries. In Karera, not more than five birds were found. Even territorial males which were seen for many years and as villagers remember, were not seen during the display season in 1993. Condition of Ghatigaon was as bad as karera. The Forest Range officer, in charge of bustard at Ghatigaon saw only two birds in six months.

Gujarat

In Gujarat, by 1980s the bustards had become extremely rare and survived only in two districts i.e. Bhatiya in Kalyanpur taluka of Jamnagar, and Abdasa and Mandvi talukas of Kutch. There was stray record from Surendranagar district. Total population in the whole of Gujarat was estimated to be between 20 and 30 bustards. During the last ten years, bustard number declined by almost fifty per cent. In 1990, a small bustards sanctuary of 100 ha was established near Lala village where breeding was noted. In the Bhatiya area of Jamnagar district, where the Gujarat Government is planning a bustard sanctuary five birds were seen in 1983. It is doubtful if the population can recover from such low numbers.

Maharashtra

Nannaj area in Solapur district of Maharashtra has now the most satisfactory increase in the bustard numbers, mainly because the Forest Department provided good measures to protect its grassland habitat. In 1981, the maximum number seen was only eight. Since then successful breeding has been observed and by August 1993, there were 37 bustards. This increase was due to successful breeding and immigration from surrounding areas.

Andhra Pradesh

Rollapadu bustard sanctuary in Kurnool district has also shown positive result, thanks to effective protection to the bustard and its grassland habitat, especially during the initial stages of establishment of the sanctuary. Territorial males and successful breeding were sighted every year.

Karnataka

In Karnataka bustards are chiefly seen around Ranebennur, Blackbuck Sanctuary and Guttal plantation in Dharwad district. However, as the sanctuary it self is not very suitable for bustards due to excessive growth of eucalyptus, the birds are seen outside the sanctuary in grazing land which suffer from over-exploitation.

Rajasthan

During 1980s, more than half of the bustards in India were present in Rajasthan, mainly in the Thar desert (Rahamani, 1989; Rahamani and Manakadan, 1990). In nine districts; Kota, Ajmer, Bhilwara, Jalore, Pali, Bikaner, Jodhpur, Jaisalmer and Barmer (Vardhan and Goriup, 1980). Now, all over the Thar desert, bustard population has drastically declined e.g. Diyatra, Bap, Sam, Sudasari, the bustard numbers have halved, or disappeared, as in Khuri, Miyajlar.

Reasons for Decline

1. Destruction of breeding areas

Destruction or alteration of grassland habitat is perhaps the most important reason of the decline of bustards in Madhya Pradesh and Gujarat. Conservation success depends on protection to adult birds and to their breeding areas.

2. Disturbance

There is no bustards area in India which is free from human disturbances. Like grazing, cultivation and their activities linked to increase in human population.

3. Shooting

At Karera, Ghatigoan, Nannaj etc., poaching of bustards is under control but shooting appears to be playing a major role in decimation of bustards of the Thar desert. All the four species of bustards found in the India have the highest legal protection but implementing the law more stringently to eliminate poaching is crucial. The habitats of the bustards are not well protected and this constitute the biggest threat to bustards. Therefore our major conservation efforts have to be addressed to habitat protection especially breeding grounds.

Project Bustards

The grassland and desert are under represented in the protected area network in our country. The studies by Rodgers and Panwar (1988), on endangered species Project and grassland ecology project of the Bombay Natural History Society have identified many potential areas which could be developed as new sanctuaries. There is a need to have a nationally co-ordinated conservation project for the Indian bustards and their, habitats. Hence a project bustard is suggested.

Nesting Ecology of Baya Birds in the Western Ghat Regions of Karnataka

Introduction

In the evolution tree of the animal kingdom, next to insects, birds are the important animals adopted to live in all types of habitat. These are the homeothermic animals being spread in different parts of the world in different habitat conditions. There are more than 8650 genera of birds have been identified and studied on the earth. In India 1200 species of birds reported from various parts, which include 300 species of migratory birds visiting every year during winter. Late Dr. Salim Ali has classified Indian avifauna into 71 families.

The general conservation and management of birds encompasses the aspects like how do men trouble the birds, how do birds harm men and what is the possible measure to solve the problems. Though, India has not lost any bird species, but according to IUCN red data book 6 birds are endangered, 6 are vulnerable, 5 are intermediate, 4 rare and one is extinct prone in Karnataka respectively (Hosetti and Venkateswaralu, 2001).

Systematics

More than 95 species of weaver birds known to occur in the world, but India possess only four of them.

They are classified as follows:

Class	:	Aves
Sub class	:	Neornithes
Order	:	Passeriformes
Family	:	Ploceidae
Sub family	:	Ploceinae
Genus	:	Ploceus
Species	:	*philippinus, bengalensis, manyar and megharhynchus.*

Common names

Name of the Species	Common name
1. *Ploceus philippinus*	Common Indian baya
2. *Ploceus bengalensis*	Black throated baya
3. *Ploceus manyar*	Streaked baya
4. *Ploceus megharhynchus*	Finns baya

Keys for Identification

The key characteristics described in the book of Indian birds are as follows :

P. Philippinus

Male: During breeding season the crown is yellow, rest of the upper parts are dark brown streaked with yellow on the back. Breast yellow, throat and ear, dark brown and rest of the under part is cream buff.

Female : Crown and back buff streaked with dark brown, neck yellowish buff, throat is whitish yellow, Breast is yellowish buff with brown streaks. During non breeding season both male and female look more or less alike.

Ploceus bengalensis

Male: During breeding season it has dull red crown or golden yellow crown, white throat and black band separating it from the fulvous white underparts.

Female : Similar to other baya

Ploceus manyar

Male: During breeding season it differs from other baya in having with streak black on fulvous breast in both sexes. During non breeding season both sexes look alike.

Ploceus megharhynchus

Male: This bird was not noticed in the Western Ghats of Karnataka. During breeding season, its head and breast are bright yellow, and dark brown streaked on the above.

Female: Head and nape pale brown or yellowish brown with dark streaks. During non breeding sexes are alike.

Among the three species recorded in our study area *P. philippinus* was most common bird and found in a variety of habitats and tree species. *P. bengalensis* was found nesting in grasses, reed beds and sugar cane fields and remained away from human settlements. The *P. manyar,* a single colony was found in a most protective habitat in the mid of wetland called Navile tank in the Shimoga city. The size is around 15 cm (6 inch) of the size of sparrow. Breeding season varies according to the monsoon season of the area.

Study Area

The study area is located in the vicinity of foothills of Western Ghats, receiving an average rainfall of 850 mm. The area is divided into three imaginary zones.

Zone I : Northern side

Zone II : Northeast

Zone III : Southwest

Methods

In the present study we have enumerated only nesting colonies of *P. philippinus*. We have also encountered the colonies of other species like *bengalensis and manyar* species. The data on *philippinus* species is used for discussion in this paper. The colonies were located by following their songs and some time we have also taken the help of localities.

We have visited a total of 969 colonies in the year 1999-2000. As each male constructs 3-4 nests and each female pairs 3-4 times in a season, based on this particular reproductive behaviour, it is calculated that the total number philippinus birds may be around 300-350.

Objectives

To Know the following :

a. Orientation of nests
b. Nesting habitats
c. Host plants and their diversity
d. Breeding activity
e. Feeding activity

Orientation

Our study on orientation of nests by the baya *P. philippinus* revealed that the first year male or the yearlings without breeding dress and the sexually matured adult males have constructed their nests in all possible orientations. They are mainly lateral, central and peripheral sides of the host plants and other combinations.

In zone I lateral colonies were maximum (81.8%). Such colonies can face different directions in lateral positions. Among these again the colonies facing exactly eastern direction were maximum in number. In Zone II again lateral oriented colonies were maximum (80.7%). Among these the eastern side facing colonies were maximum. A similar trend was also noticed in Zone III. There was a definite orientation pattern observed in these birds with respect to stem of the host plants. It is generally the entrance tube which directed towards outside while egg chamber was positioned towards the stem. Due to this particular orientation the entrance tube always remained oriented towards outside. The flying routes of the birds are so managed that the host stem or other vegetation do not pose any hindrance. Orientation of these nest was different in vegetated and non vegetated areas, dug wells and irrigation channels, wherein they oriented in relation to their nearest surfaces. The *P. philippinus* hangs its nests according to the direction of prevailing wind.

Selection of Twigs

The selection of twig is also one of the important factors for nest weaving. Twigs should be generally thin pliable and pendent to horizontal branches (leaves in case of palms and grasses) are selected by the *P. philippinus*. Twig having thickness more than a thumb are usually avoided, possibly they can not be accommodated in the grip of claws during knitting of fibers in the nest initiation. Strongly upwarding branches are preferred. Twigs which are sufficiently tough to support the weight of the nest are chosen.

The onset of rainy season of the area is mainly known to influence the activity of the nesting by the baya birds. It operates for four months usually from June to September. Any variation in south west monsoon is found to influence directly on the nesting behaviour and orientation pattern of the nests.

Nesting Patterns

A typical normal nest of *P. philippinus* is a bottle shaped nest called as normal nest and can be divided into three parts viz., a stalk, middle body and the entrance tube. The entrance tube is slightly shifted towards the egg chamber to form a bulging chamber.

Abnormal Nests

Besides normal nests, various other types of abnormal nests were constructed by the sexually matured males and yearlings. Many variations can be seen in the nests of baya, which make the nest abnormal. The category of abnormality is in the structure of the nest or any part of it. Abnormal structure may appear due to deposition of parts, formations of additional parts elaboration of parts and abolition of normal parts (Sharma, 1995). The second category of abnormality is regarding the position of nests or any part of it.

In our investigations in the Karnataka a total of 969 nests were visited from the three Zones I, II and III.

Though bistoreyed nests were common, four storeyed nests were also encountered in our study area. Multistoreyed nests are noticed in those areas where comparatively long monsoon periods prevailed. In Karnataka monsoon period is too short to fabricate 6-10 storeyed nests. Multi stalked nests were also common in this area. Presence of two or more stalks provide extra attachment and strength to the nest which make a hanging nest prone to simple harmonic motions on windy days. To minimise the windy movements of the nests more than one support are used for stalking the nest. Additionally the length of the stalk is also kept short. It is mainly based on the habitat conditions . If the tip is harder then in order to have more security more than one stalks are weaved. Stalkless nests are also seen in the area and such nests are equally liked by the females.

The other type of abnormal nests are Bell Jar shaped, buttressed, blind or closed, chained, meshed, branched, fused and stomach shaped nests. In addition to this some normal nests possessed double openings in order to facilitate the quick visits by the mother.

Host Plants

It was observed that in our study area the common baya constructed its nests on 31 species of plants belonging to 18 families. Among all these trees most preferred species of plants were *Cocos nucifera, Acacia planiformes and Pheonix sylvestris.*

Summary

1. The population of common baya was maximum in zone II is due to rich and diverse type habitat and food resources.
2. More than 50% of the nests were directed towards the eastern side to avoid S.W. monsoon. In contrast the nest located on the banks of irrigation channels, open wells and sloppy areas showed different orientations.
3. Among abnormal nests bistoreyed abnormality was maximum and it was followed by multi-stalked nesting.
4. For nesting 32 different plants species were selected, among those coconut, acacia and pheonix tree species were most preferred.
5. *P. philippinus* is not specific in hanging its nests, unlike other species of baya.
6. Population density of *philippinus* was maximum in the study area and it was followed by *bengalensis* species. A single colony with 14 nests of *manyar* species was also recorded.

Future Scope and Significance

Among the three species of baya birds, *P. manyar* may considered as endangered species in Karnataka. Before saying this, there is need to undertake a detailed survey in other parts of our country. There is a need to study the pest status of these baya birds in relation to different crops in different areas of our country.

Establishment of Orchidaria in the Western Ghats of Karnataka State

Preamble

The orchids are distributed every where in the Western Ghats. These produce fascinating beautiful flowers that mimic female bees, wasps and beetles. Some of them are attractive visually and olfactorily to insects and birds, which inturn visit them and try to copulate them. In this process of pseudo mating they apparently pollinate the flowers. The persons interested on these flowers collect them indiscriminately which has lead to their virtual disappearance. There is an need to be conserved these orchids not only in the wild, but also in orchidaria for research and enjoyment of their ornamental beauty and medicinal importance. Accordingly an attempt is made to set up orchidaria corresponding to specific locations like Town Park, Kudremukh, Bhagavathi Nature Camp and Bannerghatta National Park. These are the three recognised hot spots of Orchids in the Western Ghats of the state. In orchidaria these may become commercially potential genotypes for propagation and trade.

In Karnataka state one could see lack of organized orchidarium at state level holding wild orchids as germplasm stock under natural settings. Partially this lacuna has bean met by establishing an orchidarium at Cauvery Nisargadhama, Kushalnagar, Kodagu District. To date, this orchidarium holds 85 species mostly collected from distinct vegetational types of the Karnataka part of the Western Ghats. Further two illustrated books on the orchids of Kodagu have been published (Rao, 1998, 1999). Similarly, the orchid survey was conducted at the Kudremukh National Park during 1998-2000 to collect 59 orchids from distinct forest divisions. So far, this has yielded a few rare and unreported orchid species, which have been planted at the Bhagavathi Nature Camp Orchidarium.

Kudremukh National Park in Karnataka is one of the best examples of Shola forests found in the Western Ghat belt of Karnataka and possess a rich assemblage of wild orchids of which a few of them are unreported to orchid flora of Karnataka. Further some of them previously thought to be restricted to a few areas in Karnataka state have been recorded in this chapter with a note on their occurrence in the Kudremukh National Park and also in other parts of Karnataka. The present collection in the Western Ghats has found to be rich and diverse in respect of orchids due to the fact that there are suitable niches for the quild system of orchids. The intensive collection includes unreported endemics which have been recorded earlier only in the states of Kerala and Tamil Nadu. They have a great potential also as earners of foreign exchange, besides having medical and commercial importance in the country it self.

Culture of orchids in orchidaria

Epiphytic orchids are local specific but not host specific. In view of this, the point of consideration is how best they could be planted in the orchidarium. Among the wild orchids some of them thrive best in pots, wooden or bamboo baskets while others on trees under natural settings depending on the locations of the orchidarium. When pot culture is practiced big pots with sufficient space should be used for sympodial orchids. When epiphytic orchids are grown in pots should be filled up with burnt bricks of assorted sizes, chips of old bark, charcoal, bone chips and moss clumps. For ground orchids, decomposed cattle manure and organic manure with turfy loam forms a good mixture for rapid growth resulting in bright flowers. In earlier stages watering is necessary which in course of time should be kept in a shady place alternately subjecting the pots to sunlight and rain water during non-flowering seasons. Finally one should have or learn a sound knowledge of their growth features over the months by careful observations to have an ideal eye catching orchidarium. The orchids listed below were collected from Kudremukh National Park have been planted at Bhagavathi Nature camp, Kudremukh, Town park, Kudremukh and Bannerghatta National Park.

Orchidarium at the Town Park Kudremukh and Bannerghatta National Park

1. *Aerides crispum* Lindl.
2. *Coelogyne breviscapa* Lindl.
3. *Cymbidium aloeifolium* (L). Sw.
4. *Dendrobium harbatulum* Lindl.
5. *Dendrobium barbatulum* Lindl.
6. *Dendrobium macrostachyum* Lindl.
7. *Eria reticosa* Wt.
8. *Oberonia verticiliata* Wt.
9. *Rhynchostylis retusa* (L.) BL
10. *Cleisostoma tenuifolium* (L.) Garay
11. *Trias stocksii* Benth. ex. Hook.
12. *Luisia macrantha* Blatt. & McC.
13. *Cocologyne nervosa* A. Reich.
14. *Dendrobium aqueum* Lindl.

15. *Dendrobium ovatum* Lindl.
16. *Dendrobium crispa* Lindl.
17. *Vanda tessllata* (Roxb.) Hook ex Don.
18. *Eria microchilos* (Dalz.) Lindl.
19. *Acampe praemrosa* (Roxb.). Blatt. & McC.
20. *Pholidota pallida* Lindl.
21. *Robiquetia josephiana* Manilal & Sathish.
22. *Liparis viridiflora* (BL.) Lindl.
23. *Cymbidium discolor* Lindl.
24. *Satyrium nepalense* Don.
25. *Taprobarea spathulata* (L) Christ.

Species present in the orchidarium of Kudremukh National Park, Bhagavathi Nature camp, Kudremukh

1. *Aerides crispum* Lindl.
2. *Aphyllorchis montana* Reichb.
3. *Bulbophyllum proudlockii* (K & P)
4. *Calnathe triplicate* (Willem) Ames
5. *Coelogyne breviscapa* Lindl
6. *Cymbidium aloeifolium* (L) Sw
7. *Dendrobium lawianum* Lindl.
8. *Dendrobium barbatulum* Lindl.
9. *Dendrobium macrostachyum* Lindl.
10. *Eria microchilos* Wt.
11. *Eria Retieosa* Wt.
12. *Habeneria longicorniculata* Grash. (Ground orchid)
13. *Malaxis versicolor rheedi* Sw. (Ground Orchid)
14. *Habeneria marginata* Coleb (Ground Orchid)
15. *Nervilia crociformis* Zoll. and Mor. (Ground Orchid)
16. *Oberonia verticillata* Wt.
17. *Peristylus gooderoides* Lindl.
18. *Porpax reticulata* Lindl.
19. *Rhynchostylis retusa* BL.
20. *Sarcanthus pauciflorus* Wt.
21. *Trios stocksii* (Benth) Hook. F.
22. *Luisia birchea* B.P.
23. *Luisia macrantha* Blatt & McC
24. *Coclogyne nervosa* A. Reich
25. *Epipogium roseum* (D. Don) Lindl.
26. *Dendrobium aqueum* Lindl.
27. *Nervilia argona* Grandl (Ground orchid)
28. *Dendrobium ovatum* Lindl.
29. *Dendrobium crispa* Lindl.
30. *Vanda tessellata* (Roxb) Hook.
31. *Dendrobium erepidatum* Lindl.
32. *Gastrochilus acaulis* Lindl.
33. *Luisia zeylanic* Lindl.
34. *Acanthe philippensis* bicolor Lindl.
35. *Malaxis rheedei* Sw (Ground orchid)
36. *Trias stocksii* Benth ex Hooker.
37. *Acampe praemrosa* (Roxb) Blatt & McC.
38. *Gastrochilus flabliformis* (Blkatt & McC). Saldhana
39. *Luisia birchea*
40. *Rhynchostylis retusa* Bl.
41. *Dendrobium soni* (Hybrid)
42. *Pholidota pallida* Lindl.
43. *Chiloschrista glandulosa* Schlter
44. *Liparis viridifolia* Lindl.
45. *Diplocentrum congestum* Lindl.
46. *Tainia bicornis* (Lindl) Reich.
47. *Rhynostylis retusa* (L) Bl.
48. *Cymbidium discolor* Lindl.
49. *Satyrium nepalense* DON
50. *Luisia macrantha* Blatt & McC.

Importance of Bats

Endemic Bat (Otomops wroughtonii)

The endemic and rare bats are found only in the caves of Talewadi village of Belgaum district. *O. wroughtonii* is considered as one of the rarest insectivorous bats in the world. It is endemic to Belgaum district and not found any where on the earth. The cave is large dense with large number of millipedes and some other insects. The rocks in the caves have crevices in which the bats reside. Fecal dropping are comparatively less in this area compared to other types of bats. It is very difficult to locate these bats because they fly very swiftly. Early reports by Prater (1960) revealed that these bats lived in groups and he killed around 40 of them. It is believed that the construction of dams and mining activities may not harm the bats in the future. But how far it is true and this can adjudicated only by the bat biologists.

Painted Bats

The painted bat *Kerivoula picta* is widely distributed in India, there very few sight records of this bat from Maharashtra. Systematic studies on this bat is not undertaken in India. Banana plantation is the prime habitat for this bat species. These bats are facing a serious threat due to the clearing of banana plants to make a way for Sapota orchard *(Achras zopota).*

Fruit Bats

Bats belong to order Chiroptera which can be further subdivided into mega and micro chiropterans. Mega-chiropteran bats are confined to old world tropics and are exclusively phytophagus whereas micro-chiropteran bats are mostly insectivorous. In Indian subcontinent out of the 12 species of fruit bats three species are most commonly known. Those are the flying fox (*Pteropus giganteus*), the fulvous fruit bat (*Rousettus leschenavlti)* and short nosed fruit bat (*Cynopterus sphinx)*. About 70-100 species on insect eating bats species found in different parts of India. Their diversity is more in the Eastern and Western Ghats, Himalayas, Andaman and Nicobar Islands.

The frugivorous bats and their pest control measures revealed that these are eco-friendly rather than thinking them as pest on orchards. Approximately, 250 species of bats of the paleo-tropical family Pteropodidae and the neotropical family, Phyllostomidae, regularly visit plants for either fleshy fruits of floral resources.

The foraging activities of a frugivorous bats can be simply described on the following way: At dusk fruit bats level their roosting site and commit to a particular resource patch which is often fruiting tree. This bat then selects a fruit from those available and either eats it or carries it down to its roosting place. During feeding it might feed on an entire fruit or sometimes a part of it leaving the remains as masticated or rejected pellets or uneaten parts, usually found under the roosting site. The bat may carry a gut load during its foraging activities. At dawn, bats return to their roosts and a final gut load is either eaten or regurgitated.

In the eyes of a frugivorous bat, not all plant species are equally attractive. Within and between fruit trees, there is considerable variation in ripeness, size and other physical or chemical attributes. Fruits eaten by bats are usually fleshy or have a musty odor and high sugar content. The role played by bats as seed dispersers helps in promoting gene flow, making seeds less obvious to seed predators and moving seeds to more favorable sites for colonizing. Movement of pollen between individuals thus promotes plants to produce smaller amount of nectar so that bats would visit more than one individual flower to satisfy foraging requirements. Thus in case of keystone species such as *Ficus* spp, bats help in the propagation of seeds, promote dispersion of seeds to various patches far away from the parent plant. Thus there exists a kind of mutual relationship between fruit bats and the plants served by them for pollination and dispersal.

Most bats are seasonal at various patches. The phenologies of fruiting correspond directly to the reproductive status of these bats. As frugivores, both megabats and man have similar tastes and both share many fruits. So man through the tending of wild fruit trees and planting to enhances the chances of expreriencing heavy damage to fruits in farm lands and orchards. One possible reason for the increasing menace of bats in fruit orchards is the dwindling population of wild fruit trees and a general reduction of forest cover. The rapid loss the ficus trees has lead bats to forage in orchards is perhaps a strong factor. Thus certain management techniques could be considered in managing this menace.

The techniques can be classified into two main categories:

(a) The roost management and

(b) Crop management.

(a) *Roost management*

1. *Physical methods* : Making of sounds at their roost site which are present in window caves or palm trees. The making of loud sounds could keep bats away for some time.
2. *Chemical repellents* : chemicals having a bad odour drive them away. This should be sprayed when bats leave their roost during the late evening hours. For example fresh cow dung mixed with water or goat dung or paint.

(b) *Crop management*

The success of crop protection methods depends on the availability of alternative food resources in the nearby vicinity. Roost control methods are effective only if alternative foraging habitats are available closely otherwise they would continue to forage in orchards although their roosts may be shifted.

All Fruit Bats are not Nuisance

The suborder megachiroptera consists of 175 species of fruit bats. Apart from food habits and size the mega bats differ from micro bats in having a small claw on the second digit of fore limbs. Fruit bats have relatively large eyes, and are well adopted for nocturnal vision. The olfactory system plays an important role in the location of food. The sensitive nose is useful in detecting the aromatic scent of ripe fruits in darkness. The fruit bats play a beneficial role in pollinating flowers and dispersing seeds. They do damage commercial fruits while visiting orchards for foraging. There is a necessary to investigate the quantification of damage caused by fruit bats. However the Wildlife Protection Act 1972 has already classified bats as vermin.

The Indian Sarus Crane (*Grus antigone*)

The Indian Sarus Crane, *(Grus antigone)* is distributed mainly in the northern part of India. Gole (1989) recently carried out a survey to estimate the distribution of the species in the range mentioned by Ali and Ripley (1969). Though it is a wetland species, it also developed flexibility in its habitat requirements. Ecology of the sarus crane is not studied extensively and the previous work includes qualitative description of their distribution, diet, habitat, some aspects of its morphology and eggs. Among the four species of cranes reported from the Indian continent, the sarus crane is resident of Koeldeo National Park and the rest are migratory. Among the migratory cranes, Siberian crane was a permanent visitor to India, which is becoming rare during the past 4-5 years. The Common crane and Demoiselle crane are occasional visitors. These are regularly sighted in the Western ghat region of Karnataka state. These frequently visit the wetlands and irrigated croplands of Shimoga district, Hidakal reservoir of Belgaum, Haveri district and other water tanks.

An in-depth study on the Sarus crane was carried out by Ramachandran and Vijayan during 1984 to 1988 at Koeldeo National Park located 2 km away from Bharatapur city in Rajasthan on the south west of Mathura and 50 km away from Agra city. The total area of Koeldeo Park is 29 sq. km., out of which 8.5 km are inundated and the area dries up during summer leaving some pools in deeper areas. The Park is surrounded by agricultural lands and settlements. Some of the area forms good grasslands and the rest is deciduous thorny scrub jungle (Fig 9.6).

Population and Distribution

The population of the Sarus crane remain almost same throughout all parts of the year except during spring and summer. During summer most of the birds leave the park for feeding outside, while some pairs remain inside throughout the year as resident group. These may be breeding pairs of the park. During breeding period most of the cranes are confined to the waterspread areas of the Park. In 1986 with the onset of monsoon and subsequent flooding of the Park, the massive summer congregation of Sarus dispersed leaving only around 24-25 birds inside the park which were breeding in the park.

Breeding Season

The main breeding season during 1982-1985 was August to October. But there are reports that, it can also breed starting from June to October. The breeding season of Sarus is highly variable, it may stretch upto to March.

Nesting Site

Usually the nest is constructed on a slightly raised area in the submerged parts of the wetland. The nesting site is invariably sourrounded by water, the depth of which varied between 15 cm to 20 metres. The nesting site was selected in such a way that one or two sides of the nest have open expansions of grasses which may help in detecting the predators at a distance.

The nesting material varied from nest to nest and was dependent on the nesting site. They seldom bring materials to the nest and construct it. Of the eight nests intensively studied by Ramachandran and Vijayan (1992). The nest was usually a pile of grass on a mat of *Paspalum distichum* with nest radius of 1 to 1.5 metres. The base of the nest is broader with a slight depression on the top for the eggs. The grass pile reaches 70 cm in height from the surface of the water. It may also be constructed by *Acacia* twigs and the *Paspalum* grass. Both male and female participate in the act of nest building for period of one month. Repair and strengthening of the nest takes place even after egg laying. There was a gradual reduction in the breeding population of the bird in the Park during the past two decades.

Territoriality

These birds maintain a well established territory during breeding. The pair remain in the same area throughout the year except in the summer months. However during non-breeding season they cover large area for foraging. The breeding territory ranged around 0.068 sq. km and the nest is built on one corner of the territory. Most of the territories had two zones, a shallow water area where parent birds feed and a deeper zone in the immediate vicinity of the nest. The deeper area is reserved for the young ones. By the time young ones hatch out, the water recedes the area becomes shallower and all the birds feed in that area.

The incubation period for the bird is 28 to 34 days. Both sexes participate in incubation. In the initial stage of incubation, while one parent incubating the other remained nearby and engaged in the strengthening the nest. It is observed that during exchange of duty of incubation the partners give duet dance or prolonged unison call and then go for incubation. It is believed that such display will strengthen the pair bonding between the two. Usually male spend more time in incubation than female. The chicks begin to move along with parent on the day of hatching itself. The parent carry the food in their beak and feed the chicks. The chick always remain with the parent for one season then move away to live independently.

Food and Feeding Habit

The food material observed include rhizomes of *Nymphaea* sp., *Scirpus tuberosus, Eleocharis plantaginea,* insects, grass seeds mollusks and snaces. During summer when they move out they feed on wheat seeds from the croplands in dry areas. These bird roost in paris or with family parties. The roosting site will be usually on grass patches of the wetlands. Some time they roost along with Siberian cranes. Since these Sarus cranes use moderate and heavily flooded grassland for feeding during breeding season, these habitats have become a critical sites for the survival of these species.

Annual Cycle of Sarus Crane

The annual biological cycle of the Sarus crane is closely interwoven with the seasonality of the climatic regime. Since the Koeldeo National Park is the last safe refuge for these cranes in India, they aggregrate in large numbers near the pools of the park. With the onset of monsoon and subsequent flooding the pools in the park, most of the Sarus cranes disperse leaving around 25-30 birds in the Sanctuary, which forms the breeding or resident population of the park. They are territorial during this season and all the chicks will be out by the end of October or November, if conditions are favourable then during January and February a few pairs breed and produce chicks again. Once the water pools dry cranes from neighbouring areas congregate and making them non territorial.

Due to heavy anthopogenic pressure around the park, it revealed that during the last two decades there was a declining trend in their population in the park. Another threat to the birds is contamination of water by industrial effluents and pesticides residues into the park waters. There is a need to develop the conservation strategy for Sarus crane by considering the above discussed factors.

Southern Crow Pheasant *Centropus sinensis parrotius* Stresemann

Relatively less information is available on the crow pheasants of India. Little information is available on the biology of Indian Crow Pheasant *(Centropus sinensis sinensis)*. The detailed study on Southern pheasant *(Centropus sinensis parrotius)* was carried out in Point Calimere Wildlife Sanctuary, in Tamil Nadu, by Dr. V. Natarajan in 1988. He has studied six different breeding territories in Kodikkarai village. Sightings of breeding pairs were marked on them and the territories are calculated. During the period season of 1986 to 1989 study was carried out and a total of 34 nests were located. Among these, 21 nests in the village area were continuously monitored and the clutch size, incubation, hatching rate, and breeding success were recorded.

Breeding Season

At point calimere the crow pheasant raises a single brood in a season which extends from November to May. The egg laying was found during December to March.

Territory

The breeding pair occupy a well defined area defended by both partners. The size of territory usually varies between 0.9 to 7.2 ha with a mean of 3.8 ha. In crow pheasant nests nest may be seen in same tree for consecutive years. Other pairs entering the territory were threatened by making a sound K-wiss call. Other birds like crow, shikra, brahminy kite were tolerated in the territory but not on the nesting tree. The coop-coop-coop-call by the pairs is an advertise to threaten the intruders and keep them away and this call is herd frequently during the breeding season.

Courtship Behaviour

The display leads to mating. For the first time the male arrives with a frog in the mouth in its bill and hop along with the female on the branches of a tree. The female then glide from the tree to the ground and run for a short distance with wings open, vibrating, uttering a harsh cry sk-e-e-aw, similar to the call of common crow pheasant. Both fly on a tree and hop again from branch to branch, they copulated two to three times. The frog in the mouth will either eaten by the male alone or it will be shared with the female also.

The Nest

While searching for a nest-site the crow pheasants hop from branch to branch, sit on a perch and produce calls, each partner respond to the calls. Both sexes usually examine the site and take part in nest building. Crow pheasant prefer thorny plants to build their nests. In Karnataka the common crow pheasant usually construct its nest in thorny lemon trees. These birds construct their nest like a globe or dome with a lateral entrance. Some time it may build nests deep cup like.

Egg Laying, Clutch Size and Incubation

Usually eggs were laid at the time of completion or about to complete the nest. Afterwards both the parents mostly become busy in completion of the dome. The eggs were oval in shape and chalky white in colour, become yellow stained as incubation proceeds. The length of the eggs was 34.8 mm and breadth 14.8 mm respectively. The clutch size varied from one to four eggs. Both sexes take part in incubation during day time but at night time only one will incubate the eggs. During first three-four day old young ones are fed with regurgitated food, especially snails. Both parents feed the nestlings. The number of feeding visits increased till 16th day declined thereafter. Parents visited the nest 4-5 times per hour. Both parents involved in cleaning the nests.

Growth of Nestlings

The just hatched nestling will be black in colour with eyes closed. Dorsally covered with long white hair like down feathers named as trichoptiles by Friedmann (1930). The upper mandible is black with pink edges. The center of the belly is pinkish, the legs and claws grayish in colour and the plumage changes with age.

Anti-Predatory Strategies by Nestlings

The nestlings were known to exhibit, escaping through the rear of the nest chamber, excretion of foul smelling obnoxious sticky fluid and hissing sound like a snake when they are threatened by a outsider was observed.

Fledging

It was known to occur in about 19-21 days. Even after fledging the young were known stay near the nesting tree for few more days and later they slowly move to adjacent trees by hopping along the other branches. The parents continue to feed them even after fledging atleast for about one month.

Dispersal of Young

The ringing experiments conducted by Natarajan (1996) revealed that even two months after fledging the young ones were seen living and feeding with their parents and dispersed thereafter.

The predators of the eggs include jungle crow, house crow, brahminiy kite and the common mongoose *Herpestis edwardsi.* These predators will be chased away by the parents when they come near the nest.

White Winged Wood Duck, *Carina scutulata* in India.

It is one of the most endangered ducks of India. It was sighted commonly in the north-eastern parts of India till the beginning of twentieth century. Later on there has been a steady decline in the population of this duck. In Dibrugarh and Lakhimur districts of Assam, its number declined from an estimated 1900 to a minimum of 44 by 1992 (Green, 1992, 1993). Dr. H. S. A. Yahya of Aligarh Muslim University carried out a pilot survey to collect on hand information on the status and habitat conditions of this duck in five key habitats of Assam and Arunachal Pradesh in 1993.

His results revealed that out 65 birds estimated in the India, Dr. Yahya sighted only 26 animals in intensive survey of about 40 days. More number of these birds were located in Doom Dooma and Dibru Saikhowa Wildlife Sanctuary. About three key sites were surveyed in Arunachal Pradesh and no bird was sighted except their calls and sound. The estimated population number 65 birds may be an underestimate. According to Yahya the total number of white winged wood ducks may be around 100 pairs in the India.

The habitats of the duck are severely depleted and no site in the north is free from human influence. Depletion and fragmentation of habitat and poaching of young/eggs appears to be the main reason for the drastic decline in the population of this bird. Poaching and selling of adult birds still continued in these key areas. Several such cases were noted by Dr. Yahya of Aligarh. Arunachal beel is a suitable habitat for these birds but the quarrying, selective logging and hunting by the locals are common problems to be encountered by these animals.

Conservation Strategies Required to be Adopted

a. There is an urgent need to protect the habitat from further degradation. A sort of fencing is required at vulnerable sites like Ubahata, Littong and Namholong.

b. There is a need to locate viable nesting pairs and protect them from poaching. Artificial nest boxes may be supplied and monitored for these birds.

c. More staff and manpower is required to check and limit all illegal activities in their habitats. The Littong and Kakopathar forest areas may be announced as a Wildlife sanctuary and may be given special protection. Updating the present captive breeding center is needed.

d. There is also an urgent need for creating awareness among the public about the wildlife status of the bird.

Conservation of Indian Marine Turtles

Turtles are referred as terrestrial animals based on their place of origin and birth, however they spend most of their life time in the water, except that they come to shores during breeding season. Turtles are given much importance in terms of their study and conservation as well. Marine turtle population is declining day by day due to rapacious turtling and over exploitation. For the past few decades their survival is threatened with extinction and they are under constant anthopogenic pressure. The marine turtle flesh has been an item of tasty table delicacy since long. As a fellow harmless creature both in water and on land, they posed never any threat either to marine ecosystems or to human beings. These poor creatures are troubled by man directly or indirectly through ages.

The sea turtles are slaughtered on the beaches, the eggs are raided by the birds, animals and humans, thereby checking the chances of their progeny to continue (Pritchard, 1980). Indians by and large have been appreciably ethical towards all types of diversity. The marine turtles should be viewed as friends of human. Out of the seven worldwide species of sea turtles the following species are found in Indian maritime.

1. Olive ridley *Lepidochelys olivacea*
2. Green turtle *Chelonia mydas*
3. Leatherback turtle *Dermochelys coriacea*
4. Hawksbill turtle *Eretemochelys imbricata*
5. Loggerhead turtle *Caretta caretta*

The turtles have been given their due place in the Hindu mythology by recognizing them as second incarnation of lord Vishnu calling as Koorm Avtar. They are the important members of our biodiversity due to their biological, socio-economic, educational and scientific values as natural resources along the national coastline of about 7500 km. Though their population is dwindling, none of them are called as endemic and these have been placed in Schedule I of the WPA 1972 and listed in the Appendix I of the CITES. Due to their peculiar way of life their protection has become difficult. The marine turtles being a migratory species recognizes no boundary.

Conservation

Fortunately the Indian coast have been supported by a number of sea turtles as their feeding and breeding grounds extending from Bay of Bengal to Andaman Nicobar islands, Gulf of Mannar, Indian Ocean, Arabian Sea upto Gujarat Coast. With reference to these animals the Worlds biggest rookery is Gahirmatha and Bhitarkanika in Orissa coast should be worth to mention. In recent times these resting beaches have become serious concern both at national and international levels. The Indian interest in sea turtles is not as old as other countries. Who have spared more time and money for conservation and management of these animals. The scientific community should be concerned about the habitat disturbance, mass killing and over exploitation of these poor creatures. The conservation strategies undertaken under the leadership of Dr. Silas (1984), CMFRI, Cochin is land mark in the management history of sea turtles in India. It has been resulted in the great awareness in the scientific community and public as well. There is need to protect apart from ridley turtle, other turtles along the coast of Andhra Pradesh, Tamil Nadu, Karnataka, Goa, Maharashtra and Gujarat states. Presently the Orissa having Olive Ridley as their prime land gained special momentum. This place is facing a number of human induced threats like mortality, predation, illumination, over exploitation and domestic and international trade.

Problems affecting the turtles

The Indian turtles are facing many fold problems and some of the recent problems are listed by Yadav (1999) as follows :

1. Anthorpogenic activities are affecting the habitat loss, inundation and poaching of eggs.
2. Nesting places in the beaches are threatened by resort developers and illegal human settlements.
3. Mechanized trawl fishing and shrimp trawling caused mortality of turtles
4. Lack of political will and administrative support, coordination of management and identification of problems of the areas etc.
5. Lacunae in publicising the negative impacts of commercial exploitation on turtles.
6. Absence of strict execution of Coastal zone regulation Act, 1992.
7. Direct and indirect impacts of tourism, beach mining and industrial pollution
8. Lack of patrolling, combating, poaching of turtles, nests and their eggs.
9. Removal of mangroves and replacing them by aquaculture ponds.
10. There is a need to establish the main breeding grounds as Protected Areas, or as wildlife sanctuaries or these turtles may be protected by launching a species project in line with the Project Tiger by involving locals and fisher folk and forest officials.

Conservation Priorities

1. Converting/encroachment of breeding grounds of turtles settlements should be banned and strictly monitored.
2. Understanding the turtle biology and protection of the fragile niches
3. Habitat degradation, coastal wall construction and beach nourishments should be addressed.
4. Discharge of effluents into the beaches should be checked.
5. Establishments of special conservation schemes for marine turtles and a ban on sales of turtles and their products;
6. All activities along the fragile areas should be assessed through the Environmental Impact Assessment studies.
7. Involvement of NGO's in the protection activities/campaign;
8. Public awareness through mass media Seminars, Symposia and Workshops.
9. Ban on ecotourism or beach activities during their breeding seasons and protection by strict watch.

10
Social Biota for Biodiversity

Soil macrofauna biodiversity was studied during the May-October periods of the years 1978-1991 in the following set of ecosystems located in the territory of the Central Chernozem Reserve and the Kursk Agriculture Station : 120 ha of natural unmown meadow steppe; 500 ha of mown meadow steppe; 60 ha of pasture, 45 ha of barley field without fertilisation, 1 ha of permanent black fallow ploughed twice every year. The mowing cycle in the mown steppe is hay-making once per year for three consecutive years followed by one year without hay-making.

The meadow steppe and pasture ecosystems are the only grassland ecosystems in the Central Russian region that have never been ploughed during human history. They were used as horse pasture or hay-making land until the 1930s, when they became part of a reserve. Other small islands of real steppe can be found in Ukraine. These ecosystems were thoroughly investigated by climatologists, soil scientists, botanists, geochemists, vertebrate zoologists and other for long periods of time, and a lot of data have been published on the chemical composition and long-term dynamics of the ecosystems.

Plant biodiversity studied have been carried out for more than 60 years while soil animal biodiversity research began about 40 years ago. The main studies on soil chemistry were carried out in 1950 - 1980, plant chemistry in 1970 - 1980 in the same ecosystems. In most of our studies we used the following method of soil macrofauna estimation. Soil samples were taken by soil corer up to 30 cm depth.

However, it was found that collection of these large samples led to disturbance of the vegetation cover around every sample of an area four to six times larger than the area of the sample. This sampling procedure could change the vegetation cover in and around places of collection and alter natural processes in the unique reserve of ecosystems. It was therefore decided to modify the sampling procedure. i.e. sample size and number of samples.

Twenty samples of 10 cm x 10 cm every 2 weeks were considered sufficient for the study of seasonal dynamics of soil macrofauna in the reserve habitats. Population number of different species of soil macrofauna obtained by this sampling procedure were practically equal to those found with the earlier method. The total number of soil samples collected during the period of investigation was near 3000. This includes all samples gathered during studies cited below.

The species number and population characters of soil macrofauna decreased in the following order : unmown steppe > mown steppe > pasture > barley field > permanent black fallow. It is notworthy that in the black fallowecosystem only one or two species from any of the macrofauna groups were found, namely the earthworm Nicodrilus caligonosus, the diplopods Brachyiulus jawlowskii and Megaphyllum rossicum, the staphylinid beetle Amischa analis, and the carabid beetles Bembidion lampros and Microlestes minutulus. None of the geophilids or lithobiids that are common in the other ecosystems were found in the black fallow ecosystem.

Practically all animals found in the field and the fallow soil are microbivorous or carnivorous. Although grazing pressure at the pasture is relatively low, and along the pasture borders there are patches of steppe vegetation, soil macrofauna biodiversity is low in the pasture compared with the steppe ecosystems. Only the carabids and elaterids show an increased specied number in the pasture soil compared with the barley field and black fallow. Amara species carabids and wireworms are new members of the groups under study in the pasture soil, in addition to the species already mentioned for the barley field and black fallow.

Staphylinid beetles found in the pasture soil were all coprofillous. Regarding geophilids, *Pachyinerium ferrugineum* replaces *G. proimus* and *S. nemorensis* found in the barley field soil. Density and biomass of the soil macrofauna population in the pasture soil is similar to that in the soil of mown steppe, but biodiversity is lower.

Biodiversity is notably greater in the steppe ecosystems compared with the other ecosystems studied and is comparable in the mown and unmown steppe plots. However, density and biomass are markedly higher in the unknown steppe than in the mown steppe. There are no differences in the species composition of earthworms, diplopods, geophilids, lithobiids and elaterid larvae identified in soils of the mown and unmown steppe. Some new species appear in these groups in addition to those mentioned above.

The differences are marked only for population density and biomass of separate species in these groups. *E. nordenskioldi* is the dominant species in the unmown steppe soil and *N. roseus* in the mown steppe soil. On the other hand, the staphylinids and carabids differ markedly in these ecosystems. The Soerensen index for carabids is 0.64 and for staphylinids is 0.52. The Jaccard index is 0.47 and 0.36, respectively. All other community parameters, such as mean species number per sample, mean density and biomass of animal populations, are correlated with the changes of biodiversity. Mean species number per sample shows a stronger correlation with the population density and biomass than does total species number. This is because mean species number per sample is a characteristic of dominant and common species in determined density and biomass.

A decrease of biodiversity of soil mesofauna groups is a common phenomenon in sets of ecosystems with increasing human influence. Among the many possible reasons, soil perturbation, and microclimatic and soil chemistry changes are considered the leading causes. Even a low-impact human activity such as hay-making is reflected in the soil macrofauna composition and differences between the mown and unmown steppe communities are considerable.

Pitfall data collected in the same period indicate that in the mown steppe the share of ubiquists among staphilinid beetles and carabid beetles is 17% and 5% respectively, in the unmown steppe it is 28% and 7%, respectively, in the pasture 32% and 44%, respectively, and in the barley field and black fallow ecosystems 100% for both groups. It is interesting that the share of ubiquists in the field increased from 72% in the 1960s to 100%. This may be a reslt of an increase in the human impact on the local environment as a whole. In our opinion the higher proportion of ubiquist species of beetles in the unmown steppe is related to the microclimatic conditions in the thick layer of, litter, where conditions are optimal for ubiquist species.

On the other hand, *E. nordenskioldi* is the only non-ubiquist species among earthworms in both ecosystems but is dominant in the unmown steppe. Two other species are found in all anthropogenic habitats and may be considered as ubiquists with *N. roseus* dominating in the mown steppe. Based on data on biodiversity, it is difficult to estimate what type of a reserve regime is closer to the initial prehistoric conditions of real steppe (mown or unmown steppe). The grass litter layer is very thick in the mown steppe but practically absent in the mown steppe.

Geobotanists consider pasture regimes to be the initial one for real steppe, and the mown steppe regime as analogous to the pasture (grazing) regime. However, it is evident then that pasture and mown steppe cannot be considered as analogues. The biodiversity and community characters prove this assumption. Grazing activity in pasture at the level of one cow per hectare is sufficiently high to disturb steppe ecosystems and change them entirely. It is very significant that this level of pasture activity does lead to the disappearance of the grass litter layer.

Many pasture soil species are ubiquists or coprofillous. Our observations showed that such coprophagous beetles as Aphodius and Onthophagus species are common in surface soil of pasture. These species are practically absent in the mown and unmown steppe. The point of view that pastures and mown steppe are analogous ecosystems was universally adopted by geobotanists. However, a study of species composition at a steppe plot after artificial fire shows that during the first 2 years soil fauna species composition is similar to that on mown steppe plots.

Then species composition is changing toward that typical of unmown steppe. It confirms our opinion that steppe ecosystems are cyclicphyrogenic ecosystems, and not pasture ecosystems. Therefore the species composition of soil macrofauna in the mown and unmown steppe denotes that these ecosystems can be regarded as separate stadia of post-fire succession series in meadow steppe, and not as an initial state of real steppe ecosystems. Earlier we showed that changes in biomass and production of soil animal populations in agroecosystems depend on available organic matter and nutrients in arable soils. Correlation between species composition and population density and biomass suggests that resources of nutrients also have an influence on biodiversity.

The differences in composition and biomass of the soil fauna in arable plots are small although the quantity of organic resouirces in the field soil are somewhat higher than in the fallow soil. In the fallow soil there are no such sources of available carbon, nitrogen and phosphorous as plant roots for animals and microorganisms. Available organic matter for animals in field and fallow soils is mainly present in the form of decomposable humic compounds and microorganisms.

The low biodiversity in field and fallow soils are probably a reflection of small pools of available organic matter, nitrogen and phosphorus sources for the fauna. Stores of available organic matter in the natural steppe are three times higher than in the field and five times higher than in the fallow. The microgen pool in the steppe is 1.5 times higher than in the field and fallow. The phosphorus pools in the different ecosystems are practically equal.

However microbial biomass in the soil and in the tits of soil animals are the main sources of essential nutrients for animals. Biomass of microorganisms in field and fallow soils is higher than in meadow steppe soil, but it is only 'potential' biomass. To be the nutrient pool for animals, in microorganisms need stores of available energy in soils. These stores are concentrated in dead plant residues and plant exudates.

Hence the biodiversity of soil fauna is possibly correlated with amounts of available organic matter and nutrients, and not only with habitat conditions. In arable soil it will be dependent on input of available carbon compounds for soil microorganisms. If we compare overall biodiversity of ecosystems we usually only consider biodiversity of multicellular organisms and not the biodiversity of microorganisms. Unfortunately, there are some problems with the comparison of biodiversity of taxa of different size levels, and of ecosystems as a whole. Over 10 years ago, we first suggested a new index for biodiversity comparisons.

Our aim was to find indicators for a more objective estimation of biodiversity. In our opinion, DNA in ecosystems could be an indicator of 'potential' biodiversity in an ecosystem as a whole. We called the suggested character the 'index of genetic saturation' (ICS) IGS is equal to the sum S(mjx kj), where S is the sum of the products of mj and *kj* for species of a separate group (S is equal to the species number in the group) in an ecosystem.

11
Biodiversity and Neotropical Primates

Traditional classifications divided New World primates into "marmosets" or "callitrichids" and "nonmarmosets" or "cebids", in which the marmoset patterns of morphology were seen as primitive and those of cebids as advanced. Most authorities recognize 16 extant platyrrhine genera, including Ateles, Brachyteles, Lagothrix, Alouatta, Pithecia, Chiropotes, Cacajao, Aotus, Callicebus, Cebus, Saimiri, Cebuella, Callithrix, Leontopithecus, Saguinus, arid Callimico.

In contrast with the general concordance of reviewers working at the genus level, considerable disagreement exists among authorities working at higher taxonomic levels. In a 1986 revision, Ford reorganized recent platyrrhine genera using a cladistic approach. In 1992, Rosenberger of the National Zoological Park, Smithsonian Institution, launched a new synthesis based on phylogenetic relationships, adaptive change, and the fossil record. He recognized four monophyletic subfamilies in that classification, Callitrichids became callitrichines, closely related to squirrel monkeys and capuchins. Many features of callitrichines are now thought to have resulted from selection for small size and thus are derived and not ancestral or primitive traits. Identification of the phylogenetic link between the Cebinae and Callitrichinae was made possible by the discovery of a new set of fossils at La Venta, Colombia, with morphological characteristics intermediate between those of the two subfamilies.

The rate of description of fossil platyrrhine genera has increased markedly in recent years. Of 16 known fossil genera, 7 were described since 1985 and 4 of these since 1990. Our understanding of phylogenetic relationships and extinction of neotropical primates certainly will continue to improve if the trend toward increased description of fossil forms continues. New advances in molecular genetic techniques now make it possible to objectively test hypotheses about the phylogenetic relationships among the platyrrhines.

The Brazilian geneticists Maria Sampaio and Horacio Schneider examined sequences of DNA in nuclear globin genes and found close relationships between Alouatta and the atelines, between Cebus and Saguinus, and among Pithecia, Chiropotes, and Cacajao, adding additional support to the phylogenetic model developed by Rosenberger. Much of what we know about the diversity of neotropical primates, at the species level is due to the painstaking descriptive detail provided in the published works of Hershkovitz.

The biological accuracy of taxonomic designation becomes particularly important because legislative protection and conservation funding for full species is significantly greater than those for subspecies. In addition, conservation objectivs of reintroduction and translocation of isolated populations are predicated on the conservation of genetic diversity within a species. If genetic management is to have imperative that we be able to distinguish conspecifics forms less closely related forms.

Revision of neotropical primates at the species level should be given high priority, both in terms of contributions to primate conservation and basic search. Recently developed noninvasive genetic techniques such as sequencing of mitochondrial DNA extracted from hair follicles of museum may make it possible to resolve some of these controversies. Living neotropical primates are found from central Mexico to northern Argentina and include some 68 species, comprising about one-third of all living species of primates. In the mid 1980s, much emphasis was put on identification of major phytogeographic regions that contained relatively large numbers of primates, particularly rare, endangered, or endemic species. These analyses resulted in maps of priority areas for primate conservation and research. The discovery of four unknown primates in such a short period of time, including one species in the backward of one of the largest cities in the world, underscores how little we really know about the diversity of even well-studied taxa such as nonhuman primates and emphasizes the need for more basic research and surveys on primates in the Neotropics. However, if conservation efforts on behalf of any species of primates are to succeed, it will be necessary to go beyond the identification of its geographic distribution and taxonomic relationships with other primates. Although biologists have been conducting censuses of neotropical primates for decades, lack of standardization of methods and the difficulty of observing primates in closed forest has made it difficult to assess the accuracy of the results of these surveys.

Nonetheless, numbers of neotropical primates appear to be decreasing. For example, in 1969, the Brazilian conservationist Alvaro Aguirre estimated that 2,000-3,000 muriquis survived in Brazil's Atlantic forest. In the mid-1080s, the minimum number of muriquis known to exist in 10 locations as published as 240. However, it was not until 3 years ago - when Kicrelff, a Brazilian graduate student, used playbacks of golden lion tamarin vocalizations in virtually every forest that might contain these primates-that an accurate estimate of the number of individuals in the wild became known.

This, the first exhaustive census for any neotropical primate, suggested that only 559 individuals in 103 groups remain in the wild and that the species is subdivided into 16 isolated subpopulations, 12 of these containing only 1 group. The accurate nature of this information removed any question about the degree of threat to this species arid prompted significant changes in the focus of conservation and related research activities.

Monitoring changes in quantity and quality of habitat

In 1986, critical regions for conservation of neotropical primates were identified as huge phytogeographic areas such as Amazon basin and "megadiversity countries" that contained a large diversity of primate taxa. Since that date, our ability to pinpoint regions for conservation has improved markedly. In 1990, approximately 100 specialists on Amazonian fauna, flora, soils, and climate met in Manaus, Brazil, to produce a map of biologically important regions in the Amazon Basin based on criteria of endemism rarity, and diversity.

Important here is the diversity of specialists contributing information and the speed with which the results were disseminated. Perhaps the most significant recent advance in our ability to quantitatively evaluate primate habitat is the combination of remote-sensing technology and geographic information system (GIS) analysis. A decade ago satellite images of regions of South America were expensive, difficult to acquire, and could be analysed in only a very few locations.

Currently, a variety of high-quality inexpensive magnetic image data are available from several sources and cover virtually every region on Earth. Recent improvements and reduction of hand-held geographic position system (GPS) equipment now allow a researcher to precisely determine latitude and longitude for samples of habitat, this simplifying the ground-truthing of satellite data. Desktop computer versions of GIS software make it possible to conduct sophisticated habitat analyses without acquiring specialized equipment.

New techniques such as gap analysis, which evaluate the protection status of plant and animal communities by GIS overlay of distributional data on maps of existing protected areas, help conservationists to be proactive rather than reactive in their efforts to preserve biodiversity. The accessibility of satellite data and new analytical techniques resulted in three large steps forward in terms of establishing conservation and research priorities in the Neotropics.

First, we now are able to quantify changes in vegetation patterns over time. For example, in tire 1980s, widespread disagreement over rates of deforestation in the Brazilian Amazon hampered efforts at international collaboration in conservation. In the 1990s, independent reports by Brazilian and American scientists agreed that previous estimates were several-fold larger, than actual rates of deforestation.

Since future treaties on biodiversity and climatic change are likely to require countries to limit their emissions of greenhouse gases, including those resulting from deforestation, estimates of deforestation based on satellite data acquire additional political and economic importance. Second, we now can use satellite data to test hypotheses about the effects of anthropogeic factors on habitat quantity and quality. For example, we can use archival data to document habitat changes that followed the construction of a road into a frosted area in the Amazon and subsequently use that information to predict how a proposed road might alter undisturbed forest in another region.

Third, based on any criteria that can be mapped, it is now possible to quantitatively evaluate areas for creation as reserves, or as corridors between reserves. Remote sensing and GIS techniques are becoming increasingly powerful research tools and should be included in the tool kit of any researcher or conservationist doing field work on neotropical primates.

Ecology and Evolution

Hundreds of scientific reports have been published in the past decade on the ecology and behaviour of neotropical primates. Field research using the comparative method, long-term demographic studies, and research addressing the effects on human disturbance factors on primate population size and structure. Although the principles of evolutionary biology should be useful ill explaining primate behaviour, primates rarely have been at the forefront in this work. Cheney *et al.* identify several reasons why primates are rarely used to test evolutionary theory.

Most species of primates are difficult to identify individually or even to observe under field conditions. Many species are so long-lived that accumulation of data on life histories may require decades of observation. In addition, the social behaviour of many species is sufficiently complex that it becomes difficult to collect information on a sufficient number of individuals to obtain a complete picture of patterns of social organization.

Seasonal and annual variation further complicates the design of field studied examining the effects of ecological variables or the adaptive significance of specific behaviours in tropical ecosystems. Finally, formulation of generalizations often is impeded by differences in researchers, methodology. One approach that has proved useful in circumventing these obstacles is to ask how several primate taxa behave under the same circumstances and when the same observational methodology is used.

Conclusion then may be inferred from differences between species. This comparative approach has the added advantage of focusing on an interesting question rather than on an interesting primate. A good example of the comparative method applied m field research on primates is Terborgh's long-term study of the comparative ecology of five New World primates in Man National Park, Peru.

Comparisons across species allowed Terborgh to make inferences about, for example, the function of territorial defence, factors that influence size of the group and the relationship between size of the group and the social system in primates. Another fruitful comparative study was conducted by Peres at a remote site in Amazonas, Brazil. Peres compared the feeding ecology and habitat use of 13 primate species for a period of 20 months.

Parallel studies of native and reintroduced tamarins allowed researchers to generalize about the effects of captivity on locomotion, communication range use foraging, and reproduction. A second approach that has proved useful both in terms of basic research and cosnervation application is long-term monitoring of derflographic variables for a single primate population.

The conservation implications of these data are discussed below. Calculation of reproductive success under field conditions can be problematic, particularly for species of primates in which more than one male copulates with a reproductive female. Under these circumstances, new molecular genetic techniques, notably DNA finger-printing, may allow paternity exclsion and thus direct determination of male reproductive success. It long has been thought that deforestation and have negative effects on most primate populations. However, the effects of these and other anthropogenic factors rarely have been quantified for neotropical species.

In recent reviews of the impact of habitat disturbance on a variety of species of primates, Johns and Johns and Skorupa reported "that body size and degree of frugivory were negatively correlated in species" probability of survival. Redford and Robinson surveyed the available literature and predicted that hunting by humans would cause primate population density to decrease as prey body mass increases. Freese *et al.* surveyed primate densities in Bolivia and Peru and suggested that predation by humans was the most important factor affecting primate densities.

Peres used transact censuses of primates in hunted and unhunted areas to test hypotheses about the effect of hunting on primate densities. Peres found that body size alone largely determined the choice of primates hunted in Amazonian forests and that large species, such as woolly monkeys, often were hunted to extinction in areas near human settlements, even in the absence of habitat disturbance. Redford reached similar conclusions for a variety of large animals in neotropical forest. Two conclusions are suggested by these studies.

- Conservation efforts need to focus on large-bodied primates with low rates of reproduction, even before habitat destruction begins.

- Past and present hunting pressure must be included in population dynamics of large-bodied primates, even in areas of apparently intact habitat.

Conservation strategies

The alarm that sounded in the mid-1980s about the immediate threat to most species of neotropical primates was answered in a number of ways. I have classified these strategies as information management and networking application of scientific method, captive breeding, conservation education and international collaboration. The advent of widespread use of electronic mail dramatically increased the speed and decreased the cost of international communication.

Transfer of data files, even from remote field stations, has become routine in the past years. Computerized bulletin boards are available on a number of relevant topics, including conservation biology, primates, ecology, the Neotropics in general, as well as several tropical countries. Many international nongovernment conservation organisations (NGOs) support regional programs to develop and conduct conservation and activities in neotropical countries.

Some of these initiatives, e.g., the Golden Lion Tamarin Conservation Program in Brazil, which has been supported for 10 years by grants from the World Wildlife hind and other NGOs, focus on conservation of primates and their habitat. The Nature Conservancy sponsors the establishment of Conservation Data Centers in 3 Latin American and Caribbean countries to aid in decision-making in conservation and sustainable development. Conservation International supports a Rapid Assessment Program in which teams of experts census diverse taxa of fauna and flora to set priorities for conservation action.

Scientific method

The past decade has seen a heartening increase in the generation of conservation theory and the application of rigorous scientific method and problem-solving techniques to conservation questions. Graduate degrees in conservation biology now are offered at many universities in the United States and Europe.

The National Science Foundation now funds grant proposals in conservation and restoration biology. Although the primary concern of conservationists is the survival of ecosystems, much of the theoretical and applied research in conservation biology is conducted at the population level. The link between population and ecosystem conservation is thought to be the appropriate selection of population of "flag ship species" that serve as the focus of conservation campaigns, "umbrella species" that have ranges large enough to include significant portions of natural ecosystems, and "keystone species" whose presence or absence determines the abundance and distribution of a number of other species.

Computer models such as VORTEX are used to estimate the probability of population extinction and time to extinction based on the effects of stochastic events such as inbreeding environmental variation and catastrophes on population demographic variables. The probability of survival of multiple subpopulations in a metapopulation also may be estimated.

Examples of PHVA applied to four species of lion tamarins are presented in Seal *et. al.* The advantages of PHVA include the simultaneous treatment of several variables, the collaborative process by which the best available information is assembled for all interested parties, the quantitative nature of the results, and the versatility in evaluating management options. The shortcomings of PHVA are related to the assumptions implicit in the computer model and the effects of estimating unknown values.

For example, the impact of environmental variation on fecundity and survivorship rarely is known for population is nature. Small changes in the specified effects of environmental variation on reproduction or mortality can produce 10-fold differences the predicted probability of extinction of the population.

Captive breeding and reintroduction

Not long ago, zoos were seen as consumers of primates and other animals, competing to assemble a collection of the rarest and most valuable examples for public exhibit. Recent changes in the objectives of modern zoological parks make these institutions important contributors to *in situ* conservation of neotropical primates. The principal objective of captive propagation now is seen as the re-establishment or reinforcement, not replacement of wild populations. With over 13,000 specimens in zoological collections, primates are overrepresented relative to other mammalian orders.

Nevertheless, only about 50% of primate taxa are represented in captivity. International management committees, organized to maximize retention of genetic diversity and minimize probability of Ion in captivity and the wild, are increasing for many vertebrate taxa, including primates. A number of advisory groups work to coordinate the *ex situ* and *in situ* primate conservation activities of zoos.

In 1992, a Taxon Advisory Group (TAG) to the American Association of Zoos and Aquaria (AAZPA) was formed to coordinate the facilitate captive breeding of New World Primates in support of *in situ* conservation efforts. Faunal Interest Groups (EIGs) coordinate conservation activities in zoos for fauna and flora from a specific region, e.g., Brazil, Studbooks, containing complete demographic records for a species ill captivity, facilitate the scientific management of captive populations, e.g., by facilitating selection of appropriate individuals for pairings reduce loss of genetic diversity. Studbooks are maintained for 15 species of neotropical primates, and have been proposed for several others. For a few species, information on native populations is sufficient for inclusion in studbooks.

The reintroduction of captive-born neotropical primates has been initiated for several species in the past few years. In addition to the reintroduction of golden lion tamarins mentioned above, reintroduction and monitoring of *Callithrix geoffroyi* in Espirito Santo state, Brazil, began in 1991. The costs and criteria for success in a primate reintroduction are reviewed by Kleiman *et. al.* The IUCN-SSC Reintroduction Specialist Group and the AAZPA Reintroduction Advisory Group both were created in the past few years to facilitate in situ conservation through integration of wild and captive population management.

Conservation Education

The role of community education in directing the attitudes and behaviours of local peoples toward sustainable use of natural resources long has been recognized as an essential component of any viable conservation strategy. This is particularly true where charismatic species such as primates can be used as flagships for conservation efforts.

Whereas 42.5% of these responses were attributed to project activities, only 6.9% reported seeing tamarins in the forest were mentioned most often included television and radio, a total effect roughly equivalent to all other project activities combined. Graduate training in conservation education for South and Central American nationals is one of the best investments that can be made to ensure the future of neotropical primates.

International Collaboration

A final example illustrates progress in international collaborations. In this case, private lands were purchased and annexed to a Brazilian federal reserve too small to guarantee the survival of an endangered primate. A PHVA for golden headed lion tamarins, conducted in the Una Biological Reserve, southern Bahia State, Brazil, suggested that the probability of extinction of that primate was relatively high. Based on the findings of the PHVA, a coalition of eight international conservation organizations conducted fund-raising efforts to buy out squatters and purchase land to annex to Una Reserve.

Funds were transferred in several instalments to Fundacao Biodiversitas, a Brazilian NGO, which purchased lands at market price from local landowners and immediately conferred title to the Brazilian agency responsible for administration of the Reserve. To date, the area of the Reserve has been increased from 5.342 hectares to 7.059 hectares.

12
Biodiversity Loss and Threat

Data from the last three decades demonstrate that on a worldwide scale the total area and woodland has remained fairly stable. Considering the conversion of natural forests in many parts of the world, this means that in general nature natural forests are being replaced by woodlands which are the result of human activity, be it planting of trees or the development of secondary forests on land which has been cultivated in the past. With the increasing demand for land, and for wood worldwide, this trend is likely to continue for decades to come.

Consequently, it can be expected that forests which bear the marks of human influence will play an increasingly important role in species conservation, particularly in tropical countries where the largest biodiversity must be preserved, but also where economic needs put considerable pressure on natural resources. This is being increasingly recognized by forestry officials and forest managers worldwide. Many countries in the temperate zones which no longer have any significant amount of mature natural forest manage biodiversity in anthropogenic forests.

In many European countries there is only little need to generate economic returns from those forests, while in the USA the National Forest Service recognizes a serious need to balance economic demands with biodiversity. In many tropical countries where forests are managed for economic interests, maintaining a diverse forest with a large number of species is often considered contradictory to the principal goal of increasing returns. This, for instance, remains the case with most of the plantation and land development programs that the Indonesian Ministry of Forestry has designed for areas of forest land which have been designated Conversion Forest. Indonesia has a total land area of 199 x 10^6 ha, including about 10^9 x 10^6 ha of forest and woodland of which 86.4 x 10^6 ha are forest cover. This is about 6% of the world's total remaining tropical forest, and the third largest area after.

Brazil and Zaire, Indonesia's forests contain abut 10% of the world's plant species, 12% of the world's mammals, 16% of the world's reptiles, and 17% of the world's bird species. Because of its unique geographic location on both sides of the Wallace Line, between the continents of Asia and Australia, a high rate of species endemism occurs. On the other hand, Indonesia's forests constitute an important economic resource for the country. Until recently, timber was the second largest generator of foreign exchange after oil, and remains an important source of development capital, notwithstanding this economic importance. Indonesia has repeatedly expressed an awareness of it's responsibility tot he world community to preserve its forests and the biodiversity they contain.

In order to safeguard both the important economic role of the timber industry and the country's biodiversity. Indonesia has functionally divided its large forest. This division was laid down, until

1996, in the Forest Land Use by Consensus which is now being replaced by Provincial Land Use Ordination Plan. Roughly one-third of the country's forest lands are set asside for conservation or protection purposes. The larger part is designated as Production Forest, while the remaining area is designated as Conversion Forest or Nonclassfied Forests. Conversion and Non-classified Forests may be used for agricultural or other purposes.

Hence, the GOI has made it a goal to eradicate all forms of swidden agriculture in the whole country. It has designed several programs to reach this goal. These programs vary according to the category of forest in which they are being conducted. So called 'social forestry' programs have been designed to stop forest encroachment by small farmers in Conservation, Protection, or Production Forest areas.

Furthermore, we argue that alternative choices for economic development of swidden agriculturists can be designed which can preserve, or even enhance the biodiversity in these same areas. In particular we will try to demonstrate : (1) that conservation of biodiversity is an integral part of resource management among the very same swidden agriculturists as an important subsistence strategy : (2) that several of the programs which are designed to eradicate swidden agriculture in Indonesia ironically will lead necessarily to a loss of this biodiversity that is maintained by swidden agriculturists : and (3) that instead of promoting large-scale timber plantation or estate plantations, developing traditional forest management is both an attractive option for regional economic development as well as a way of preserving the biodiversity that is an integral part of many swidden agricultural landscapes.

It is estimated that swidden agriculturists are responsible for about half of the estimated 1 x 106 ha land deforested in Indonesia yearly, although little accurate data exist to sustain such an estimate. Of the total area under swidden agriculture, about 25% is in Production Forests, 15% in Protection Forests, 29% in Conversion Forests, and 31% in Unclassified Forest. Even though 60% of swidden agriculture is located in forest land which is planned to be used for non-forest purposes, swidden agriculture is considered to have serious implications for the future of forestry in Indonesia.

Since the early 1970s, the GOI has designed programs to deal with swidden agriculturists. In their earlier phase these programs attempted to relocate swidden agriculturists to out of forest areas. Especially since the mid-1980s, when the issue of sustainable forest utilization has become widely discussed in Indonesia, many directives from the Ministries of Forestry, Agriculture, Transmigration, Social Affairs and Home Affairs have appeared which are meant to 'stabilize' swidden agriculture.

The social forestry, or village development programs, which are applied in Conservation, Protection, or Production Forest areas, aim to give people responsibility for managing the forests in their areas. In fact social forestry programs are still ill-defined, and subject to much discussion. Experiences with social forestry in Java, where local control over forests was replaced by government control some time ago, give little hope for similar programs for the outer islands.

Social forestry is not attempted in areas which have been designated as Conversion Forest. In this category of state forest land three different programs have been defined : (1) the establishment of industrial timber estates; (2) the nuclear estate program and (3) the Rehabilitation and Expansion of Export Crops program. The timber estates development program began in 1984. This program has a high, priority with the GOI, since in the near future, it is supposed to substitute for the expected decline in supply of timber as a result of exhaustion of the natural forests. Until 1988, some 1.4 x 10^6 ha of timber production plantation existed. The largest parts of those timber estates are still found in Java, but the second largest parts are located in the four provinces of Kalimantam with most in West Kalimantan Indonesia's goal is to reach 6 x 10^6 ha of industrial timber plantation, by the year 2000. These plantations are to be established in Conversion Forest or Production Forest areas if they do not fall under a forest concession and do not contain sufficient commercial natural forest.

In practice many state HTI projects have been established in secondary forest, scrub land or grass land. The location of HTI estates in secondary forest areas implicitly means that it is located in land that is occupied by swidden agriculturists. One of the objectives of HTI estates is also to rehabilitate degraded lands. However, often any land that is under fallow is qualified as degraded by forestry officials, although from the swidden agriculturists point of view this land still may be productive rice land.

The participation of swidden agriculturists in industrial timber estates is mainly as providers of labour and land. Land needed for the estates is obtained from local farmers who are given some compensation, but this is not higher than US$ 15.00 ha. In the original wans the timber estates were to provide work to the local population, hence generate income, and thus reduce the need for swidden farmers to grow their own food, and slash more forests.

The wages paid are rupiahs 3750 day 1, or US$ 1.70. Only when the implementation of industrial timber estates is combined with the Transmigration Program are housing infrastructure, schools, medical posts, and the like, provided to the transmigrants coming from Java, Madura, Bali, and Lombok, as well as local farmers who agree to participate. HTI/TRANS projects aim to include about 50% of 'local' transmigrants in their schemes. The NES/PIR or Nucleus Agricultural Estate Program, which began in 1979, and which is the responsibility of the Ministry of Agriculture, aims at establishing estate crops. In this program government or private companies establish an estate, which may be as large as 50,000 ha. About 60% to 80% of the land is dedicated to small holder production of the estate crop.

Participant farmers are given 2 ha of land to establish the estate crop, and 0.5 1 ha for a house lot and a home garden. The estate company provides facilities to the participating farmers to establish their 2 ha of estate crop, and it guarantees to buy the yield, as well as to provide technical assistance and credits for inputs. Farmers accept these facilities as a loan which they have to pay back. In West Kalimantan farmers must repay US$ 4900 plus an interest of 6%. Farmers who become estate members commit themselves to work as labourers on the company-owned part of the estate, for which they receive daily wages.

In many PIR/TRANS projects local swidden agriculturists are incorporated also, but since the estates rely on transmigrants for their labour, fewer local farmers can be involved which means that relatively more farmers will be resettled. Farmers usually do receive compensation for the land they give up. Reports, however, document the coercive, methods used by the state companies to obtain traditionally held land. NES/PIR or PIR/TRANS estates are entirely converted to mono-crop plantations, except for a smaller part which is converted to food crop area.

In 1993 the total area under oil palm plantation in West Kalimantan was 142, 403 ha the total area planned to be converted into nuclear estate crop is 1.2 x 106 ha, or 8% of the total surface of

West Kalimantan. The third program that affects swidden agriculturists in West Kalimantan is the Rehabilitation and Expansion of Export Crops program. This program provides credit to farmers to improve the quality of their small holder plantations. In the outer islands, support is given for 2 ha of export crop land. In West Kalimantan this program only focuses on improving the quality of rubber production. Very soon after, the first *Hevea* rubber appeared in Sarawak early this century, it also was brought to West Kalimantan by the Dutch.

Most rubber, however, has been grown as 'forest rubber', planted on swidden fields, and left to grow, together with the spontaneously, occurring secondary vegetation. As a result, most rubber gardens are diverse, species rich, anthropogenic forests. Farmers who receive credit and technical assistance under the Expansion of Export Crops Program commit themselves to plant only improved robber clones. These are high yielding rubber clones, which are much less resistant to pests than the locally used varieties.

In many cases, how ever, farmers have actually been sold the locally existing rubber as imporved high yielding rubber.

Farmers who plant improved rubber clons are supposed to conduct a thorough slash-weeding from the onset of the plantation until the trees are large enough to be tapped. As a consequence, improved rubber gardens have a characteristic mono-crop species composition which is similar to that of Pinus *Merkusii, Acacia mangium,* or oil palm plantations and not like the diverse forest rubber gardens found in Kalimantan and Sumatra.

A dominant, mono-culture plantation has a grassy weed vegetation covering the forest floor. Since currently more than 4,00,000ha of land are under traditional species-rich smallholder rubber gardens in West Kalimantan, the extension of loss of biodiversity can expected to be considerable if these areas are to be converted to monoculture improved rubber gardens.

In 1990 the whole island had 12.7 x 10^6 inhabitants, 9.1 x 10^6 of which lived in the Kalimantan provinces. About one-third of the people who live on the island are broadly classified as indigenous Dayak. Throughout the history of Borneo the island has been of interest for its natural resources. In pre-Dutch colonial periods for various forest resources and minerals and for the last several decades for its oil and timber.

Gradually various other groups came to claim their niche : the Chinese, the Buggies from Sulawesi, and later people from other islands from the archipelago, as transmigrantes, or as spontaneous migrants, continuously pushing back the native Dayak to more remote places. Together with the booming of timber and oil exploitation came the large scale opening up of large parts of West, South and, too a lesser extent East Kalimantan. This opening up was followed by large scale agricultural development projects. The native Dayak population mostly living in the remoter interiors relies for its daily staple on upland rice cultivation, produced in yearly cleared swiddens. Much controversy and confusion still exists about swidden agriculture.

The term covers agricultural practices which begin with the slashing and burning of vegetation of an area of land, using it for intensive cultivation for one or several years, followed by fallowing the same plot for a longer period. Among most swidden agriculturists, fallow fields continue to have an economic importance since the secondary vegetation contains spotaneously occurring species, or perennial species which, were planted at an earlier stage and which continue to yield useful products.

Some of these species occur mainly in fallow vegetation, and are only exceptionally found in primary forest Jessup claims that some Kenyah farmers in East Kalimantan purposely maintain a sufficient area of fallow to assure the supply of such species. Hence, as a rule, among swidden agriculturists some kind of management of the fallow vegetation occurs. Among many swidden agriculturists the dominant economic imperative is to optimize returns of labour, rather than returns of land.

Swidden agriculture is still practised by many farmers in the world, and subsequently appears in, many different forms. Even on the island of Borneo, significant variations of swidden agricultural patterns can be observed. For instance, lban farmers historically had the reputation of preferring land that was newly opened in mature natural forests, and shied away from making swiddens in secondary forest, although Jessup and Sellato question the validity of such generalizations based on ethnic affiliation. Other groups, like the *Kenyah* and *Kenyah,* cultivated the area around a new settlement for several decades, and then moved the whole village to a new site.

A reason to move the village could be the exhaustion of agricultural land, or because diseases occurred at increasing frequencies, or in some cases the location had become unsafe as a result of invasions by other groups. Recent migrations have occurred because people wanted to gain easier access to markets and government services. The *Bidayuh,* a name originally used to distinguish the non-Malay, non-Iban inhabitants of Sarawak's first division, constitute several different linguistic subgroups, who originated from the Indonesian part of West Kalimantan, and which are found as far South as the Ketapang basin.

Bidayuh farmers practice a type of swidden agriculture which is more sedentary than that of several other groups in Borneo who migrate more often. Rather than moving entire villages, Bidayuh villages which become too large, are split and only portions moved to a new area of forest to start their own settlement. The stable Bidayuh villages, however, frequently move the location of their settlement within the village territory, over distances that do not exceed about 1 km.

The more permanent character of Bidayuh settlements has certain consequences for the way swidden agriculture is practised. In order to be able to reside over the long term in a single region, a farmer must have a sufficiently large holding of land under various stages of fallow. This holding is built up gradually from when a couple become an independent farming household, until the area is large enough such that fields can be used and followed at sufficiently great intervals. Since *Bidavuh* farmers have more stable holdings, the area to which they hold tenure claims are smaller than, for instance, the ever-moving *Than.*

The landscape of a typical *Bidavuh* village is usually characterized not only by land under swidden and fallow, but also by many patches of forests of various sizes. For instance, among two different *Bidavuh* groups in the North of West Kalimantan, *Mate* Dayak and *Jangkang* Dayak, six types of managed forests have been described. These include communally held forest, or *hutan ttupan,* which are usually located close to villages and may be as large as 100ha. *Hutan tutupan* mostly function as a reserve for timber for house construction, rattans, medicinal plants, fishing and hunting sites, and to protect water source. A smaller version of the same type of forest, of between 1 and 3 ha, is forest islands or *pulau rimba.* Such patches of forest become surrounded by one household's agricultural fields or fallows and according to local customary law, became private property of this household. These forests are often kept as private forest reserves, and yield the same products as *hutan tutupan.* In the same category of protected areas of mature natural forest fall the *Sompuat,* which are patches of forest which surround honey trees, mostly *Koompassia excelsa* or *Koompassia malaccensis.*

The *sompuat* forests are kept intact in order to provide an appropriate habitat for the *Apis dorsata* honey bees that built their nests in the central tree. Obviously, preserving areas of mature natural forests also implies preserving the diversity of species that are present in them. However, that is not where *Bidayuh* biodiversity forest conservation stops. Two types of managed forests are important, riot only economically, but also for the diversity of plant species they contain. These are *tembawang* or fruit forest gardens, and rubber gardens created by individual farmers planting trees on sites previously used for agriculture or around the settlement area, or sometimes directly in mature forest.

Planting may continue for many years until eventually most of the area designated for tree planting is covered. Such *tembawang* are privately owned and inherited by descendants of the people who established the forest. They are usually about 1-2 ha in size. Since many farmers in a single village create individual *tembawang* on adjacent fields, over time these merge into a single stretch of contiguous forest. Because of the short distances over which the village is frequently moved, the *tembawang* area often trails the advancing settlement as a long band of high closed forest. Rubber gardens are usually located adjacent to *tembawang.* Once the rubber seedlings are planted, no wedding occurs until the rubber is fully grown; about 10 years.

Only then are trails established to reach trees which have to be tapped. Very often additional fruit trees are planted in rubber gardens, as a result of which they tend to become an extension of the *tembawang* area, rather than a totally different type of managed forest. The biodiversity conservation role of managed natural forests like *hutan tutupan, pulau rimba,* and sompuat are obvious. However, in this respect tembawang and rubber gardens are no less important.

In one of the villages researched, for instance, 400 plant species were recorded in five *tembawang.* These five *tembawang* had an average of 170 plant species per hectare, while one specific *tembawang* had a total of 231 plant species per hectare. In two plots of mature natural forests in the research area 256 and 258 plant species per hectare were recorded. High species diversity can also be found in rubber gardens.

From the biological conservation standpoint therefore, it would be better to leave swidden agricultural land the way it is, rather than try to implement large-scale plantations of mostly single species. However, the objectives of HTI, NES/PIR and NES/TRANS and export crop development are not primarily to conserve biodiversity, but rather to increase or sustain the state's income from timber production, or estate crops.

These incomes are indeed intended to increase living standards of local fanners, but also enhance the country's timber, oil palm and rubber industry, in order to provide income for the state and to generate foreign exchange. The programs that deal with swidden agriculturists in Production, Protection, or Conservation forests, focus largely on developing alternatives which reduce the reliance of farmers on forests as a resource to create agricultural land. Some of the strategies are to set aside forest land from which products can be obtained for sale in local markets to obtain monetary income.

The programs reserved for areas of Conversion or Unclassified forest land do not consider options of forest exploitation, although they are required to dedicate some land to other uses. For instance, the HTI estates have to dedicate 10% of the land to *tanaman kehiduppi,* which can be private vegetable or mixed fruit gardens. Similarly, PIR projects provide land to participating farmers for their own house gardens. Considering the objectives of estate development, alternative choices will have to provide the same incomes as the estate options to be socially acceptable. They have to be generated at least partly from some kind of exploitation of forests in order to be able to choose alternative approaches for developing swidden agriculture without losing its intrinsic biodiversity.

Reports exist from *Bidavuh* farmers in central West Kalimantan who gain substantial income from *durian* yields from *tembawang* forests. These fruits are being bought by traders from Kutching, Sarawak who travel for 6 h to reach the producers during the fruiting season. Expanding markets in urban Indonesia, Malaysia and Singapore only will increase the demand for such products.

Hence, the opportunity to increase incomes by producing durians in diverse forests gardens only will increase. Existing forest management practices, furthermore, provide a unique opportunity for smallholder timber production which eventually could sustain the supply of the local timber industry. Such timber can be planted in and harvested from rubber gardens, *tembawang, pulau rimba* and *hutan tutupan.* The 33% of area under forest in the village where Dayak forest management was studied suggests the potential that such traditional forest management holds for the production of high quality timber.

Some kind of tree cover creates a better environment to plant species which are naturally found in closed forests, rather than in open terrain where many seedlings die, or where fires may destroy plantations. Additionally, in diverse holdings, other faster-growing species could be produced on small lots which would be rotated, for instance, with rice production in order to supply an industry that needs faster-growing species of lower wood quality.

13

Biodiversity in Farming

Agricultural progress seems even more paradoxical when we consider biodiversity use in agriculture. Several thousand plant and animal species are known to have served as food in human history. The displacement of traditional crop varieties by newly-developed high-yielding varieties has been a major factor in this development. FAO estimates that in the next 10 years 90% of the existing genetic diversity within the main crops is at serious risk of disappearing. The vast majority of the literature dealing with biodiversity use in agriculture is characterized by a symptomatic approach, and centers on providing data that prove the continuous reduction of both inter and intra-species biodiversity used in food production on its associated risks for the sustainability of our food supply and on alternative farming techniques to halt the erosion of biodiversity use in agriculture.

A discussion of the driving forces that have generated the reduction of biodivesity use in agriculture is generally missing. Put another way, the detection of symptoms is directly coupled to a prescription of remedies without any serious attempts to diagnose that diesase. Perhaps this is because a thorough diagnosis would necessarily raise some politically unpleasant issues. The decreasing energy output/input ratio in agriculture means that the food security of humankind is increasingly dependent on rapidly dwindling oil reserves and on an increasing environmental impact per unit of crop output. The dramatic reduction of biodiversity worldwide spells serious problem for the maintenance of ecological, equilibria at the planet level. That is the path of technological in food production chosen right now by humankind is not sustainable. Not only are these unpopular issues to discuss, but it is becoming evident that most of the undesirable side-effects leading toe unsustainability are difficult, if riot impossible, to incorporate in conventional sets of measures and indicators used to assess risks, costs and benefits of technological change. In spite of these difficulties, it is crucial to understand the driving forces that are shaping current trends of technological development in otherwise, we seriously risk caveats in discussions of future agricultural policies. It is common knowledge that traditional techniques achieved high output/input energy ratios and used much more biodiversity than modern high-input agriculture.

Individual scientific disciplines that study change in socio-economic systems tend to focus on the effect of one factor at te time. Although this approach provides a high resolution for the factor in question, it does not shed much light on the bigger picture. In this respect the agricultural sciences are no exception.

Ecologists tend to focus on the on-site and off-site effects of agriculture on the environment without considering agriculture as a livelihood. Economists are concerned with the economic return of the labour and capital resources invested in the farming process, and tend to more the ecological impacts. Agronomists occupy themselves basically with the physical performance of the production system. Rural sociologists and anthropologists are mainly interested in the effects of technological changes on the lives of individual farmers and the identity of rural communities. This specialization or *reductionism* of established scientific discplines is at times necessary, but often creates confusion for decision-makers that have to decide about the future directions of technical development in agriculture.

All of the above-mentioned aspects of agriculture must be addressed since each one is fundamental for the sustainability of agriculture. The problem lies in the fact that a detailed discussion of any one of these aspects becomes useless if not put into relation to the functioning of the system as a For this reason, it is of utmost importance to put the different disciplines in perspective within a common frame of analysis.

The existence of different perspectives for assessment of sustainability of agricultural development has been pointed out by many workers. In spite of minor differences in definitions and emphasis among the different authors, there is substantial agreement that at least three fundamental perspectives should be considered when attempting to define sustainable agriculture:

— The ecological view: agricultural techniques must be environmentally sound:

— The economic view: agricultural techniques must be economically viable:

— The social view: agricultural techniques must be acceptable to farmers and society given culture, ethics and religon.

Assessing only agronomic performance or ecological compatibility of various agricultural techniques is not sufficient to evaluate their sustainability. We need a parallel check on economic viability and social acceptability. The farming system as a sub-system of society: constraints imposed by population density, income level and labour time allocation in society. In order to verify whether certain agricultural production techniques are convenient for a given society, we should first analyse how the organization and technological development of society influence the way in which food is produced in its agricultural sector. Two main characteristics of society that I will examine in this paper are:

1. Population density, or demographic pressure. One of the most obvious aspects of demographic pressure that directly influences the agricultural sector is the amount of arable land per capita available for agricultural production.
2. The level of technological development of a society, or socio-economic pressure. This can be conveniently quantified by the following two indicators:

— The average per capita income or average per capita commercial energy consumption ill society.

— The fraction of total human time available that is allocated to work in the primary sectors of the economy. Of the primary economic sectors, agriculture is one of the most affected in terms of contraction of human time allocation. A cross-section analysis over a sample of 60 countries for the years 1990-1991 shows that the fraction of the labour force engaged in agriculture is inversely correlated with the GDP per capita. Countries with GDP per capita of more than US$ 10000 per year have less than 7% of the working force in agriculture, and the richest countries even as little as 2%.

The demographic pressure or population density of society is simply defined as the ratio. It is well known that population density has been an important factor in shaping the development of agricultural production techniques. Apart from scientific analyses, common sense also suggests that a high demographic pressure in society will favour farming techniques that yield a high food production per unit area.

The effect of demographic pressure on agricultural production can be expressed by the indicator AP_{DP} (=agricultural productivity imposed by demographic pressure), which is defined as the productivity of land (yield of food energy per hectare) that would be needed to obtain a situation of complete food sef-sufficiency in society (matching the aggregate food requirement of society). Thus defined numerical assessments of AP_{DP} require knowledge of the aggregate requirement of food in society (considering the food system under analysis as closed), which is in turn determined by the population size of society, the food consumption pattern, and post-harvest losses: The amount of land available for food production , which depends on availability of arable land, characteristics of arable land, and alternative land-uses (dependent on population size and level of technical development).

The effect of socio-economic pressure on agriculture can be expressed by the indicator AP_{SEP} agricultural productivity imposed by socio-economic pressure, which is defined as the productivity of labour in agriculture (yield of food energy per hour of labour in the agricultural sector) that would be needed to obtain food self-sufficiency for society.

Considering society's food system as closed, AP_{SEP} is obtained by dividing the yearly aggregate requirement of ood in society by the labour time available in a year for generating food supply. The labour time available in a year for generating the food supply depends on the size of the labour force, the unemployment rate, tile fraction of the labour force obsorbed by the non-agricultural sectors, and the average world load. The US economy, with only 2% of its workers engaged in agriculture, has an APSEP of 270 MJ of food energy per hour of agricultural labour. This value is well out of the range of possible productivity of fanners in developing countries (less than 10 W 1 and is not even reached by frmers in the European Union. The requirement for food in society is proportional to population size, and therefore to the total human time available to society, whereas internal food production is proportional to the labour time available in the agricultural sector. Hence, in order to achieve and maintain food self-sufficiency, any reduction in the fraction of total labour time allocated to the sector must be compensated by an increase in the amount of food produced per hour of labour in the agricultural sector. According to this rationale, a socioeconomic structure that allocates only a small fraction of its labour force to the agricultural sector will select farming techniques that result in a high food production per unit of labour in the agricultural sector.

The indicators AP_{DP} and AP_{SEP} represent the minimum density of biophysical throughput, per hectare and per hour, respectively, that would be needed on average in the agricultural sector to achieve a state of food self-sufficiency in a defined society. Hence, they are based on characteristics of the socioeconomic system as a whole, and their value can be calculated without any knowledge of what is going on at thelevel of the fanning system in the society. How, then, can these indicators be related to the actual technological changes occurring at the farm level? Are the values of the indicators AP_{DP} and AP_{SEP} that characterize the pressure that society exerts on its fanning systems, effectively related to the techniques of food production realized at the farm level? to answer these questions it is useful to compare, for several types of societies, the values of AP_{DP} and AP_{SEP} with actual values achieved in agricultural production in these societies. For instance, the minimum productivity of land imposed by demographic characteristics (AP_{DP} can be compared with the acutal agricultural productivity per hectare (AP_{ha}) for the country as a whole (average), or at the regional level, or for individual farms. The actual productivity of land or yield of food energy per hectare achieved by a country, or farming system is obtained by dividing the aggregate production of food within the defined territory by the amount of land used in the production process. Similarly, expected minimum value of labour productivity in agriculture according to socio-economic characteristics can be compared with the actual agricultural productivity (throughput of food energy) achieved per hour of labour in agriculture for the country as a whole (average) or for selected farming systems. Such a comparative analysis, at the national level has been carried out for a large sample of countries, to see whethr the demographic and socio-economic pressure are indeed related to the way in which food is actually produced in different socio-economic contexts.

The analysis shows that even if the two indicators do not indicate mandatory threshold values, they should in any case be considered attractor values, toward which the agricultural system feuds to move in practice. If a country has a high demographic pressure it is also likely to have a large production of food energy per hectare in the agricultural sector.

Countries not achieving self-sufficiency must rely on imports or food aid to guarantee food secutiry for their population. The analysis described above indicates the existence of a link between the variables AP_{DP} and AP_{SEP} on the other hand, that are defined at the hierarchical level of society as a whole, and the variables AP_{DP} and AP_{hour} on the other hand, that are defined and measured at the hierarchical level of farming systems. This link suggests that is is possible to describe technical changes in food production on both levels on the same geometric plane. This plane is defined by two axes, the density of agricultural throughput per hour of labour and the density of agricultural throughput per hectare of land.

Considering the process of agricultural production, changes in AP_{ha} and AP_{hour} reflect changes in technical coefficients of technologies adopted at the farming system level. On the higher level, society as a whole, changes in AP_{DP} and AP_{hour} reflect alterations in the characteristics of the socio-economic system to which the agricultural sector belongs. Being part of both levels, agriculture must operate in such a way that the combination of values of AP_{DP} and AP_{hour} is compatible with the constraints imposed by society, that is the combination of the values of AP_{DP} and AP_{SEP}. Clearly, farming systems of subsistence societies occupy in area of the plane that is quite distant from that occupied by farming systems of industralized countries. In general, farming systems tend to have an actual agricultural performance that is closely related to the set of socio-economic constraints corresponding to the socio-economic context in which they are operating. Describing changes in agricultural production on the plane formed by the agricultural throughput per hour labour and the throughput per hectare of land makes it possible to define trajectories of changes for both socio-economic variables and technological coefficients in agriculture, which are expected to accompany trends in technological development. For instance, current trends in economic development suggest an economic growth of developing countries that is characterized by increases in both population size and per capita consumption. This will be reflected in a parallel increase in APDP and APSEP. Sooner or later, these changes in demographic and socio-economic pressure (APDP and APSEP) will be reflected into altered technical coefficients of farming techniques. That is, technological development in agriculture in this situation tends to evolve toward higher values of AP_{DP} and AP_{hour}.

General trends in the evolution of food production techniques for different types of socio-economic systems can be represented on the two-dimensional plane. Four main types of socio-economic systems, having different combinations of demographic and socio-economic pressure.

— *Societies with low demographic and socio-economic pressure:* The population is growing faster than the GNP per capita, which means that AP_{DP} will grow faster than APSEP. Hence, they will move toward a situation typical of PR China.

— *Societies with low demographic and high socio-economc pressure:* Economic development is expected to be maintained and population growth will be relatively slow but steady. On the plane, this means a slow movement toward higher values of AP_{DP}.

— *Societies with high demographic and low socio-economic pressure:* They tend to show economic growth and maintain if not expand their already huge population size. At national level, an increasing GNP per capita will result in an accelerated absorption of the labour force currently engaged in agriculture by other sectors of the economy. This will inevitably require a major increase in agricultural labour productivity to maintain security. Hence, a movement toward the West European conditions of agricultural production is to be expected.

— *Societies with high demographic and socio-economic pressure:* These societies have no alternative but to try to maintain a high material standard of living and keep population growth to a minimum. This means a more or less stable and high level of AP_{SEP} and a very slowly increasing value of AP_{DP}.

For these societies, trying to reduce the environmental impact of their food production becomes a major factor. Note that food import from the international market, a must for countries where $AP_{DP}>AP_{ha}$ and/or $AP_{SEP}>AP_{hour}$ is based on the existence of surpluses produced by countries where the relation between these parameters is inverse. At present, the USA, Canada, Australia and Argentina combined 80% of the net export of cereals on the world market.

However, at their present rate of population growth and because of an increasing concern for the environment this surplus may be seriously eroded ill the near future. for instance, the USA is expected to double its population in 60 years. Food import problems will be accelerated if a general concern for the environment will lead to a less intensive agriculture all over the world. At world level, demographic and socio-econimic pressure are certainly expected to increase. It is often overlooked that, at world level, there is no option to import food from elsewhere. When increases in demographic and socio-economic pressure are not matched by an adequate increase in productivity of land and labour, food imports of the rich will be based on starvation of the poor. Worldwide, arable land per capita is already less than 0.27 ha per capita at present, and is expected to continue to shrink because of population growth.

In addition, arable land is being lost. For instance, during the past 40 years nearly one third of the world's cropland has been abandoned because of soil erosion and other types of degradation. Most of the added land that replaces this loss has come from marginal land made available by deforestation.

The increasing dependence of the food security of poor densely-populated countries on oil may become problematic in the near future. In the last few years, PR China and other Asian countries have been among the countries with the fastest growing economics in terms of GNP. At world level, fuelling the economic development of densely-populated countries represents a major drain on natural resources, notably fresh water and fossil energy stocks. Not only do we risk that a dramatic surge in oil demand will augment oil prices, but we may soon face the bottom of the oil barrels.

The goal of achieving sustainability of food production in the future is closely related to the challenge of mediating between the socio-economic constraints 'imposed' by society, that push for more concentrated and faster agricultural throughputs, and ecological constraints 'imposed' by natural processes that push in the opposite direction.

Natural processes provide the services on which all agricultural production relies, such as water and nutrient supply, oil fertility, pollination, stabilization of atmospheric composition, and regulation of biogeochemical cycles. Respecting ecological constraints means keeping the flows of biomass produced int he agroecosystem by human management within a range of values that is compatible with regulation mechanisms typical of natural ecosystems.

A large discrepancy between the actual density of agricultural throughput and the densities imposed by either socio-economic or ecological constraints indicates unsustainability of agricultural production. Agriculture is ecologically unsustainable when the acutal density of agricultural throughput exceeds the critical environmental loading, and socioeconomically unsustainable when the actual density of agricultural throughput is too low to provide food security and/or economic viability for farmers.

Agriculture can be defined as a human activity that exploits natural processes and natural resources in order to obtain food and other products considered useful by society. The verb 'exploits' suggests that we deal with an alternation. Indeed, within a defined area humans alter the natural distribution of animal and plant populations in order to selectively increase the density of certain flows of biomass that they consider more useful for the socio-economic system.

Adopting this definition, we implicity assume that agriculture has to do with manipulation of biodiversity. Following the schema of ecosystem structure proposed by Odum, a natural ecosystem can be seen as a network of matter and energy flows in which nutrients are mainly recycled within the system and solar energy is used to sustain this matter cycling.

The amount of solar energy used for self-organization by the ecosystem is proportional to the speed of its matter cycles (in terrestrial ecosystems this also has to do with availability and circulation of water). Agriculture then involves a boosting of only those matter and energy flows in the network that humans consider beneficial, and an elimination or reduction of the flows that they consider detrimental to their purposes. In (Icing so, humans amplify some of the existing genetic information (that is contained in populations representing 'useful' biomass), and reduce or eliminate some other genetic information. In other words, changes in the hierarchical structure of matter and enregy flows in the ecosystem must be coupled to changes in the hierarchical structure of genetic information clustered in species and distributed over taxa in the geographical space altered for agricultural purposes. The hierarchical structure of genetic information can be stabilized only by a quasi-steady-state in nutrient flows, which in turn depends on the ability to use available flows of free energy.

The resonance between coded information in DNA and the metabolism of living systems implies that the reverse must also be true. The particular hierarchical structure of matter and energy flows is determined by the rate of input/output of individual components of the network and their spatial and quantitative distribution; that is, on a large scale, it depends on the hierarchical organization of genetic information. Profound changes in the densities of harvested biomass can have dramatic consequences for the ecosystem structure. The elimination of ecosystem diversity can affects the stability of the biosphere when the scale of human alteration is increased to reach global relevance.

For example, in order to significantly increase yields of biomass (the amount taken away from the agroecosystem per unit of area and time), it is necessary to fight competing species (pests) and to bypass physiological rates of nutrient cycling in the ecosystem (fertilization) when these become limiting. This calls for selection of crop species and varieties that perform well tinder human-managed conditions.

The interaction of the agricultural sector with both ecosystem and society can be studied by following inputs and outputs of the four main steps of the food system cycle:

- Food production, where food is defined as energy and matter compatible with human metabolism.
- Making the food accessible to consumers, where accessible food is defined as meals ready to be consumed according to defined food consumption patterns.
- Consuming food and generating wastes, where wastes are defined as energy and matter no longer compatible with human needs.
- Recycling wastes to agricultural inputs, where agricultural inputs are defined as energy and matter compatible with the agricultural process.

All steps are interconnected, and hence the definition of a flow as an input, an available resource, an accessible resource, or waste, is arbitrary. First of all, the definition depends on the point of view from which the system is analysed. For example, the introduction of trees leads to increased evapotranspiration, which may be negative in terms of less accessible water in the soil , but positive in terms of more available water in the form of rain clouds. Second, the definition depends on the compatibility of the throughput density with the processes regulating the particular step in question.

For example, human excrement of a small Chinese village are valuable input for agriculture, but the sewage of a big city is a major pollution problem. Whether or not the speed of a throughput at any particular step is compatible with the system as a whole depends on the internal organization of the system. Nobody expects science to make it possible to feed a person 30000 kcal of food per day, about 10 times the normal amount, while keeping that person alive for a long period of time. Why then do many believe it to be possible to increase th productivity of agroecosystems by several times without generating any, negative side-effects on agroecosystem health?

Indeed, the increasing intensity of agricultural throughputs over increasing areas of our planet should be considered a major threat to the conservation of biodiversity. The more the amount of agricultural biomass harvested from a defined population (considering a single species at the time) in an altered ecosystem differs in density from the biomass turnover of a similar species in naturally occurring ecosystems, the larger canbe expected to be the environmental impact on the agroecosystem and the energetic cost of the agricultural production.

Extremely high densities of agricultural throughput (per hectare and/or per hour) necessarily require production techniques that ignore the functional mechanism of natural ecosystems, that is the nutrient cycles powered by a linear flow of solar energy. Clearly, individual species, taken one at the time, cannot generate cycles in matter, since they perform only a defined 'job' (occupy a certain niche) in the ecosystem.

Technical progress sooner or later implies a switch from low-input agriculture based on nutrient cycling within the agroecosystem to high-input agriculture in which the throughput density of harvested biomass is directly controlled and maintained at elevated levels through reliance on external inputs that provide linear flows of both nutrients and energy. In low-input agriculture, where the harvested flow of biomass remains within the range of natural turnover values, the activity of production interferes only to a limited extent with the system of controls regulating matter and energy flows in the ecosystem. This form of agriculture requires that at least several distinct species are used in the process of production so as to maintain the internal cycling of natural inputs as a pillar of the agricultural production process.

For humans this implies a poor control over the flow of produced biomass in the agroecosystem, because of the low productivity per hectare when assessing the yield of a particular crop at the time and the variability in productivity in case of perturbations. Especially serious is the problem of facing dramatic increases of demographic pressure. This type of farming cannot work properly with levels of throughput per hectare too high.

On the positive side, direct and indirect biophysical costs of production are low. Few technical inputs are required per unit of output produced and, when population pressure is not too high, the environmental impact of this form of agricultural prodcution is modest. The contrary is true for high-input agriculture. In this case, the harvested flow of biomass is well out of the range of values that is compatible with regulation processes typical of natural ecosystems (when considering yields of a single cultivated species at the time). In high-input agriculture, human management is based on an eradication of the natural structure of controls in the ecosystem.

In fact, where several tonnes of grain have to be produced per hectare and hundreds of kilograms of grain per hour of agricultural labour, natural rates of nutrient cycling and a natural structure of biological communities are unacceptable. In high-input agriculture, not even the genetic material used for agricultural production is related to the original ecosystem in which production takes place.

In this way, humans keep the flow of produced biomass tightly under control and adjust yields to match increasing demographic and socio-economic pressure (e.g. the green revolution). However, this control is paid for with a high energetic cost of food production, because humans must continuously provide artificial regulation in the agroecosystem in the form of inputs and defense against undesired, competitive species.

The environmental impact of high-input food production is necessarily large, and involves a dramatic reduction of biodiversity in the altered area, destruction of the entire set of mechanisms that regulated ecosystem functioning before alteration, and several negative side-effects such as on-site and off-site pesticide and fertilizer pollution, soil erosion, and salinization.

How do changes in technical coefficients of farming techniques, aimed at an increase of both land and labour productivity, affect biodiversity use in agriculture.' What is the nature of trade-off between biodiversity use M agriculture on the one hand and productivity of land and/or labour on the other hand? At the level of the farming system the parameters land productivity and labour productivity are interdependent. The existence of this link is illustrated by the possibility to express the parameter labour productivity as, the product of two terms, one of which is the productivity of the land itself :

$$\text{labour productivity (kg hr}^{-1}) = \text{land productivity (kg ha}^{-1}) \times \text{land managed per hour (ha hr}^{-1})$$

Hence labour productivity can be augmented by either increasing the land area managed per hour of labour, increasing the land productivity or a combination of these two solutions. Usually, technical progress induces a combination of the two solutions which are mutually reinforcing.

The land managed per hour of labour can be augmented, while maintaining the same yield per hectare, by a substitution of machine power for human power. This solution pushes for monocultures, because a monocultures facilitates mechanization by, synchronizing and concentrating in time and space the requirement of inputs and the harvest of outputs. With this solution, labour productivity can be raised by using more land per farmer, while avoiding all increase in energy costs per hectare in the form of land-saving inputs such as fertilizers and irrigation. This solution is typical of farming systems in Canada and Australia, where the demographic pressure is low.

Since capital substitution for labour (human power) must have a positive economic return to pay back its own investment, economic feasibility is a crucial factor for this solution. There are two obvious ways to increase farm revenues in this situation. First, by trying to get the most out of the local conditions by focusing on the crop(s) that show(s) the best comparative advantage in the area.

Second, by trying to get the most out of the capital investment by expanding the area of the farm, carefully concentrating cultivation in favourable spots and avoiding marginal lands. To facilitate mechanization, fanners generally tend to increase the dimension of their fields, by eliminating patchworks and edges, despite the fact that those landscape patterns and elements are known to be important in the maintenance of beneficial insect species for pest control and conservation of biodiversity in general.

Thus, the use of machine power to increase labour productivity will drive technological development in agriculture further toward the exploitation of comparative advantages and economies of scale (e.g. enlargement of farms). The building of specialized technical infrastructures in the area will reinforce the trend of product specialization and generate a mode-locking effect.

While increasing yields per hectare without enlarging the land area managed per unit of labour can be a (temporary) solution to match an increasing demographic pressure, it has limited power to meet the demand for a higher labour productivity. The problems with this solution are evident in PR China, where massive fertilization and irrigation are practiced to augment yields coupled to labour-intensive farming techniques. Although in this way agriculture will feed a densely-population country, it will provide little support for its socio-economic development.

The equation of labour productivity indicates that both an increase in land productivity and an increase in land managed per hour of labour can have a positive marginal return in terms of an increased labour productivity and economic return. Therefore, the particular choice of technical inputs in agricultural production–machinery to boost the land managed per unit of labour or fertilizers and irrigation to boost land productivity–will depend on the expected marginal return of capital investments in these two factors, Sparsely-population developed countries, such as Canada, the USA and Australia tend to invest more in machinery, even though a certain investment in land-saving inputs is made when the marginal return is high.

Densely-populated developing countries, such as PR China, Egypt and Bangladesh, tend to predominantly invest in fertilizers and irrigation, and invest only little in machinery. Densely-populated wealthy countries, such as Japan, Italy and the Netherlands, are in a peculiar situation. In these countries, farming can become a 'mission impossible', because of the simultaneous high demographic (arable land per farmer cannot be significantly expanded) and socio-economic pressure (the need for high labour productivity and economic return in the agricultural sector).

Despite heavy investments in both machinery and land-saving inputs, farmers in these countries still face the unpleasant reality of a land and labour productivity that do not match the minimum outputs imposed by their demographic and socio-economic pressure.

The need for high labour productivity (socio-economic pressure) and eventually land productivity (demographic pressure) at the level of society, reflected into changes in technical coefficients at the farm level, is intricately linked to changes occurring in the energy budget of socio-economic systems to which the agricultural system belongs. Economic development, assessed at the socio-economic level, inevitably results in a continuously increasing opportunity cost of human labour and capital at the farm level, and pushes technical development of the farming system towards mechanization, specialization, economies of scale, lineraization of nutrient inputs in the agroecosystem, and dependence on price support and other subsidies.

This chain of changes, well-known in the field of agricultural economics though generally described from a different perspective, is the main cause of the abandonment of traditional farming strategies, and the consequent loss of biodiversity in agriculture. Most notably, two important trends follow from this process of technological change : adoption of monocultures, and reduced use of by-products, such as crop residues and manure, in the agricultural process.

Natural ecosystems (e.g. mature natural forest) are able to sustain a large amount of biomass, but with a net primary productivity that is relatively low compared to agroecosystems. Traditional farming systems based on gardening in natural forests can produce large quantities of biomass per hectare on the cleared areas, but only with a long cycle of rotation. When the entire area involved in the rotation is considered, overall crop productivity per hectare is quite low.

Other systems of traditional farming based on inter-cropping, multiple cropping and animal husbandry have proved to be highly productive in terms of food supply per hectare whenever ample fresh water is available. However, the combined exploitation of diverse natural resources generally requires labour-intensive manage merit. Rural communities in China and elsewhere in East Asia, masters in the integrated use of natural resources, have been able to feed themselves and pay taxes in form of part of the harvest in spite of a limited endowment of land, for centuries.

However, the particular techniques of multiple cropping, animal husbandry and integrated use of natural resources implied that more than 80% of the population was engaged in agriculture. Moreover, note that the population of China did not surpass 600 million until synthetic fertilizers were made available to China m the late 1970s. Whenever socio-economic development calls for a reduction of the labour force in agriculture, the society must confront the need for an increase in farm labour productivity.

Recycling of natural inputs within the farming system is abandoned as soon as this becomes too costly in terms of land and/or human labour requirement, provided of course that alternative technical inputs for use in agriculture are available on the market. For example, the trend of abandoning cultivation of green manure in favour of food grain when artificial fertilizer is accessible is discussed in detail by Smil. In general, the decision whether or not to abandon the recycling of natural inputs will depend on the amount of energy or nutrients gained by recycling, divided by the extra hours of labour required, the opportunity-cost of labour, and the price of technical inputs used as alternatives.

When the inputs made available per hour of labour by recycling are significantly less than the inputs that could be bought on the market with the economic return of one hour of labour allocated to off-farm activities, then recycling will generally be abandoned. In this case, linearization of nutrient inputs is more convenient in terms of economic cost/benefit. Within a market economy, the choice to specialize and rely on monocultures often leaves no alternative but to also switch to linearization of nutrient inputs in the form of fertilization.

Criteria adopted for choosing and developing farming techniques have generally been based on socioeconomic constraints only (costs/benefits referring to economic growth) and therefore have been biased in favour of continuously increasing densities of harvested biomass.

Ecological constraints imposed by the characteristics of natural ecosystems, that would put an upper limit to the densities of harvested biomass, tend to be ignored. This is for the simple reason that socioeconomic constraints to agriculture leading to technological 'progress' have universal application (technical coefficients refer to unit operations in artificial environments and can easily be generalized in terms of monetary costs/benefits), whereas ecological constraints are location-specific and extremely difficult to define and quantify. The consequent imbalance between the short-term and long-term (ecological) perspective when deciding how to mange agricultural production may have disastrous consequences given the immense scale of agricultural activities worldwide approximately 50% of the terrestrial area is devoted to agriculture, while in temperate ecosystems agriculture occupies 70% of the land.

As Hollings has pointed out, in biological systems a general homogenization of activities over a large scale is a disaster waiting to happen: it dramatically increases the risk that perturbation damaging to that pattern will occur and be amplified over the whole system.

In conclusion, strict respect the pattern of natural process in the agroecocystem to preserve biodiversity would slow down the throughput of food production per hectare of land and per hour of labour. When these throughputs become so low so as to become incompatible with socioeconomic characteristics this solution becomes economically non-viable.

On the other hand, matching demographic and socioeconomic pressure above certain ecological thresholds leads to ecological unsustainability. For example, the production of tonnes of grain per hectare per year and hundreds of kilograms of grain per hour of labour with less than 0.2 hectare of arable land per capita can only be obtained in an ecologically unsustainable way through linearzation of nutrient flows in the agroecosystem, that is continuous harvests (no rotation) and heavily reliance on technical inputs such as machinery, fertilizer, pesticides, and irrigation.

Discussions of sustainability of world agricultural production should carefully consider both ecological constraints such as the possible limiting role of land water, and other natural resources, and socioeconomic constraints to food production.

Current strategies of technological development in agriculture not only appear unable to solve problems of food security in the long term, but also have been increasingly malfunctioning in the last decade, both in developed and in many developing countries.

In developed countries this is because excessive socioeconomic pressure has caused farms and fanners to disappear, while remaining farming systems employ production techniques that are detrimental to ecological equilibria, and in developing countries it is because excessive demographic pressure has caused poverty in rural areas to increase and food production per capita to decrease.

Paradoxically, this evident malfunctioning is perhaps also a reason for optimism. After all, when dealing with human systems, cultural mode-locking is the most persistent and important cause of lack of adaptability in response to changing boundary conditions.

As concerns agriculture, however, there is an increasing, though not yet general consensus among farmers, scientists, politicians and the general public that we somehow have to change the way we look at and practice agriculture. In terms of opportunities for future changes in farming strategies, the situation is dramatically different in developed and developing countries.

This risk-free environment, that was and still is financially supported by the rest of society, has been realized through guaranteed prices, input subsidies, disaster, relief, etc. Indeed, modem high-input agriculture depends economically on the rest o society, that mobilizes economic resources to maintain its agricultural sector. We may thus conclude that in developed countries the economic resources are available to support the agricultural sector, but that they should be spent in a different way.

Developing countries have never disposed over sufficient economic resources to generate a risk-free environment for farmers and investors in high-input agriculture. On the contrary, the agricultural sector is usually taxed to support other sectors of the economy.

In the absence of government support, risks implied by excessive simplification of ecosystems typical of high-input agriculture are transferred to individual farmer families. This makes such a solution less attractive, especially for poor farmers operating in marginal areas. Nevertheless, the continuous increase in demographic and/or socioeconomic pressure as well as access to Western technologies, has pushed for the adoption of high-input techniques in local attempts to augment agricultural throughputs. Given the rapidly growing population in developing countries (actual increases in demographic pressure) and the gap still to fill before reaching desired income levels (potential socioeconomic pressure), we can expect the density of agricultural throughputs to increase in most, if not all, developing countries, even though this could be destructive for the environment in many cases.

In spite of this wide difference, there is a striking similarity between high-input and low-input agriculture. In both cases, food security totally depends on the availability of fresh water, fertile soil, pollination, natural recycling of nutrients and pollutants, biological pest control and nitrogen fixation. These are servies that cannot be substituted by injection of technical capital, and that are hevaily dependent on biodiversity.

However, this view implicitly implies the adoption of the entire world as the hierarchical level at which the economic process is assessed and optimized. This level is not necessarily the best possible choice, especially when dealing with agricultural production. In fact, the use of a smaller space-time scale for defining strategies of technological development also has positive effects. First, a smaller scale makes it easier to involve local people in the decision-making process and increase the responsibility of local communities for resource management.

Thus, decisions related to sustainability should be made as close as possible to local people, and with the participation of all major stake-holders. Apart from the ethical aspect, this would make it possible to benefit from indigenous knowledge of the nature of the problem of sustainability dealt with. Second, internalizing most of the external services on which economic systems rely (by reducing trade) makes it easier for the system to respond quickly to changes and signals coming from surrounding ecosystems, inspite of their intricate and varying nature. Societies heavily relying on trade often cannot see the importance of the ecological services on which they directly and indirectly rely. Therefore, energy expenditure in transportation from distant ecosystems should be justified only in cases of substantial comparative advantages. Third, the goal of harmonizing energy and material circulation with the characteristics of local ecosystems (decreasing the environmental impact of economic activities) can be more easily achieved it' the space-time closure of the activities of production and consumption in the society is similar to the space-time closure of the matter cycles occurring in local ecosystems. These points suggest that the enlargement of scale of economic systems obtained by heavy trading can also induce negative effects, both on the stability of the biosphere and on the integrity of local socioeconomic systems. There is therefore also a rationale for policies aiming at increasing the level of self-sufficiency and addressing the issue of social equity at the level of national and regional economic systems.

When defining the terms of reference for sustainable agriculture, biodiversity should be considered as an output rather than as an input in agriculture. Agriculture should not only produce food, but also preserve a healthy environment. The latter task, grossly neglected, involves guaranteeing a higher level of cycling of nutrients, better control of pollution of fresh water supplies, and a sound management of the landscape in general. For all these tasks preservation of biodiversity is essential.

Moreover, the perception of the trade-off between long-term ecosystem health and short-term maximization of production is affected by the level of development. In countries with a high material standard of living, citizens often consider it acceptable to reduce economic returns for the benefit of the environment, while developing countries do not have room to reduce economic returns. They have no other option but to boost short-term returns at the cost of long-term sustainability.

At the moment of discussing agricultural policies in view of sustainability, the small space-time scale typical of economic analysis should always be balanced by a long-term and worldwide view. But is this a feasible idea? Developed countries might be prepared to sacrifice part of their own economic return for the benefit of their immediate surroundings, but doing this for the benefit of other, distant societies and ecosystems is something quite different. The fact that all socioeconomic systems, ecosystems and agroecosystems form part of a larger whole, the functioning of which is affected by the health of all its individual components, often appears to be beyond our capacity and willingness of understanding. The 'biodiversity revolution in agriculture' should result in a definition of a set of new optimizing parameters for agricultural techniques.

14

Nature and Natural Resources Conservation

Kingdom of Member States of the Association of South East Asian Nations (ASEAN) have *Recognizing* the importance of natural resources for present and future generations :

Conscious of their ever-growing value from a scientific, cultural, social and economic point of view:

Conscious also that the interrelationship between conservation and socio-economic development implies both that conservation is necessary to ensure sustainability of development, and that socio-economic development is necessary for the achievement of conservation on a lasting basis:

Recognizing the interdependence of living resources, between them and with other natural resources, within ecosystems of which they are part;

Wishing to undertake individual and joint action for the conservation and management of their living resources and the other natural elements on which they depend;

Recognizing that international co-operation is essential to attain many of these goals;

Convinced that an essential means to achieve such concerted action is the conclusion and implementation of an Agreement agreed as follows:

I. Conservation and Development

Article 1

Fundamental Principle

1. The Contracting Parties, within the framework of their respective national laws, undertake to adopt singly, or where necessary and appropriate through concerted action, the measures necessary to maintain essential ecological processes and life-support systems, to preserve genetic diversity, and to ensure the sustainable utilization of harvested natural resources under their jurisdiction in accordance with scientific principles and with a view to attaining the goal of sustainable development.
2. To this end they shall develop national conservation strategies, and shall co-ordinate such strategies within the framework of a conservation strategy for the Region.

Article 2

Development Planning

1. The Contracting Parties shall take all necessary measure, within the framework of their respective national law, to ensure that conservation and management of natural resources are treated as an integral part of development planning at all stages and at all levels.
2. To that effect they shall, in the formulation of all development plants, give as full consideration to ecological factors as to economic and social ones.
3. The Contracting Parties shall, where necessary, take appropriate action with a view to conserving and managing natural resources of significant importance for two or several Contracting Parties.

II. Conservation of Species and Ecosystems

Article 3

Species - genetic diversity

1. The Contracting Parties shall, wherever possible, maintain maximum genetic diversity by taking action aimed at ensuring the survival and promoting the conservation of all species under their jurisdiction and control.
2. To that end, they shall adopt appropriate measures to conserve animal and plant species whether terrestrial, marine and freshwater, and more specifically
   (a) Conserve natural, terrestrial, freshwater and marine habitats;
   (b) ensure sustainable use of harvested species;
   (c) protect endangered species;
   (d) conserve endemic species; and
   (e) take all the measures in their power to prevent the extinction of any species or sub-species.
3. In order to fulfil the aims of the preceding paragraph of this Article the Contracting Parties shall, in particular, endeavour to
   (a) create and maintain protected areas:
   (b) regulate the taking of species and prohibit unselective taking methods;
   (c) regulate and, where necessary, prohibit the introduction of exotic species;
   (d) promote and establish gene banks and other documented collections of animal and plant genetic resources.

Article 4

Species - Sustainable use

The Contracting Parties shall pay special attention to harvested species, and, to that effect, shall endeavour to

1. develop, adopt and implement management plans for those species, based on scientific studies and aiming at

(a) preventing decrease in the size of any harvested population to levels below those which ensure its stable recruitment and the stable recruitment of those species which are dependent upon, or related to them;

(b) maintaining the ecological relationship between harvested, dependent and related populations of living resources of the ecosystem considered;

Article 5

Species - endangered and endemic

1. Appendix 1 to this Agreement shall list endangered species recognized by the Contracting Parties as of prime importance to the Region and Parties as of prime importance to the Region and deserving special attention. The Appendix shall be deserving special attention. The Appendix shall be adopted by a meting of the Contracting Parties.

Accordingly, Contracting Parties shall, wherever possible

(a) prohibit the taking of these species, except for exceptional circumstances by special allowance from the designated authorities of the Contracting Parties.

(b) regulate the trade in and possession of specimens and products of those species accordingly;

(c) especially protect habitat of those species by ensuring that sufficient portions are included in protected areas;

(d) take all other necessary measures to improve their conservation status, and restore their populations to the highest possible level.

2. Each Contracting Party shall, wherever possible, apply the above measures to species endangered at national level.

3. The Contracting Parties recognize their special responsibility in respect of species that are endemic to areas under their jurisdiction and shall undertake accordingly to take wherever possible, all the necessary measures to maintain the population of such species at the highest possible level.

   (a) restoring depleted populations to at least the levels referred to in sub-paragraph (1) of this paragraph;

   (b) preventing changes or minimizing risk of changes in the ecosystem considered which are not reversible over a reasonable period of time.

4. Take the appropriate and necessary legislative and administrative measures on harvesting activities in the light of their national interest whereby

   (a) such activities must conform to the management plans referred to above;

   (b) the conduct of such activities is controlled by a permit system;

   (c) all indiscriminate means of taking and the use of all means capable of causing local extinction of, or serious disturbance to, populations of a species or related species are prohibited;

   (d) such activities are prohibited or strictly regulated at certain period, seasons or places of importance in the life cycle of the species;

(e) such activities may be regulated more strictly, temporarily or locally in order to assist restoration of population levels or counterbalance any threat caused by special circumstances;

(f) special measures, such as restocking, are provided for whenever the conservation status of a species so warrants;

(g) trade and possession of specimens or products of specimens are regulated whenever such regulations meaningfully contribute to the implementation of the harvesting regulations.

Article 6

Vegetation Cover and Forest Resources

1. The Contracting Parties shall, in view of the role of vegetation and forest covers in the functioning of natural ecosystems, take all necessary measures to ensure the conservation of the vegetation cover and in particular of the forest cover on lands under their jurisdiction.
2. They shall, in particular, endeavour to

(a) control clearance of vegetation; endeavour to prevent bush and forest fires; prevent overgrazing by, inter alia, limiting grazing activities to periods and intensities that will not prevent regeneration of the vegetation;

(b) regulate mining and mineral exploration operations with a view to minimizing disturbance of vegetation and to requiring the rehabilitation of vegetation after such operations;

(c) set aside areas as forest reserves, inter alia, with a view to conserve the natural forest genetic resources;

(d) in reforestation and afforestation planning avoid as far as possible monoculture causing ecological imbalance;

(e) designate areas whose primary function shall be the maintenance of soil quality in the catchment considered and the regulation of the quantity and quality of the water delivered from it;

(f) ensure, to the maximum extent possible the conservation of their natural forests, particularly mangroves with a view, inter alia, to maintaining maximum forest species diversity;

(g) develop their forestry management plans on the basis of ecological principles with a view to maintaining potential for optimum sustained yield and avoiding depletion of the resource capital.

Article 7

Soil

1. The Contracting Parties shall, in view of the role of soil in the functioning of natural ecosystems, take measures, wherever possible towards soil conservation, improvement and rehabilitation; they shall in particular, endeavour to take steps to prevent soil erosion and

other forms of degradation, and promote measures which safeguard the processes of organic decomposition and thereby its continuing fertility.

2. To that effect, they shall, in particular, endeavour to
   (a) establish land use policies aimed at avoiding losses, of vegetation cover, substantial soil losses, and damage to the structure of the soil;
   (b) take all necessary measures to control erosion, especially as it may affect coastal or freshwater ecosystems lead to siltation of downstream areas such as lakes or vulnerable ecosystems such as coral reefs, or damage critical habitats, in particular that of endangered or endemic species;
   (c) take appropriate measures to rehabilitate eroded or degraded soils including rehabilitation of soil affected by mineral exploitation.

Article 8

Water

1. The Contracting Parties shall, in view of the role of water in the functioning of natural ecosystems, take all appropriate measures towards the conservation of their underground the surface water resources.
2. They shall to that effect, in particular, endeavour to
   (a) undertake and promote the necessary hydrological research especially with a view to ascertaining the characteristics of each water shed;
   (b) regulate and control water utilization with a view of achieving sufficient and continuous supply of water for, inter alia, the maintenance of natural life supporting systems and aquatic fauna and flora;
   (c) when planning and carrying out water resource development projects take fully into account possible effects of such projects on natural processes or on other renewable natural resources and prevent or minimize such effects.

Article 9

Air

The Contracting Parties shall, in view of the role of air in the functioning of natural ecosystems, endeavour to take all appropriate measures towards air quality management compatible with sustainable development.

III. Conservation of Ecological Processes

Article 10

Environment Degradation

The Contracting Parties, with a view to maintaining the proper functioning of ecological processes, undertake, wherever possible, to prevent, reduce and control degradation of the natural environment and to this end, such endeavour to undertake, in addition to specific measures referred to in the following article;

(a) to promote environmentally sound agricultural practice by inter alia, controlling the application of pesticides, fertilizers and other chemical products for agricultural development shcemes, in particular for wetland drainage or forest clearance, pay due regard to the need to protect critical habitats as well as endangered and economically important species;

(b) to promote pollution control and the development of environmentally sound industrial processes and products;

(c) to promote adequate economic or fiscal incentives for the purposes of subparagraphs (a) and (b) above;

(d) as far as possible to consider the originator of the activity which may lead to environmental degradation responsible for its prevention, for rehabilitation and remedial measures required;

(e) to take into consideration, when authorizing activities likely to affect the natural environment, the foreseeable interactions between the new activities proposed and those already taking place in the same area, and the result of such interactions on the air, waters and soils of the area;

(f) to pay particular attention to the regulation of activities which may have adverse affects on processes which are ecologically essential or on areas which are particularly important or sensitive form an ecological point of view, such as the breeding and feeding grounds of harvested species.

Article 11

Pollution

The Contracting Parties, recognizing the adverse effect that polluting discharges or emissions may have on natural processes and the functioning of natural ecosystems as well as on each of the individual ecosystem components, especially animal and plants species, shall endeavour to prevent, reduce and control such discharges, emissions or applications, in particular by

(a) submitting activities likely to cause pollution of the air, soil, freshwater, or the marine environment, to controls which shall take into consideration both the cumulative effects of the pollutants concerned and the self-purificating aptitude of the recipient natural environment.

(b) making such controls conditional on, inter alia, appropriate treatment of polluting emissions; and

(c) establishing national environmental quality monitoring programmes, particular attention being paid to the effects of pollution on natural ecosystems, and co-operation in such programmes for the Region as a whole.

IV. Environmental Planning Measures

Article 12

Land Use Planning

1. The Contracting Parties shall, wherever possible in the implementation of their development planning, give particular attention to the national allocation of land usage. They shall endeavour to take the necessary measures to ensure the integration of natural resources conservation into the land use planning process and shall, in the preparation and implementation of specific land use plans at all levels, give as full consideration as possible to ecological factors as to economic and social ones. In order to achieve optimum sustainable land use, they undertake to base their land use plans as far as possible on the ecological capacity of the land.
2. The Contracting Parties, shall, in carrying out the provisions of paragraph (1) above, particularly consider the importance of retaining the natural high productivity of areas such as coastal zones and wetlands.
3. They shall, where appropriate, coordinate their land use planning with a view to conserving and managing natural resources of significant importance for two or several Contracting Parties.

Article 13

Protected Areas

1. The Contracting Parties shall as appropriate establish, in areas under their jurisdiction, terrestrial, freshwater, coastal or marine protection areas for the purpose of safeguarding
   - the ecological and biological processes essential to the functioning of the ecosystems of the Region;
   - representative samples of all types of ecosystem of the Region;
   - satisfactory population levels for the largest possible number of species of fauna and flora belonging to those ecosystems;
   - areas of particular importance because of their scientific, educational, aesthetic, or cultural interest;

   and taking into account their importance in particular as;
   - the natural habit of species of fauna and flora, particularly rare or endangered or endemic species;
   - zones necessary for the maintenance of exploitable stocks of economically important species;
   - pools of genetic material and safe refuges for species, especially endangered ones;
   - sites of ecological, aesthetic or cultural interest;
   - reference sources of scientific research;

They shall, in particular, take all measures possible in their power to preserve those areas which are of an exceptional character and are peculiar to their country or the Region as well as those which constitute the critical habitats of endangered or rare species, of species that are endemic to a small areas and of species that migrate between countries of Contracting Parties.

2. Protected areas established pursuant to this Agreement shall be regulated and managed in such a way as to further the objectives of the propose of which they have been created. Contracting Parties shall, wherever possible, prohibit within such protected areas activities which are inconsistent with such objectives.
3. Protected areas shall include

(a) *National Parks*

(i) This expression denotes natural areas that are sufficiently large to allow for ecological self-regulation of one or several ecosystems, and which have not been substantially altered by human occupation or exploitation.

(ii) National Parks shall be placed under public control, their boundaries shall not be altered nor shall any portion of any National Park be alienated except by the highest competent authority.

(iii) National Parks shall be dedicated to conservation and to scientific, educational and recreational uses and the common welfare of the people.

(b) *Reserves*

(i) This expression denotes areas set aside for the purpose of preserving a specific ecosystem, the critical habitat of certain species of fauna or flora, a water catchment area or for any other specific purpose relating to the conservation of natural resources or objects or areas of scientific, aesthetic, cultural, educational or recreational interest.

(ii) After reserves have been established their boundaries shall not be altered nor shall any portion of such reserves be altered nor shall any portion of such reserves be alienated except by the authority establishing them or by higher authority.

(iii) Reserves shall be dedicated to the purposes for which they have been created and, in the light of the national interests of the Contracting Parties, any activity inconsistent with such purposes shall be prohibited.

4. Contracting Parties shall, in respect of any protected areas established pursuant to this Agreement

(a) prepare a management plan and manage the area on the basis of this plan;

(b) establish, wherever appropriate, terrestrial or aquatic buffer zones that shall be located around protected areas and which, in the case of marine areas, may include coastal land areas or watersheds of rivers flowing into the protected area; in such buffer zones all activities that may have harmful consequences on the ecosystems that such areas purport to protect shall be prohibited or regulated and activities which are consistent with the purpose of the protected area shall be promoted.

5. Contracting Parties shall, in respect of any protected area established pursuant to this Agreement, endeavour to

(a) prohibit the introduction of exotic animal or plant species;

(b) prohibit the use or release of toxic substances or pollutants which could cause disturbance or damage to protected ecosystems or to the species they contain;

(c) to the maximum extent possible, prohibit or control any activity exercised outside protected areas when such an activity is likely to cause disturbance or damage to the ecosystems or species that such protected areas purport to protect.

6. Contracting Parties shall co-operate in the development of principles, objectives, criteria and guidelines for the selection, establishment and management of protected areas in the Region with a view to establishing a co-ordinated network of protected areas throughout the Region giving particular attention to those of regional importance. An Appendix containing such principles, objectives, criteria and guidelines shall be drawn up in the light of the best scientific evidence as adapted to the conservation requirements of the Region and shall be adopted by a meeting of Contracting Parties.

7. In addition to the establishment of the protected areas referred to in paragraph 3 of this Article. Contracting Parts shall promote, through the adoption of appropriate measures, the conservation of natural areas by private owners, community or local authorities.

Article 14

Impact Assessment

1. The Contracting Parties undertake that proposals for any activity which may significantly affect the natural environment shall as far as possible be subjected to an assessment of their consequences before they are adopted, and they shall take into consideration the results of this assessment in their decision-making process.

2. In those cases where any such activities are undertaken, the Contracting Parties shall plan and carry them out so as to overcome or minimize any assessed adverse effects and shall monitor such effects with a view to taking remedial action as appropriate.

V. National Supporting Measures

Article 15

Scientific Research

The Contracting Parties shall individually or in co-operation with other Contracting Parties or appropriate international organizations, promote and, whenever, possible, support scientific and technical programmes of relevance to the conservation and management of natural resources, including monitoring, research, the exchange of technical information and the evaluation of results.

Article 16

Education, Information and Participation of the Public, Training

1. The Contracting Parties shall endeavour to promote adequate coverage of conservation and management of natural resources in education programmes at all levels.

2. They shall circulate as widely as possible information on the significance of conservation measures and their relationship with sustainable development objectives, and shall, as far as possible, organize participation of the public in the planning and implementation of conservation measures.

3. Contracting Parties shall endeavour to, individually or in co-operation with other Contracting Parties or appropriate international organizations, develop the programmes and facilities necessary to train adequate and sufficient scientific and technical personnel to fulfil the aims of this Agreement.

Article 17

Administrative Machinery

1. The Contracting Parties shall identify or maintain the administrative machinery necessary to implement the provisions of this Agreement, and, where several governmental institutions are involved create the necessary co-ordinating mechanism for the authorities dealing with designated aspects of the environment.
2. They shall endeavour to allocate sufficient funds to the task necessary for the implementation of this Agreement, as well as sufficient qualified personnel with adequate enforcement powers.

VI. International Co-operation

Article 18

Co-operative Activities

1. The Contracting Parties shall co-operate together and with the competent international organizations, with a view to co-ordinating their activities in the field of conservation of nature and management of natural resources and assisting each other in fulfilling their obligations under this Agreement.
2. To that effect, they shall endeavour
   (a) to collaborate in monitoring activities;
   (b) to the greatest extent possible, co-ordinate their research activities;
   (c) to use comparable or standardized research techniques and procedures with a view to obtaining comparable data;
   (d) to exchange appropriate scientific and technical data, information and experience, on a regular basis;
   (e) whenever appropriate, to consult and assist each other with regard to measures for the implementation of this Agreement.
3. In applying the principles of co-operation and co-ordination set for the above, the Contracting Parties shall forward to the Secretariat
   (a) information of assistance in the monitoring of the biological status of the natural living resources of the Region;
   (b) Information, including reports and publications of a scientific, administrative or legal nature and, in particular, information on

- measures taken by the Parties in pursuance of the provisions of this Agreement;
- the status of species included in Appendix 1;
- any other matter to which the Conference of the Parties may give special priority.

Article 19

Shared Resources

1. Contracting Parties that share natural resources shall co-operate concerning their conservation and harmonious utilization, taking into account the sovereignty, rights and interests of the Contracting Parties concerned in accordance with generally accepted principles of international law.
2. To that end, they shall, in particular
   (a) co-operate with a view to controlling, preventing, reducing or eliminating adverse environmental effects which may result in one Contracting Party from the utilization of such resources in another Party;
   (b) endeavour to conclude bilateral or multilateral agreements in order to secure specific regulation of their conduct in respect of the resources concerned;
   (c) as far as possible make environmental assessments prior to engaging in activities with respect of shared natural resources which may create a risk of significantly affecting the environment of another sharing Contracting Party or other sharing Contracting Parties;
   (d) notify in advance the other sharing Contracting Party or the other sharing Contracting Parties of pertinent details of plans to initiate, or make a change in, the conservation or utilization of the resource which can reasonably be expected to affect significantly the environment in the territory of the other Contracting Party or Contracting Parties;
   (e) upon request of the other sharing Contracting Party or sharing Contracting Parties, enter into consultation concerning the above mentioned Plans;
   (f) inform the other sharing Contracting Party or other sharing Contracting Parties of emergency situations or sudden grave natural events which may have repercussions on their environment;
   (g) whenever appropriate, engage in joint scientific studies and assessments, with a view to facilitating co-operation with regard to environmental problems related to a shared resource, on the basis of agreed data.
3. Contracting Parties shall especially co-operate together and, where appropriate, shall endeavour to co-operate with other Contracting Parties, with a view to
   (a) the conservation and management of
   - border or contiguous protected areas;
   - shared habitats of species listed in Appendix 1;
   - shared habitats of any other species of common concern;

   (b) the conservation, management and, where applicable, regulation of the harvesting of species which constitute shared resources;

- by virtue of their migratory character, or
- because they inhabit shared habitats.

Article 20

Transfrontier Environmental Effects

1. Contracting Parties have in accordance with general accepted principles of international law the responsibility of ensuring that activities under their jurisdiction or control do not cause damage to the environment of the natural resources under the jurisdiction of other Contracting Parties or of areas beyond the limits of national jurisdiction.
2. In order to fulfil this responsibility, Contracting Parties shall avoid to the maximum extent possible and reduce to the minimum extent possible adverse environmental effects of activities under their jurisdiction or control, including effects on natural resources, beyond the limits of their national jurisdiction.
3. To that effect, they shall endeavour
   (a) to make environmental impact assessment before engaging in any activity that may create a risk of significantly affecting the environment or the natural resources of another Contracting Party or the environment or natural resources beyond national jurisdiction;
   (b) to notify in advance the other Contracting Party or Contracting Parties concerned of pertinent details of plans to initiate, or make a change in, activities which can reasonably be expected to have significant effects beyond the limits of national jurisdiction;
   (c) to enter into consultation concerning the above-mentioned plans upon request of the Contracting Party or Contracting Parties in question;
   (d) to inform the Contracting Party or Contracting Parties in question of emergency situations or sudden grave natural events which may have repercussion beyond national jurisdiction.
4. Contracting Parties shall, in particular, endeavour to refrain from actions which might directly or indirectly adversely affect wildlife habitats situated beyond the limits of national jurisdiction, especially habitats of species listed in Appendix 1 or habitats included in protected areas.

VII. International Supporting Measures

Article 21

Meeting of the Contracting Parties

1. Ordinary meetings of the Contracting Parties shall be held at least once in three years, in as far as possible in conjunction with appropriate meetings of ASEAN, and extraordinary meetings shall be held at any other time, upon the request of one Contracting Party provided that such request is supported by at least one other Party.

2. It shall be the function of the meetings of the Contracting Parties, in particular

(a) to keep under review the implementation of this Agreement and the need for other measures, in particular the Appendices;

(b) to adopt review and amend as required any Appendix to this Agreement;

(c) to consider reports submitted by the Contracting Parties in accordance with Article 28 or any other information which may be submitted by a Party, directly or through the Secretariat;

(d) to make recommendations regarding the adoption of any Protocol or any amendment to this Agreement;

(e) to establish working groups or any other subsidiary body as required to consider any matter related to this Agreement;

(f) to consider and undertake any additional action, including the adoption of financial rules, that may be required for the achievement of the purposes of this Agreement.

Article 22

Secretariat

On the coming into force of this Agreement the Contracting Parties shall designate the Secretariat responsible for carrying out the following functions;

(a) to convene and prepare the meetings of Contracting Parties;

(b) to convene diplomatic conferences for the the purpose of adopting Protocol;

(c) to transmit to Contracting Parties notification, reports and other information received in accordance with this Agreement;

(d) to consider inquiries by, and information from, the Contracting parties, and to consult with them on questions relating to this Agreement.

(e) to perform such other functions as may be assigned to it by the Contracting Parties;

(f) to ensure the necessary co-ordination with other competent international bodies and in particular to enter into such administrative arrangements as may be required for the effective discharge of the secretariat functions.

Article 23

National Focal Points

In order to facilitate communications with other Parties and the Secretariat, the Contracting Parties shall designate an appropriate national agency or institution responsible for co-ordinating matters arising from consultations and channelling communications between Contracting Parties or with the Secretariat.

VIII. Final Clauses

Article 24

Adoption of Protocols

1. The Contracting Parties shall co-operate in the formulation and adoption of Protocols to this Agreement, prescribing agreed measures, procedures and standards for the implementation of this Agreement.
2. The Contracting Parties, at a diplomatic conference, may adopt Protocols to this Agreement.
3. The Protocols of this Agreement shall be subject to acceptance and shall enter into force on the thirtieth day after the deposit with he Depositary of the Instruments of Acceptance of all the Contracting Parties.

Article 25

Amendment of the Agreement

1. Any Contracting Party to this Agreement may propose amendments to the Agreement. Amendments shall be adopted by a diplomatic conference which shall be convened at the request of the majority of the Contracting Parties.
2. Amendments to this Agreement shall be adopted by a consensus of the Contracting Parties.
3. Acceptance of amendments shall be notified to the Depositary in writing and shall enter into force on the thirtieth day following the receipt by the Depositary of notification of the acceptance by all the Contracting Parties.
4. After the entry into force of an amendment to this Agreement any new Contracting Party to this Agreement shall become a Contracting Party to this Agreement as amended.

Article 26

Appendices and Amendments to Appendices

1. Appendices to this Agreement shall form an integral part of the Agreement.
2. Amendments to an Appendix:
   (a) Any Contracting Party may propose amendments to an Appendix at a meeting of the Contracting Parties;
   (b) Such amendments shall be adopted by a consensus of the Contracting Parties;
   (c) The Depositary shall without delay communicate the amendment so adopted to all Contracting Parties.
3. The adoption and entry into force of a new Appendix to this Agreement shall be subject to the same procedure as for the adoption and entry into force of an amendment to an Appendix as provided for in paragraph (2) of this Article. Provided that the new Appendix shall not enter into force until such time as the amendment to the Agreement enters into force.

Article 27

Rules of Procedure

The Contracting Parties shall adopt rules of procedure for their meetings.

Article 28

Reports

The Contracting Parties shall transmit to the Secretariat reports on the measures adopted in implementation of this Agreement in such form and at such intervals as the meetings of Contracting Parties may determine.

Article 29

Relationship with Other Agreements

The provisions of this Agreement shall in no way affect the rights and obligations of any Contracting Party with regard to any existing treaty, convention or agreement.

Article 30

Settlement of Disputes

Any dispute between the Contracting Parties arising out of the interpretation or implementation of this Agreement shall be settled amicably by consultation or negotiation.

Article 31

Ratification

This Agreement shall be subject to ratification by the Contracting Parties. The instruments of Ratification shall be deposited with the Secretary-General of the ASEAN Secretariat, who shall assume the functions of Depositary.

Article 32

Accession

1. After the entry into force of the Agreement, any Member State may accede to this Agreement, subject to prior approval by the Contracting Parties to this Agreement.
2. Instrument of accession shall be deposited with the Depositary.

Article 33

Entry into Force

1. This Agreement shall enter into force on the thirtieth day after the deposit of the sixth Instruments of Ratification.

2. Thereafter, this Agreement shall enter into force with respect to any Contracting Party on the thirtieth day following the date of deposit of the instrument of accession by that Contracting Party.

Article 34

Responsibility of the Depositary

The Depositary shall inform the Governments which have signed this Agreement;

(a) of the deposit of instruments of ratification, acceptance or accession;

(b) of the date on which the Agreement will come into force.

Article 35

Deposit and Regulation

1. The present Agreement shall be deposited with the Depositary who shall transmit certified true copies thereof to the Governments of all Contracting Parties which have signed the present Agreement or acceded to it.
2. As soon as the present Agreement enter into force, the text shall be transmitted by the Depositary to the Secretary-General of the United Nations for registration and publication, in accordance with Article 102 of the Charter of the United Nations.

15

Plant Protection International Convention

Preamble

The Contracting Governments, recognizing the usefulness of international co-operation in controlling pests and diseases of plants and plant products and in preventing their introduction and spread across national boundaries, and desiring to ensure close co-ordination of measures directed to these ends, have agreed as follows :

Article I

Purpose and Responsibility

1. With the purpose of securing common and effective action to prevent the introduction and spread of pests and diseases of plants and plant products and to promote measures for their control, the contracting Governments undertake to adopt the legislative, technical and administrative measures specified in this Convention and in supplementary agreements pursuant to Article III.
2. Each contracting Government shall assume responsibility for the fulfilment within its territories of all requirements under this Convention.

Article II

Scope

1. For the purposes of this Convention the term "plant" shall comprise living plants and parts thereof, including seeds in so far as the supervision of their importation under Article VI of the Convention or the issue of phytosanitary certificates in respect of them under Articles IV(1), (a) (iv) and V of this Convention may be deemed necessary by contracting Governments; and the term "plant products" shall comprise unmanufactured and milled material of plant origin, including seeds in so far as they are not included in the term "plants".
2. The provisions of this Convention may be deemed by contracting Governments to extend to storage places, containers, conveyances, packing material and accompanying media of all sorts including soil involved in the international transportation of plants and plant products.
3. This Convention shall have particular reference to pests and diseases of importance to international trade.

Article III

Suplementary Agreements

1. Supplementary agreements applicable to specific regions, to specific pests and diseases, to specific plants and plant products, to specific methods of international transportation of plants and plant products, or otherwise supplementing the provisions of this Convention, may be proposed by the Food and Agriculture Organization of the United Nations (hereinafter referred to as "FAO") on the recommendation of contracting Government or on its own initiative, to meet special problems of plant protection which need particular attention or action.
2. Any such supplementary agreements shall come into force for each contracting Government after acceptance in accordance with the provisions of the FAO Constitution and Rules of Procedure.

Article IV

National Organization for Plant Protection

1. Each contracting Government shall make provision, as soon as possible and to the best of its ability, for

(a) an official plant protection organization, with the following main functions:

(i) the inspection of growing plant, of areas under cultivation (including fields, plantations, nurseries, gardens and greenhouses), and of plants and plant products in storage and in transportation particularly with the object of reporting the existence, outbreak and spread of plant diseases and pests and of controlling those pests and diseases;

(ii) the inspection of consignments of plants and plant products moving in international traffic, and, as far as practicable, the inspection of consignments of other articles or commodities moving in international traffic under conditions where they may act incidentally as carriers of pests and diseases of plants and plant products, and the inspection and supervision of storage and transportation facilities of all kinds involved in international traffic whether of plants and plant products or of other commodities, particularly with the object of preventing the dissemination across national boundaries of pests and diseases of plants and plant products;

(iii) the disinfestation or disinfection of consignments of plants and plant products moving in international traffic, and their containers, storage places, or transportation facilities of all kinds employed;

(iv) the issue of certificates relating to phytosanitary conditions and origin of consignments of plants and plant products (hereinafter referred to as "phytosanitary certificates");

(b) the distribution of information within the country regarding the pests and diseases of plants and plant products and the means of their prevention and control;

(c) research and investigation in the field of plant protection.

2. Each contracting Government shall submit a description of the scope of its national organization for plant protection and of changes in such organization to the Director-General of FAO, who shall circulate such information to all contracting Governments.

Article V

Phytosanitary Certificates

1. Each contracting Government shall make arrangements for the issue of phytosanitary certificates to accord with the plant protection regulations of other contracting Governments and in conformity with the following provisions:
   (a) Inspection shall be carried out and certificates issued only by or under the authority of technically qualified and duly authorized officers and in such circumstances and with such knowledge and information available to those officers that the authorities of importing countries may accept such certificates with confidence as dependable documents.
   (b) Each certificate covering materials intended for planting or propagation shall be as worded in the Annex to this Convention and shall include such additional declarations as may be required by the importing country. The model certificate may also be used for other plants or plant products where appropriate and not inconsistent with the requirements of the importing country.
   (c) The certificates shall bear no alterations or erasures.
2. Each contracting Government undertakes not to require consignment of plants intended for planting or propagation imported into this territories to be accompanied by phytosanitary certificates inconsistent with the model set out in the Annex to this Convention.

Article VI

Requirements in Relation to Imports

1. With the aim of preventing the introduction of diseases and pests of plants into their territories, contracting Governments shall have full authority to regulate the entry of plants and plant products, and to this end, may :
   (a) prescribe restrictions or requirements concerning the importation of plants or plant products;
   (b) prohibit the importation of particular plants or plant products, or of particular consignments of plants or plant products;
   (c) inspect or detain particular consignments of plants or plant products;
   (d) treat, destroy or refuse entry to particular consignments of plants or plant products, or require such consignments to be treated or destroyed.
2. In order to minimize interference with international trade, each contracting Government undertakes to carry out the provisions referred to in paragraph 1 of this Article in conformity with the following:

(a) Contracting Governments shall not, under this plant protection legislation, take any of the measures specified in paragraph 1 of this Article unless such measures are made necessary by phytosanitary considerations.

(b) If a contracting Governments prescribes any restrictions or requirements concerning the importation of plants and plant products into its territories, it shall publish the restrictions or requirements and communicate them immediately to the plant protection services of other contracting Governments and to FAO.

(c) If a contracting Government prohibits, under the provisions of its plant protection legislation, the importation of any plant or plant products, it shall publish its decision with reasons and shall immediately inform the plant protection services of other contracting Governments and FAO.

(d) If a contracting Government requires consignment of particular products to be imported only through specified points of entry, such points shall be so selected as not unnecessarily to impede international commerce. The contracting Government shall publish a list of such points of entry and communicate it to the plant protection services of other contracting Governments and to FAO, such restrictions on points of entry shall not be made unless the plants or plant products concerned are required to be accompanied by phytosanitary certificates or to be submitted to inspection or treatment.

(e) Any inspection by the plant protection service of a contracting Government of consignments of plants offered for importation shall take place as promptly as possible with due regard to the perishability of the plants concerned. If any consignment is found not to conform to the requirements of the plant protection legislation of the importing country, the plant protection service of the exporting country shall be informed. If the consignment is destroyed, in whole or in part, and official report shall be forwarded immediately to the plant protection service of the exporting country.

(f) Contracting Governments shall make provisions which, without endangering their own plant production, will reduce to a minimum the number of cases in which a phytosanitary certificate is required on the entry of plants or plant products not intended for planting, such as cereals, fruits, vegetables and cut flowers.

(g) Contracting Governments may make provision for the importation for purposes of scientific research of plants and plant products and of specimens of plant pests and disease-causing organisms under conditions affording ample precaution against the risk of spreading plant diseass and pests.

3. The measures specified in this Article shall not be applied to goods in transit throughout the territories of contracting Governments unless such measures are necessary for the protection of their own plants.

Article VII

International Co-operation

The contracting Governments shall co-operate with one another to the fullest practicable extent in achieving the aims of this Convention, in particular as follows:

(a) Each contracting Government agrees to co-operate with FAO in the establishment of a world reporting service on plant diseases and pests, making full use of the facilities and services of existing organizations for this purpose, and, when this is established, to furnish to FAO periodically the following information.

(i) reports on the occurrence, outbreak and spread of economically important pests and diseases of plants and plant products which may be of immediate or potential danger;

(ii) information on means found to be effective in controlling the pests and diseases of plants and plant products.

(b) Each contracting Government shall, as far as is practicable participate in any spcial campaigns for combating particular destructive pests or diseases which may seriously threaten crop production and need international action to meet the emergencies.

Article VIII

Regional Plant Protection Organization

1. The contracting Governments undertake to co-operate with one another in establishing regional plant protection organizations in appropriate areas.
2. The regional plant protection organizations shall function as the co-ordinating bodies in the areas covered and shall participate in various activities to achieve the objectives of this Convention.

Article IX

Settlement of Disputes

1. If there is any dispute regarding the interpretation or application of this Convention, or if a contracting Government considers that any action by another contracting Government is in conflict with the obligations of the latter under Articles V and VI of this Convention, especially regarding the basis of prohibiting or restricting the imports of plants or plant products coming from its territories, the Government or Governments concerned may request the Director-General of FAO to appoint a committee to consider the question in dispute.
2. The Director-General of FAO shall thereupon, after consultation with the Governments concerned, appoint a committee of experts which shall include representatives of those Governments. This committee shall consider the question in dispute, taking into account all documents and other forms of evidence submitted by the Governments concerned. This committee shall submit a report to the Director-General of FAO who shall transmit it to the Governments concerned, and to other contracting Governments.
3. The contracting Governments agree that the recommendations of such a committee, while not binding in character, will become the basis for renewed consideration by the Governments concerned of the matter out of which the disagreement arose.
4. The Governments concerned shall share equally the expenses of the experts.

Article X

Substitution of Prior Agreements

This Convention shall terminate and replace, between contracting Governments, the International Convention respecting measures to be taken against the Phylloxera vastarix of 3 November 1881, the additional Convention signed at Berne of 15 April 1889 and the Interntional Convention for the Protection of Plants signed at Rome on 16 April 1929.

Article XI

Territorial Application

1. Any Government may at the time of ratification or adherence or at any time thereafter communicate to the Director-General of FAO a declaration that this Convention shall extend to all or any of the territories for the international relations of which it is responsible, and this Convention shall be applicable to all territories specified in the declaration as from the thirtieth day after the receipt of the declaration by the Director-General.
2. Any Government which has communicated to the Director-General of FAO a declaration in accordance with paragraph 1 of this Article may at any time communicate a further declaration modifying the scope of any former declaration or determinating the application of the provisions of the present Convention in respect of any territory. Such modification or termination shall take effect as from the thirtieth day after the receipt of the declaration by the Director-General.
3. The Director-General of FAO shall inform all signatory and adhering Governments of any declaration received under this Article.

Article XII

Ratification and Adherence

1. This Convention shall be open for signature by all Governments until 1 May 1952 and shall be ratified at the earliest possible date. The instruments of ratification shall be deposited with the Director-General of FAO, who shall give notice of the date of deposit to each of the signatory Governments.
2. As soon as this Convention has come into force in acordance with Article XIV, it shall be open for adherence by non-signatory Governments. Adherence shall be effected by the deposit of an instrument of adherence with the Director-General of FAO, who shall notify all signatory and adhering Governments.

Article XIII

Amendment

1. Any proposal by a contracting Government for the amendment of this Convention shall be communicated to the Director-General of FAO.
2. Any proposal amendment of this Convention received by the Director-General of FAO from a contracting Government shall be presented to a regular or special session of the

Conference of FAO for approval and, if the amendment involves important technical changes or imposes additional obligations on the contracting Governments, it shall be considered by an advisory committee of specialists convened by FAO prior to the Conference.

3. Notice of any proposed amendment of this Convention shall be transmitted tot he contracting Governments by the Director-General of FAO not later than the time when the agenda of the session of the Conference at which the matter is to be considered is dispatched.
4. Any such proposed amendment of this Convention shall require the approval of the Conference of FAO and shall come into force as from the thirtieth day after acceptance by two-thirds of the contracting Governments.
5. The instruments of acceptable of amendments involving new obligations shall be deposited witht eh Director-General of FAO, who shall inform all contracting Governments of the receipt of acceptances and the entry into force of amendments.

Article XV

Denunciation

1. Any contracting Government may at any time give notice of denunciation of this Convention by notification addressed to the Director-General of FAO. The Director-General shall at once inform all signatory and adhering Governments.
2. Denunciation shall take effect one year from the date of receipt of the notification by the Director-General of FAO.

16
Biological Diversity Convention

Biological diversity is being significantly reduced by certain human activities. Aware of the general lack of information and knowledge regarding biological diversity and of the urgent need to develop scientific, technical and institutional capacities to provide the basic understanding upon which to plan and implement appropriate measures.

Noting further that *ex situ* measures, preferably in the country of origin, also have an important role to play. Recognising the close and traditional dependence of any indigenous and local communities embodying traditional lifestyles on biological resources, and the desirability of sharing equitably benefits arising from the use of traditional knowledge, innovations and practices relevant to the conservation of biological diversity and the sustainable use of its components. Recognising also the vital role that women play in the conservation and sustainable use of biological diversity and affirming the need for the full participation of women at all levels of policy-making and implementation for biological diversity conservation.

Stressing the importance of, and the need to promote, international, regional and global cooperation among States and intergovernmental organisations and the non-governmental sector for the conservation of biological diversity and the sustainable use of its components. Acknowledging further that special provision of new and additional financial resources and appropriate access to relevant technologies can be expected to make a substantial difference in the world's ability to address the loss of biological diversity.

Noting in this regard the special conditions of the least developed countries and small island States. Acknowledging that substantial investments are required to conserve biological diversity and that there is the expectation of a broad range of environmental, economic and social benefits from those investments. Recognising that economic and social development and poverty eradication are the first and overriding priorities of developing countries.

Aware that conservation and sustainable use of biological diversity is of critical importance for meeting the food, health and other needs of the growing world population, for which purpose access to and sharing of both genetic resources and technologies are essential. Noting that, ultimately, the conservation and sustainable use of biological diversity will strengthen friendly relations among States and contribute to peace for humankind.

Desiring to enhance and complement existing international arrangement for the conservation of biological diversity and sustainable use of its components, and determined to conserve and sustainable use biological diversity for the benefit of present and future generations.

Have agreed as follows :

Article 1 : Objectives

The objectives of this Convention, to be pursued in accordance with its relevant provisions, are the conservation of biological diversity, the sustainable use of its components and the fair and equitable sharing of the benefits arising out of the utilisation of genetic resources, including by appropriate access to genetic resources and by appropriate transfer of relevant technologies, taking into account all rights over those resources and to technologies, and by appropriate funding.

Article 2 : Use of Terms

For the purposes of this Convention :

"Biological diversity" means the variability among living organisms from all sources, including, inter alia, terrestrial, marine and other aquatic ecosystems and the ecological complexes of which they are part : this includes diversity within, species, between species and of ecosystems.

"Biological resources" includes genetic resources, organisms or parts thereof, populations, or any other biotic component of ecosystems with actual or potential use or value for humanity. "Biotechnology" means any technological application that uses biological systems, living organisms, so derivatives therefore, to make or modify products or process for specific use.

"Country of origin of genetic resources" means the country which possesses those genetic resources is in situ conditions. "Country providing genetic resources" means the country supplying genetic resources collected from in situ sources, including populations of both wild and domesticated species, or taken from ex situ sources, which may or may not have originated in that country "Domesticated or cultivated species" means species in which the evolutionary process has been influenced by humans to meet their needs.

"Ecosystem" means a dynamic complex of plant, animal and microorganism communities and their nonliving environment interacting as a functional unit.

"*Ex situ*, conservation" means the conservation of components of biological diversity outside their natural habitats.

"Genetic Material" means any material of plant, animal, microbial or other origin containing functional units of heredity.

"Genetic resources" means genetic material of actual or potential value.

"Habitat" means the place or type of site where an organism or population naturally occurs.

"*In situ* conditions" means conditions where genetic resources exist within ecosystems and natural habitats and the maintenance and recovery of viable populations of species in their natural; surrounding and, in the case, of domesticated or cultivated species, in the surrounding where they have developed their distinctive properties.

"Protected area" means a geographically defined area which is designated or regulated and managed to achieve specific conservation objectives.

"Regional economic integration organisation" means an organisation constituted by sovereign States of a given region, to which its member States have transferred competence in respect of matters governed by this Convention and which has been duly authorised, an accordance with its internal procedures, to sign, ratify, accept, approve or accede to it.

"Sustainable use" the use of components of biological diversity in a way and at a rate that does not lead to long term decline of biological diversity, thereby maintaining its potential to meet the needs and aspirations of present and future generations.

"Technology" includes biotechnology.

Article 3 : Principle

States have, in accordance with the Charter of the United Nations and the principles of international law, the sovereign right to exploit their own resources pursuant to their own environmental policies, and the responsibility to ensure that activities within their jurisdiction or control for not cause damage to the environment of other States or of areas beyond the limits of national Jurisdiction.

Article 4 : Jurisdictional Scope

Subject to the rights of other States, and except as otherwise expressly provided in this Convention, the provisions of this Convention apply, in relation to each Contracting Party :

(a) In the case of components of biological diversity, in areas within the limits of its national jurisdiction; and

(b) In the case of processes and activities, regardless of where their effects occur, carried out under its jurisdiction or control, within the area of its national jurisdiction or beyond the limits of national jurisdiction.

Article 5 : Cooperation

Each Contracting Party shall, as far as possible and as appropriate, cooperate with other contracting parties, directly or, where appropriate, through competent international organisations in respect of areas beyond national jurisdiction and on other matters of mutual interest, for the conservation and sustainable use of biological diversity.

Article 6 : General Measures for Conservation and Sustainable Use

Each Contracting Party shall, in accordance with its particular conditions and capabilities :

(a) Develop national strategies, plans or programmes for the conservation and sustainable use of biological diversity or adapt for this purpose existing strategies, plans or programmes which shall reflect, inter alia, the measures set out in this Convention relevant to the Contracting Party concerned; and

(b) Integrate as far as possible and as appropriate, the conservation and sustainable use of biological diversity into relevant sectoral or cross-sectoral plans, programmes and policies.

Article 7 : Identification and Monitoring

Each Contracting Party shall, as far as possible and as appropriate, in particular for the purposes of Articles 8 to 10.

(a) Identify components of biological diversity important for its conservation and sustainable use having regard to the indicative list of categories set down in Annex I;

(b) Monitor, through sampling and other techniques, the components of biological diversity identified pursuant to subparagraph (a) above, paying particular attention to those requiring urgent conservation measures and those which offer the greatest potential for sustainable use;

Identify process and categories of activities which have or are likely to have significant adverse impacts on the conservation and sustainable use of biological diversity, and monitor their effects through sampling and other techniques; and maintain and organise by any mechanism data derived from identification and monitoring activities pursuant to subparagraphs (a), (b) and (c) above.

Article 8 : In situ Conservation

Each Contracting Party shall, as far as possible and as appropriate :

(a) Establish a system of protected areas or areas where special measures need to be taken to conserve biological diversity;

(b) Develop, where necessary guidelines for the selection, establishment and management of protected areas where special measures need to be taken to conserve biological diversity;

(c) Regulate or manage biological resources for the conservation of biological diversity within or outside protected areas with a view to ensuring their conservation and sustainable use;

(d) Promote the protection of ecosystems, natural habitats and maintenance of viable populations of species in natural surroundings;

(e) Promote environmentally sound and sustainable development in areas adjacent to protected areas with a view to furthering protection of these areas;

(f) Rehabilitate and restore degraded ecosystems and promote the recovery of threatened species, inter alia, through the development and implementation of plans or other management strategies;

(g) Establish or maintain means to regulate, manage or control the risks associated with the use and release of living modified organisms resulting from biotechnology which are likely to have adverse environmental impacts that could affect the conservation and sustainable use of biological diversity, taking also into account the risks to human health;

(h) Prevent the introduction of control or eradicate those alien species which threaten ecosystems, habitats or species;

(i) Endeavour to provide the conditions needed or compatibility between present uses and the conservation of biological diversity and the sustainable use of its components;

(j) Subject to its national legislation, respect, preserve and maintain knowledge, innovations and practices of indigenous and local communities embodying traditional lifestyles relevant for the conservation and sustainable use of biological diversity and promote their wider application with the approval and involvement of the holders of such knowledge, innovations and practices and encourage the equitable sharing of the benefits arising from the utilisation of such knowledge, innovations and practices;

(k) Develop or maintain necessary legislation and/or other regulatory provisions for the protection of threatened species and populations;

(l) Where a significant adverse effect on biological diversity has been determined pursuant to Article 7, regulate or manage the relevant processes and categories of activities; and

(m) Cooperate in providing financial and other support for in situ conservation outlined in subparagraphs (a) to (l) above, particularly to developing countries.

Article 9 : Ex Situ Conservation

Each Contracting Party shall, as far as possible and as appropriate, and predominantly for the purpose of complementing in situ measures.

(a) Components of biological diversity, of such components;

(b) Conservation of and research on plants, animals and microorganisms;

(c) Adopt measures for the recovery and rehabilitation of threatened species and for their reintroduction into appropriate conditions;

(d) Regulate and manage collection of biological resources from purposes so as not to threaten ecosystem and in situ population of species except where special temporary *ex situ* measures are required under subparagraph (c) above; and

(e) Cooperate in providing for *ex situ* conservation outlined in subparagraphs (a) to (d) above and in the establishment and maintenance of *ex situ* conservation facilities in developing countries.

Article 10 : Sustainable use of components of biological diversity

Each Contracting Party shall, as far as possible and as appropriate

(a) Integrate consideration of the conservation and sustainable use of biological resources;

(b) Adopt measures relating to the use of biological resources to avoid or minimise adverse impacts on biological diveristy;

(c) Protect and encourage customary use of biological resources to avoid or minimise adverse impacts on biological diversity;

(d) Support local populations to develop and implement remedial action in degraded areas where biological diversity has been reduced; and

(e) Encourage cooperation between its governmental authorities and its private sector in developing methods for sustainable use of biological resources.

Article 11 : Incentive Measures

Each Contracting Party shall, as far as possible and as appropriate, adopt economically and socially sound measures that act as incentives for the conservation and sustainable use of components of biological diversity.

Article 12 : Research and Training

The Contracting Parties, taking into account the special needs of developing countries, shall :

(a) Establish and maintain programmes for scientific and technical education and training in measures for the identification, conservation and sustainable use of biological diversity and its components and provide support for such education and training for the specific needs of developing countries;

(b) Promote and encourage research which contributes to the conservation and sustainable use of biological diversity, particularly in developing countries inter alia, in accordance with decisions of the Conference of the Parties taken consequence of recommendations of the subsidiary Body on Scientific, Technical and Advice ; and

(c) In keeping with the provisions of Articles 10, 18 and 20, promote and cooperate in the use of scientific advances in biological diversity research in developing methods for conservation and sustainable use of biological resources.

Article 13 : Public Education and Awareness

The Contracting Parties shall :

(a) Promote and encourage understanding of the importance of, and the measures required for, the conservation of biological diversity, as well as its propagation through media, and the inclusion of these topics in educational programmes; and

(b) Cooperate, as appropriate, with other States and international organisations in developing educational and public awareness programmes, with respect to conservation and sustainable use of biological diversity.

Article 14 : Impact Assessment and Minimising Adverse Impacts

1. Each Contracting Party, as far as possible and as appropriate, shall :

(a) Introduce appropriate procedures requiring environmental impact assessment of its proposed projects that are likely to have significant adverse effects on biological diversity with a view to avoiding or minimising such effects and, where appropriate, allow for public participation in such procedures.

(b) Introduce appropriate arrangements to ensure that the environmental consequence, of its programmes and policies that are likely to have significant adverse impacts on biological diversity are duly taken into account;

(c) Promote, on the basis of reciprocity, notification, exchange of information and consultation on activities under their jurisdiction or control which are likely significantly to affect adversely the biological diversity of other States or areas beyond the limits of national jurisdiction, by encouraging the conclusion of bilaterla, regional or multilateral arrangements as appropriate;

(d) In the case of imminent or grave danger or damage originating under its jurisdiction or control to biological diversity within the area under jurisdiction of other States or in areas beyond the limits of national jurisdiction of other States or in areas beyond the limits of national jurisdiction, notify immediately the potentially affected States of such danger or damage, as well as initiate action to prevent or minimise such danger or damage;

(e) Promote national arrangements for emergency responses to activities or events, where caused naturally or otherwise, which present a grave and imminent danger to biological diversity and encourage international cooperation to supplement such national efforts and, where appropriate and agreed by the States or regional economic integration organisations concerned, to establish joint contingency plans.

2. The Conference of the Parties shall examine, on the basis of studies to be carried out, the issue of liability and redress, including restoration and compensation, for damage to biological diversity, except where such liability is a purely internal matter.

Article 15 : Access to Genetic Resources

1. Recognising the sovereign rights of States over their natural resources, the authority to determine access to genetic resources rests with the natural governments and is subject to national legislation.
2. Each Contracting Party shall endeavour to create conditions to facilitate access to genetic resources for environmentally sound uses by other Contracting Parties and to impose restrictions that run counter to the objectives of this Convention.
3. For the purpose of this Convention, the genetic resources being provided by a Contracting Party, as referred to in this Article and Article 16 and 19, origin of such resources or by the Parties that have acquired the genetic resources in complete with this Convention.
4. Access, where granted, shall be on mutually agreed terms and subject to the provisions of the Article.
5. Access to genetic resources shall be subject to prior informed consent of the Contracting Party providing such resources, unless otherwise determined by that Party.
6. Each Contracting Party shall endeavour to develop and caryy out scientific research based on genetic resources provided by other Contracting Parties with the full participation of, and where possible in, such Contracting Parties.
7. Each Contracting Party shall take legislative, administrative or policy measures, as appropriate, and in accordance with Articles 16 and 19 and where necessary through the financial mechanism established by Articles 20 and 21, with the aim of starting in a fair and equitable way the results of research and development and the benefits arising from the commercial and other utilisation of genetic resources with the Contracting Party providing such resources. Such sharing shall be upon mutually agreed terms.

Article 16 : Access to and Transfer of Technology

1. Each Contracting Party, recognising that technology includes biotechnology, and that both access to and transfer of technology among Contracting Parties are essential elements for the attainment of the objectives of this Convention, undertakes subjects to the provisions of this Article to provide and/or facilitate access for an transfer to other Contracting Parties of technologies that are relevant to the conservation and sustainable use of biological diversity or make use of genetic resources and do not cause significant damage to the environment.
2. Access to and transfer of technology referred to in paragraph 1 above to developing countries shall be provided and/or facilitated under fair and most favourable terms,

including on concessional and preferential terms where agreed, and where necessary in accordance with the financial mechanism established by Articles 20 and 21. In the case of technology to patents and other intellectual property rights, such access and transfer shall be provided on terms with recognise and are consistent with the adequate and effective protection of intellectual property rights. The application of this paragraph shall be consistent with paragraphs 3, 4 and 5 below.

3. Each Contracting Party shall take legislative, administrative or policy measures, as appropriate, with the aim that Contracting Parties, in particular those that are developing countries, which provide genetic resources are provided access to and transfer of technology which makes use of those resources... on mutually agreed terms, including technology protected by patents and other intellectual property rights, where necessary through the provisions of Articles 20 and 21 and in accordance with international law and consistent with paragraphs 4 and 5 below.
4. Each Contracting Party shall take legislative, administrative or policy measures, as appropriate, with the aim that the private sector facilitates access to, joint development and transfer of technology referred to in paragraph 1 above for the benefit of both governmental institutions and the private sector of developing countries and in this regard shall abide by the obligations included in paragraphs 1, 2 and 3 above.
5. The Contracting Parties, recognising that patents and other intellectual property rights may have an influence on the implementation of this Convention, shall cooperate in this regard subject to national legislation and international law in order to ensure that such rights are supportive of and do not run counter to its objectives.

Article 17 : Exchange of Information

1. The Contracting Parties shall facilitate the exchange of information from all publicly available source, relevant to the conservation and sustainable use of biological diversity, taking into account the special needs of developing countries.
2. Such exchange of information shall include exchange of results of technical, scientific and socioeconomic research, as well as information on training and surveying programmes, specialised knowledged, indigenous and traditional knowledge as such and in combination with the technologies referred to in Artcle 16, paragraph 1. It shall also, where feasible, include repatriation of information.

Article 18 : Technical and Scientific Cooperation

1. The Contracting Parties shall promote international technical and scientific cooperation in the field of conservation and sustainable use of biological diversity, where necessary through the appropriate international and national institutions.
2. Each Contracting Party shall promote technical in particular developing countries, in implementing this Convention, inter alia, through the development and implementation of national policies. In promoting such cooperation, special attention should be given to the development and strengthening of national capabilities by means of human resources development and institution building.

3. The Conference of the Parties, at its first meeting, shall determine how to establish a clearinghouse mechanism to promote and facilitate technical and scientific cooperation.
4. The Contracting Parties shall, in accordance with national legislation and policies, encourage and develop methods of cooperation for the development and use of technologies, including indigenous and traditional technologies, in pursuance of the objectives of this Convention. For this purpose, the Contracting Parties shall also promote cooperation in the training of personnel and exchange of experts.
5. The Contracting Parties shall, subject to mutual agreement, promote the establishment of joint research programmes and joint ventures for the development of technologies relevant to the objectives of this Convention.

Article 19 : Handling of Biotechnology and Distribution of Its Benefits

1. Each Contracting Party shall take legislative, administrative or policy measures, as appropriate, to provide for the effective participation in biotechnological research activities by those Contracting Parties, especially developing countries, which provide the genetic resources for such research, and where feasible in such Contracting Parties.
2. Each Contracting Party shall take all practicable measures to promote and advance priority access on a fair and equitable basis by Contracting Parties, especially developing countries, to the results and benefits arising from biotechnologies based upon genetic resources provided by those Contracting Parties. Such access shall be on mutually agreed terms.
3. The Parties shall consider the need for and modalities of a protocol setting out appropriate procedures, including, in particular, advance informed agreement, in the field of the safe transfer, handling and use of any living modified organism resulting from biotechnology that may have adverse effect on the conservation and sustainable use of biological diversity.
4. Each Contracting Party shall, directly or by requiring any natural or legal person under its jurisdiction providing the organisms referred to in paragraph 3 above, provide any available information about the use and safety regulations required by that Contracting Party in handling such organisms as well as any available information on the potential adverse impact of the specific organisms concerned to the Contracting Party into which those organisms are to be introduced.

Article 20 : Financial Resources

1. Each Contracting Party undertakes to provide, in accordance with its capabilities, financial support and incentives in respect of those national activities which are intended to achieve the objectives of this Convention, in accordance with its national plans, priorities and programmes.
2. The developed country Parties shall provide new and additional financial resources to enable developing country Parties to meet the agreed full incremental costs to them of implementing measures which fulfil the obligations of this Convention and to benefit from its provisions and which costs are agreed between a developing country Party by the institutional structure, referred to in Article 21, in accordance with policy, strategy, programme priorities and eligibility criteria and an indicate list of incremental costs

established by the Conference of teh Parties. Other Parties, including countries undergoing the process of transition to a market economy, may voluntarily assume the obligations of the developed country Parties. For the purpose of this Article, the Conference of teh Parties shall at its first meeting establish a list of developed country Parties and other Parties which voluntarily assume the obligations of the developed country Parties.

The Conference of the Parties shall periodically review and if necessary amend the list. Contributions from other countries and source on a voluntary basis would also be encouraged. The implementation of these commitments shall take into account the need for adequacy, predictability and timely flow of funds and the importance of burden-sharing among the contributing Parties included in the list.

3. The developed country Parties may also provide, and developing country Parties avail themselves of, financial resources related to the implementation of this Convention through bilateral, regional and other multilateral channels.
4. The extent to which developing country Parties will effectively implement their commitments under the Convention will depend on the effective implementation by developed country Parties of their commitments under the Convention related to financial resources and transfer of technology and will take fully into account the fact that economic and social development and eradication of poverty are the first and overriding priorities of the developing country Parties.
5. The Parties shall take full account of the specific needs and special situation of least developed countries in their actions with regard to funding and transfer of technology.
6. The Contracting Parties shall also take into consideration the special conditions resulting from the dependence on distribution and location of biological diversity within developing country Parties, in particular small island States.
7. Consideration shall also be given to the special situation of developing countries, including those that are most environmentally vulnerable, such as those with arid and semi-arid zones, coastal and mountainous areas.

Article 21 : Financial Mechanism

1. There shall be a mechanism for the provision of financial resources to developing country Parties for purposes of this Convention on a grant or concessional basis the essential elements of which are described in this Article. The mechanism shall function under the authority and guidance of, and be accountable to, the Conference of the Parties for purposes of this Convention. The operations of mechanism shall be carried out by such institutional structure as may be decided upon by the Conference of the Parties at its first meeting. For purposes of this Convention, the Conference of the Parties shall determine the policy, strategy, programme priorities and eligibility criteria relating to the access to and utilisation of such resources. The contributions shall be such as to take into account the need for predictability, adequacy and timely flow of funds referred to in Article 20 in accordance with the amout of resources needed to be decided periodically by the Conference of the Parties included in the list referred to in Article 20, paragraph 2. Voluntary contributions may also be made by the developed country parties and by other

countries and sources. The mechanism shall operate within a democratic and transparent system of governance.

2. Pursuant to the objectives of this Convention, the Conference of the Parties shall at its first meeting determine the policy, strategy and programme priorities, as well as detailed criteria and guidelines for eligibility for access to and utilisation of the financial resources including monitoring and evaluation on a regular of such utilisation. The Conference of the Parties shall decide on the arrangements to give effect to paragraph 1 above after consultation with the institutional structure entrusted with the operation of the financial mechanism.
3. The Conference of the Parties shall review the effectiveness of the mechanism established under this Article, including the criteria and guidelines referred to in paragraph 2 above, not less than two years after the entry into force of this Convention and thereafter on a regular basis. Based on such review, it shall take appropriate action to improve the effectiveness of the mechanism if necessary.
4. The Contracting Parties shall consider strengthening existing financial institutions to provide financial resources for the conservation and sustainable use of biological diversity.

Article 22 : Relationship with other International Conventions

1. The provisions of the present Convention shall not affect the rights and obligations of any Contracting Party deriving from any existing international agreement, except where the exercise of those rights and obligations would cause serious damage or a threat to biological diversity.
2. Contracting Parties shall implement this Convention with respect to the marine environment consistently with the rights and obligations of States under the law of the sea.

Article 23 : Conference of the Parties

1. A Conference of the Parties is hereby established. The first meeting of the Conference of the Parties shall be convened by the Executive Director of the United Nations Environment Programme not later than one year after the entry into force of this Convention. Thereafter, ordinary meetings of the Conference of the Parties shall be held at regular intervals to be determined by the Conference at its first meeting.
2. Extraordinary meetings of the Conference of the Parties shall be held at such other times as may be deemed necessary by the Conference or at the written request of any Party, provided that within six months of the request being communicated to them by the Secretariat, it is supported by at least one-third of the Parties.
3. The Conference of the Parties shall by consensus agree upon and adopt rules of procedure for itself and for any subsidiary body it may establish, as well as financial rules governing the funding of the Secretariat. At each ordinary meeting, it shall adopt a budget for the financial period until the next ordinary meeting.
4. The Conference of the Parties shall keep under review the implementation of the present Convention, and, for this purpose, shall :

(a) Establish the form and the intervals for transmitting the information to be submitted in accordance with Article 26 and consider such information as well as reports submitted by any subsidiary body;

(b) Review scientific, technical and technological advice on biological diversity provided in accordance with Article 25;

(c) Consider and adopt, as required, protocols in accordance with Article 28.

(d) Consider and adopt, as required, in accordance with Articles 29 and 30, amendments to this Convention and its annexes;

(e) Consider amendments to any protocol, as well as to any annexes thereto, and, if so decided, recommend their adoption to the parties to the protocol concerned;

(f) Consider and adopt, as required, in accordance with Article 30, additional annexes to this Convention;

(g) Establish such subsidiary bodies, particularly to provide scientific and technical advice, as are deemed necessary for the implementation of this Convention;

(h) Contact, through the Secretariat, the executive bodies of conventions dealing with matters covered by this Convention with a view to establishing appropriate forms of cooperation with them; and

(i) Consider and undertake any additional action that may be required for the achievement of the purposes of this Convention in the light of experience gained in its operation.

5. The United Nations, its specialised agencies and the International Atomic Energy Agency, as well as any State not party to this Convention, may be represented as observers at meetings of the Conference of the Parties. Any other body or agency, whether governmental or non-governmental, qualified in fields relating to conservation and substainable use of biological diversity which has informed the Secretariat of its wish to be represented as an observer at a meeting of the Conference of the Parties present object.

Article 24 : Secretariat

1. A secretariat is hereby established. Its functions shall be :

   (a) To arrange for and service meetings of the Conference of the Parties provided for in Article 23;

   (b) To perform the functions assigned to it by any protocol;

   (c) To prepare report on the execution of its function under this Convention and present them to the Conference of the Parties;

   (d) To coordinate with other relevant international bodies, and in particular to enter into such administrative and contractual arrangements as may be required for the effective discharge of its functions; and

   (e) To perform such other functions as may be determined by the Conference of the Parties.

2. At its first ordinary meeting, the Conference of the Parties shall designate the secretariat from amongst those existing competent international organisations which have signified their willingness to carry out the secretariat functions under this Convention.

Article 25 : Subsidiary body on scientific, technical and technological advice

1. A subsidiary body for the provision of scientific, technical and technological advice is hereby established to provid the Conference of the Parties and, as appropriate, its other subsidiary bodies with timely advice relating to the implementation of this Convention. This body shall be open to participation by all Parties and shall be multi disciplinary. It shall comprise government representatives competent in the relevant field of expertise. It shall report regularly to the Conference of the Parties on all aspects of its work.
2. Under the authority of and in accordance with guidelines laid down by the Conference of the parties, and upon its request, this body shall :
   (a) Provide scientific and technical assessments of the status of biological diversity;
   (b) Prepare scientific and technical assessments of the effects of types of measures taken in accordance with the provisions of this Convention.
   (c) Identity innovative, efficient and state of art technologies and know-how relating to the conservation and sustainable use of biological diversity and advise on the ways and means of promoting development and/or transfering such technologies;
   (d) Provide advice on scientific programmes and international cooperation in research and development related to conservation and sustainable use of biological diversity; and
   (e) Respond to scientific, technical, technological and methodological questions that the Conference of the Parties and its subsidiary bodies may put to the body.

The functions terms of reference, organisation and operation of this body may be further elaborated by the conference of the parties.

Article 26 : Reports

Each Contracting Party shall, at intervals to be determined by the Conference of the Parties, present to the Conference of the Parties reports on measures which it has taken for the implementation of the provisions of this Convention and their effectiveness in meeting the objectives of this Convention.

Article 27 : Settlement of Disputes

1. In the event of a dispute between Contracting Parties concerning the interpretation or application of this Convention, the Parties concerned shall seek solution by negotiation.
2. If the parties concerned cannot reach agreement by negotiation, they may jointly seek the good offices of, or request mediation by, a third party.
3. When ratifying, accepting, approving or aceding to this Convention, or at any time thereafter, a State or regional economic integration organisation may declare in writing to the Depositary that for a dispute not resolved in accordance with paragraph 1 or paragraph 2 above, it accepts one or both of the following means of dispute, settlement as compulsory;
4. If the parties to the dispute have not in accordance with paragraph 3 above, accepted the same or any procedure, the dispute shall be submitted to conciliation in accordance with Part 2 of Annex 11 unless the parties otherwise agree.

5. The provisions of this Article shall apply with respect to any protocol except as otherwise provided in the protocol concerned.

Article 28 : Adoption of Protocols

1. The Contracting Parties shall cooperate in the formulation and adoption of protocols to this Convention.
2. Protocols shall be adopted at a meeting of the Conference of the Parties.
3. The text of any proposed protocol shall be communicated to the Contracting Parties by the Secretariat at least six months before such a meeting.

Article 29 : Amendment of the Convention or Protocols

1. Amendments to this Convention may be proposed by any Contracting Party. Amendments to any protocol may be proposed by any Party to that protocol.
2. Amendments to this Convention shall be adopted at a meeting of the Conference of the Parties. Amendments to any protocol shall be adopted at meeting of the Parties to the Protocol in question. The text of any proposed amendment to this Convention or to any protocol, except as may otherwise be provided in such protocol, shall be communicated to the parties to the instrument in question by the secretariat at least six months before the meeting at which it is proposed, for adoption. The secretariat shall also communicate proposed amendments to the signatories to this Convention for information.
3. The Parties shall make every effort to reach agreement on any proposed amendment to this Convention or to any protocol by consensus. If all efforts at consensus have been exhausted, and no agreement reached, the amendment shall as a last resort be adopted by a two-thirds majority vote of the Parties to the instrument in question present and voting at the meeting, and shall be submitted by the Depositary to all Parties for ratification, acceptance or approval.
4. Ratification, acceptance or approval of amendments shall be notified to the Depositary in writing. Amendments adopted in accordance with paragraph 3 above shall enter into force among Parties having accepted them on the ninetieth day after the deposit of instruments of ratification, acceptance or approval by at least two-thirds of the Contracting Parties to this Convention or of the Parties to the protocol concerned, except as may otherwise be provided in such protocol. Thereafter the amendments shall enter into force for any other Party on the ninetieth day after that Party deposits its instruments of ratification, acceptance or approval of the amendments.
5. For the purposes of the Article, "Parties present and voting" means parties present and casting an affirmative or negative vote.

Article 30 : Adoption and Amendment of Annexes

1. The annexes to this Convention or to any protocol shall form an integral part of this Convention or of such protocol, as the case may be, and unless expressly provided otherwise, a reference to this Convention or its protocols constitutes at the same time a reference to any annexes thereto. Such annexes shall be restricted to procedural, scietific, technical and administrative matters.

2. Except as may be otherwise provided in any protocol with respect to its annexes, the following procedure shall apply to the proposal, adoption any entry into force of additional annexes to this Convention or of annexes to any protocol :
   a. Annexes to this Convention or to any protocol shall be proposed and adopted according to the procedure laid down in Article 29
   b. Any Party that is unable to approve an additional annex to this Convention or an annex to any protocol to which it is party shall so notify the Depositary, in writing, within one year from the date of the communication of the adoption by the Depositary. The Depositary shall without delay notify all Parties of any such notification received. A Party may at any time withdraw a previous declaration of objection and the annexes shall thereupon enter into force for that Party subject to subparagraph (c) below :

   On the expiry of one year from the date of the communication of the adoption by the Depositary, the annex shall enter into force for all parties to this Convention or to any protocol concerned which have not submitted a notification in accordance with the provisions of subparagraph (b) above.
3. The proposal, adoption and entry into force of amendments to annexes to this Convention or to any protocol shall be subject to the same procedure as for the proposal, adoption and entry into force annexes to the Convention or annexes to any protocol.
4. If an additional annex or an amendment to an annex is related an amendment to this Convention or to any protocol, the additional annex or amendment shall not enter into force until such time as the amendment to this Convention or to the protocol concerned enters into force.

Article 31 : Right to Vote

1. Except as provided for in paragraph 2 below, each Contracting Party to this Convention or to any protocol shall have one vote.
2. Regional economic integration organisations, in matters within them competence, shall exercise their right to vote if their member States which are Contracting Parties to the Convention of the relevant protocol. Such organisations shall not exercise their right to vote in their member State exercise theirs, and rice versa.

Article 32 : Relationship between this Convention and its Protocols

1. A State or a regional economic integration organisation may not become a party to a protocol unless it is, or becomes at the same time a Contracting Party to this Convention.
2. Decisions under any protocol shall be taken only by the Parties to the protocol concerned. Any Contracting Party that has not ratified accepted or approved a protocol may participate as an observer in any meeting of the parties to that protocol.

Article 33 : Signature

This Conventionshall be open for signature at Rio de Janeiro by all States and any regional economic integration organisation from the fifth day of June 1992 until.

Article 34 : Ratification, Acceptance or Approval

1. This Convention and any protocol shall be subject to ratification acceptance or approval by States and by regional economic integration organisations. Instruments of ratification, acceptance or approval shall be deposited with the Depository.
2. Any organisation referred to in paragraph 1 above which becomes a Contracting Party to this Convention or any protocol without any or its member States being a Contracting Party shall be bound by all the obligations under the Convention or the protocol, as the case may be. In the case of such organisations, one or more of whose member States is a Contracting Party to the Convention or relevant protocol the organisation and its member States shall decide on their respective responsibilities for the performance of their obligations under the Convention or protocol, as the case may be in such cases, the organisation and the member States shall not be entitled to exercise rights under the Convention or relevant protocol concurrently.
3. In their instruments of ratification, acceptance or approval, the organisations, referred to in paragraph 1 above shall declare the extent of their competence with respect to the matters governed by the Convention or the relevant protocol. These organisations shall also inform the Depositary of any relevant modification in the extent of their competence.

Article 35 : Accession

1. This Convention and any protocol shall be open for accession by States and by regional economic integration organisations from the date on which the Convention or the protocol concerned is closed for signature. The instruments of accessions shall be deposited with the Depositary.
2. In their instruments of accession, the organisations referred to in paragraph 1 shall declare the extent of their competence with respect to the matters governed by the Convention or the relevant protocol. These organisations shall also inform the Depositary of any relevant modification in the extent of their competence.
3. The provisions of Article 34, paragraph 2, shall apply to regional economic integration organisations which accede to this Convention or any protocol.

Article 36 : Entry into Force

1. This Convention shall enter into force on the ninetieth day after the date of deposit of the thirtieth instrument of ratification, acceptance approval or accession.
2. Any protocol shall enter into force on the ninetieth day after the date of deposit of the number of instruments of ratification, acceptance, approval or accesion, speccified in that protocol, has been deposited.
3. For each Contracting Party which ratifies, accepts or approves this Convention or accedes thereto after the deposit of the thirtieth instrument of ratificcation, acceptance, approval or accession, it shall enter into force on the ninetieth day after the date of deposit by such Contracting Party of its instrumentof ratification, acceptance, approval or accession.
4. Any protocol, except as otherwise provided in such protocol, shall enter into force for a Contracting Party that ratifies, accepts or approves that protocol or accedes thereto after

its entry into force pursuant to pargraph 2 above, on the ninetieth day after the date on which that Contracting Party deposits its instrument of ratification, acceptance, approval or accession, or on the date on which the Convention enters into force for that Contracting Party, whichever shall be the later.

5. For the purposes of paragraphs 1 and 2 above, any instrument deposited by a regional economic integration organisation shall not be counted as additional to those deposited by member States of such organisation.

Article 37 : Reservations

No reservations may be made to this Convention.

Article 38 : Withdrawals

1. At any time after two years from the date on which the present Convention has entered into force for a Contracting Party, that Contracting Party may withdraw from the Convention by giving written notification to the Depositary.
2. Any such withdrawal shall take place upon expiry of one Year after the date of its receipt by the Depositary, or on such later date as may be specified in the notification of the withdrawal.
3. Any Contracting Party which withdraws from this Convention shall be considered as also having withdraw from any protocol to which it is party.

Article 39 : Financial Interim Arrangement

Provided that it has been fully restructured in accordance with the requirements of Article 21, the Global Environment Facility of the United Nations Development Programme, the United Nations Environment Programme and the International Bank for Reconstruction and Development shall be the institutional structure referred to in Article 21 on an interim basis, for the period between the entry into force of this Convention and the first meeting of the Conference of the Parties or until the Conference of the Parties decides which institutional structure will be designated in accordance with Article 21.

Article 40 : Secretariat interim arrangements

The secretariat to be provided by the Executive Director of the United Nations Environment Programme shall be the secretariat referred to in Article 24, paragraph 2, on an interim basis for the period between the entry into force of this Convention and the first meeting of the Conference of the Parties.

Article 41 : Depository

The Secretary-General of the United Nations shall assume the functions of Depositary of this Convention and any protocols.

Article 42 : Authentic Texts

1. The original of this Convention, of which the Arabic, Chinese, English, French, Russian and Spanish texts are equally authentic, shall be deposited with the Secretary-General of the United Nations.
2. In witness whereof the undersigned, being duly authorised to that effect, have signed this Convention.
3. Done at Rio de Janeiro on this fifth day of June, one thousand nine hundred and ninety-two.

17

Natural Biological Capital of the Earth

This universe is the creation of the supreme power meant for the benefit of all His creations. Individual species must, therefore, learn to enjoy its benefits by forming a part of the system in close relation with other species. Let not only one species encroach upon the other's rights.

(Source : **Ishopanishad**)

Native biodiversities a source of pride for each country, composing as it does a shining part of the national heritage.

(Anon.)

Our National Food Security depends on our ability to conserve all out biological wealth.

(Dr. M.S. Swaminathan)

Variation is the law of nature. It occurs everywhere and every moment. The variations take place at micro levels at short space and small time period, but these become apparent only over a large space and big a time gap. The variations may be linear or cyclic. The variety and variability of organisms and ecosystems is referred to as biological diversity or biodiversity. Similarly, the biological variations initiate at the micro level (bio-molecular level or genes) and become apparent at species and ecosystem level. The biological variations in nature over time and space form the basis of evolutionary processes. Thus, *biodiversity is the degree of variety in nature and not nature itself.* Similarly, the biological diversity is not the same as biological resource although mutually, they form part of each other. The conservation of biological diversity is distinct but related to biological resources.

The World Commission on Environment and Development (WCED) constituted by the General Assembly of the United Nations published its report in 1987 which provided a major boost and endorsement to the need for conserving the world's rich biodiversity, particularly that of the tropical areas. Despite conflicting views among nations, a broad consensus was reached after bitter negotiations, and 170 countries signed the Biodiversity Convention, which is now ratified by 104 countries (42 Articles were adopted). One of the prerequisite tasks as expressed by Article 7 of the Convention is the identification and monitoring the components of biological diversity. Article 12 calls for Research and Training and suggest programmes for identification, conservation and sustainable use of biological diversity.

Biodiversity at Different Levels

Global Level

It is estimated that there exists 5-30 million species of living forms on our earth and of these, only 1.5 million have been identified and include 3,00,000 species of green plants and fungi, 8,00,000 species of insects, 40,000 species of vertebrates and 3,60,000 species of microorganisms. Recently it has been estimated that the number of insects alone may be as high as 10 million, but many believe it to be around 5 million.

Table 17.1 : Estimated number of species worldwide

Taxnomic group	No. of spp.
Bacteria	3600
Blue green algae	1700
Fungi	46983
Bryophytes	17000
Gymnosperms	750
Angiosperms	250000
Insects	750000
Sponges	5000
Custaceans	9000
Molluscs	38000
Star fishes	50000
Fishes	6100
Amphibians	19056
Reptiles	6300
Birds	9036
Mammals	4008

The tropical forests are regarded as the riches in biodiversity. According to the opinion of the scientists more than half of the species on the earth live in moist tropical forests, which is only 7% of the total land surface. Insects (80%) and primates (90%) make up most of the species.

The species diversity in tropics is high as:

- In tropics, as the conditions for evolution were optimum and for extinction fewer.
- In tropics, species diversity was conserved over geological time. Due to low rates of extinction prevailing there; and
- Biological diversity is the result of interaction between climate, organisms, topography, parent soil materials, time and heredity.

However, these explanations need experimental observations and confirmation.

Country Level

India is located in South Asia, between latitudes 6° and 38° N and longitudes 69° and 97° E. The Indian landmass extending over a total geographical area of about 3029 million hectares, is bounded by Himalayas in the north, the Bay of Bengal in the east, the Arabian sea in the west, and Indian Ocean in the South. The Wide variety in physical features and climatic situation have resulted in a diversity of ecological habitats. These vary from the humid tropical Western Ghats to the hot desert of Rajasthan, from the cold desert of Ladakh and the icy mountain of Himalayas to the warm costs of peninsular India.

There are no clear estimates about the marine biota though the coastline is 7,000 km long with a shelf zone of 4,52,460 sq km and extended economic zone of 20,13,410 sq km. There is an abundance of seaweeds, fish, crustaceans, molluscs, corals, reptiles and mammals.

Table 17.2 : Number of recorded biota in India

	Taxon	No. of Species
Flora	Bacteria	850
	Algae	2500
	Fungi	23000
	Lichens	1600
	Bryophyta	2700
	Pteridophyta	1022
	Gymnosperms	64
	Angiosperms	17000
	Total	**48736**
Fauna	Protozoans	2577
	Porifera	519
	Cnidaria	237
	Ctenophora	10
	Platyhelminthes	1622
	Nematoda	2350
	Rotifera	310
	Kinoryncha	10
	Gastroticha	88
	Acanthocephala	110
	Sipuncula	38
	Mollusca	5042
	Echiura	33
	Annelida	1093
	Onychophora	1
	Arthropoda	57525
	Phoronida	3
	Bryozoa	170
	Entoprocta	10
	Brachiopoda	3
	Chaetognatha	30
	Echinodermata	765
	Hemichordata	12
	Protochordata	116
	Fishes	2546
	Amphibians	204
	Reptiles	428
	Birds	1228
	Mammals	372
	Total	126188

(Based on available data)

Information regarding other flora and fauna are patchy. Hundreds of new species may be present in the country awaiting discovery. The Western Ghats in Peninsular India, which extend in the southern states, are a treasure house of species diversity and has about 5,000 species. It is estimated that almost one-third of the animal varieties found in India have taken refuge in Western Ghats of Kerala alone.

The Indian Gene Centre is among the twelve megadiversity regions of the world. More than 20 crop species were domesticated here. It is known to have more than 49,000 species of plants, 18,000 species of higher plants, including major and minor crop (166) and their wild relatives (326). Around 1,000 wild edible plant species are widely exploited by native tribes. These include 145 species of roots and tubers, 521 of leafy vegetables/greens, 101 of buds and flowers, 647 of fruits and 118 of seeds and nuts. In addition, nearly 9,500 plant species of ethno-botanical uses have been reported from the country of which around 7,500 are the ethno-medicinal importance and 3,900 are multipurpose, edible species.

Table 17.3 : Endemic species in India

	Group	No. of Species
Plants	Pteridophyta	200
	Angiosperms	4950
Animals		
Protozoa	Parasitic	500
	Free living	90
	Lepidoptera	9
Mollusca	Land & Freshwater	967
Pisces	Freshwater	64
	Marine	14
	Amphibia	123
	Reptilia	182
	Aves	60
	Mammalia	44

Wilk Plant Wealth

More than 75 per cent of Indian diversity (in the number of species in genera such as *Dendrobium, Bulbophyllum, Liparis, Coelogyne, Paphiopedilum, Vitis, Citrus, Musa, Rhododendron, Hedychium, Elaeacarpus,* and *Elaeagnus* occurs in this region. Over 30 species of legumes and 45 of grasses mainly occur in the temperate belt of the northeast Himalayan region. The legume flora is represented by species of *Astragalus, Caragana, Medicago, Melilotus, Praochetus, Trifolium, Trigonella* and *Vicia.* Among grasses, a large diversity is noticed in genera such as *Agrostis, alopecurus, Bromus, Calamogrostis, Dactylis, Festuca, Glyceria, Lolium, Muehlenbergia, Phleum, Poa, Stipa* and *Trisetum.*

Local inhabitants gather a wide range of plants from wild habitats, which provide edible tubers, green leafy vegetables, edible fruits and nuts etc. The more promising of these have been domesticated and/or protected under homestead management.

Ecosystem Diversity in India

These ecosystems harbour sustain the immense biodiversity. India is one of the 12 megabiodiversity centres in the world. This attributed to the immense variety in physiography and climatic situations resulting in diversity of ecological habitats ranging from tropical subtropical, temperate alpine to desert.

- Desert
- Forests
- Grasslands
- Wetlands
- Mangroves

Desert Ecosystem

According to the world bio-geography classification, India represents 2 major realms and 3 basic biomes including 12 bio-geographical regions. However, the Wildlife Institute of India has suggested that India be divided into following ten bio-geographic regions.

1. Trans Himalayan
2. Himalayan
3. Indian desert
4. Semi-arid
5. Western Ghats
6. Deccan Peninsula
7. Eastern Himalayan
8. Gangetic Plains
9. North East India
10. Islands and Coasts

Desert ecosystem is characterized by low precipitation, arid lands, which are largely barren except for sparse or seasonal vegetation cover. Species in this habitat are adapted to a extremely harsh, water-scarce environment. Covering 2% of the landmass, deserts in India re classified into three distinct types :

1. The sandy thar desert of Western Rajasthan and adjoining areas of other states.

2. The salt desert of Kutch in Gujarat.
3. The high altitude cold desert of Jammu and Kashmir, Leh Laddakh and Himachal Pradesh.

The sandy Thar desert in India covers about 2,78,330 sq. km. of which 1,96,150 sq. km., is in Rajasthan, 62,180 sq. km. in Gujarat and 20,000 sq. km. in Punjab and Haryana. The sparse vegetation is mainly composed of *Prosopis cineraria, Capparis, decidua, Ziziphus nummularia, Acacia nilotica, Clotropis procera* and *Prosopis juliflora.*

The salt desert of Rann of Kutch is spread over 9000 sq. km. It is characterized by a typical salt bush plants community of halophytes.

Extending over the North Himalayan ranges, the cold desert is characterized by extremely low temperature going down below-45°C and low rain fall ranging from 500-800 mm annually. The vegetation is sparse alpine steppe and mostly herbaceous and shrubby. Some common species are *Salix daphnoides, Myricaria elegan, Cicer microphyllum, Polygonum affine* and *Potentilla bifurea* etc.

Forest Ecosystems

The forests cover of the country is about 639,600 sq. km according to the Forest Survey of India Assessment (1995). This represents 19.46% of India's total geographical area India is endowed within diverse forest types rangign from the tropical wet evergreen forests in north-east to the tropical thorn froests in the central and western India. The forests of the country provide several essential services to the mankind. Forests are the source of a number of food items, fuel-wood, fodder and timber. Other economic uses indeed providing raw materials for forest based industries. Some of the minor forest produce including gum, resin, honey, etc. forests perform improtant ecological functions such as maintaining delicate ecological and hydrobiological blances, conserving soil, controlling floods, drought and pollution. Forests provide habitats for innumerable plants animals and microorganisms. Forests are the source of recreation and religious inspirations.

Grasslands

Grasslands, which are also called as steppes, prairies, pampas and savannas in various parts of the world, are vegetation types with predominance of grass and grass like species. In India the total area under grasslands is about 3.9% or 12 millions hectares. Grasslands in the country also exhibit as diversity ranging from semi-arid pastures in Daccan Peninsula, Humid Semi waterlogged grass land of Tarai belt, rolling shola grassland on the hilltops of Western Ghats and the high attitude alpine pasture of Himalayas. Five distinct types of grasslands have been recognised in India.

- *Sehima - Dichanthium* type
- Dichanthium - Cenchrus - Lasiurus type
- Phragmites - Saccharum - Inperata type
- Themeda - Arundinella type
- Temperate alpine type

The significance of about 10,000 odd known grass species in the world is enormous in leading vegetation cover to about 17% of the earth land surface, besides providing a number of cereals and fodder to mankind. About 15% of world grasses are represented in India (Vates et. al., 1999).

The significance of tropical grasslands has been well realized for their highest productive capacity. Grassland ecosystems are also expected to play crucial role in the global biospheric responses to climatic change and in augmenting vegetation in forests facing plunder. The W.W.F.'s Biodiversity Conservation Prioritisation (BCP) report on diwindling biodiversity of grasslands is a timely reminder on the extent and ferocity of environmental deterioration that has degraded nearly half of the earth's vegetative surface (Daioy 1995; Vates et. al. 1999). It is estimated that the Indian grasslands harbour about 1256 species belonging to 245 genera.

Wetlands

Wetlands are transitional zones that occupy intermediate position between dry land and open water. These ecosystems are dominated by the influence of water. They encompass diverse and heterogeneous habitats ranging from rivers, flood plains and rainfed lakes to swamps, estuaries and salt marshes.

India by virtue of its extensive stretch and varied terrain and climate, supports a rich diversity of inland and coastal wetland habitats. It is estimated that India has about 4.1 million ha.

Mangroves

Mangroves constitute an important economic resource providing fodder, fuel, wood, tannin, edible fished, hides, honey, wax, various chemicals and medicines. Mangroves play important role in stabilizing shorelines and protecting them from cyclones. Mangrove also harbours a variety of plants and animals.

Mangroves are salt-tolerant ecosystems in tropical and sub-tropical regions. These ecosystems are lagely characterized by assemblage of unrelated tree genea, that share the common ability to grow in saline and dual zones.

India harbours some of the best mangroves swamps in the world, located in the alluvial deltas of Ganga, Mahandi, Godavari, Krishna and Cauvery rivers and on the Andaman and Nicobar group of Islands. The total area covered by mangroves in India is about 6,700 sq. km. amounting to the 7% of the world mangrove. The largest stretch of mangroves in the country lies in the Sunderbans in West Bengal covering an area of about 4,200 sq. km. The predominant mangrove species are *Avicennia officinalis. Exoecaria agallocha, Heritiera fomes, Bruguiera parviflora, Ceriops decandra, Rhizophora mucronata and Xylocarpus granatum.*

Species Diversity

Biogeographically, India is situated at the tri-junction of three realms namely Afro-tropical, Indo Malayan and Paleoarctic relams, and therefore has characterstic element forms in each of them. This assemblage of three distinct realms makes the country rich and unique in biological diversity. Based on the available data, India ranks tenth in the world and fourth in Asia in plant diversity and ranks tenth in the number of mammalian species an eleventh in the number of endemic species of higher vertebrates in the world.

Status of Survey

at present, 1.75 million species have been recorded so far in the world (Global Biodiversity Assessment, 1995). India's contribution to this record stands at 7%. Surveys conducted so far have inventoried over 49,000 species of plants and 81,000 species of animals. As until now, only 70% of the area have been surveyed, it is estimated tha the flora and fauna already identified are only a part of what actually occur in India. The list is being constantly added to especially in the case of lower plants and invertebrate animals. Survey and inventorisation of India's biodiversity is still far from competing, especially for the lower groups of plants and invertebrate animals.

Endemic Species

Endemism of Indian biodivesity is significant. About 4,900 species of flowering plants are 33% of the recorded floras are endemic to the country. These are distributed over 141 genera belonging to 47 families. These are concentrated in the floristically rich areas of North East India, the Western Ghats, North West Himalayas and the Andaman and Nicobar Islands.

The Western Ghats and the Eastern Himalayas are reported to have 1,600 and 3,500 endemic species of flowering plants, respectively. These constitute two of 18 hot sports identified in the world. It is estimated that 62% of the known amphibian species are endemic to India of which a majority occur Western occur in Western Ghats.

Endemic species are the plants, which are limited in their distribution i.e. they are restricted to a small area and are not found elsewhere in the world. It may be due to:

- Poor adaptability of a species in a wide range of ecological
- Presence of some geographical barrier, e.g. Sea, Mountains etc.
- Failure of dispersal of reproductive organs (propagules, seeds, runners etc.).
- The species might have been comparatively young and not have enough time to spread.

Some examples of endemics include *Metasequoic* living gymnosperm endemic in China, *Sequoia* (red wood tree) endemic in coastal valleys of California, USA, *Primulla* and *Potentilla,* at high altitudes of Himalayas, *Ginkgo biloba* endemic in Japan and China.

Cultivated Plants/Agro Biodiversity

The agricultural biological diversity is a strong, evolutionary divergent line from the biodiversity, which deals with the life forms at large. Several multifaceted, scientifically established, threats to agro-biodiversity have emerged, particularly in many technologically advanced countries, in which agriculture plays key role in national economy and the **Gross Domestic Product** (GDP), the more important and relevant issues are :

1. Safe
2. Their management under a sustained system
3. Providing for further opportunities for agro-evolutionary processes.
4. The efficient, judicious and equitable use of this diversity.
5. To attain progressive increases in yielding ability and wider adaptability of improved varieties by appropriate use of conserved germplasm.

Two contrasting dimensions of agro-biodiversity may be clearly viewed in the present context, namely:

1. Its occurrence in the traditional ago-ecosystem in which cultivated plant diversity is maintained, more or less in equilibrium and
2. Its manipulation in the intensive agro cropping system, which are externally guided for their varietal components and packages of cultivation practices.

India is an acknowledge centre of crop diversity. It is considered to be the homeland of 167 important cultivated plants species and 320 species of their wild relatives. India is considered to be the centre of origin or 30,000-50,000 varieties of rice, pigeon pea, mango, turmeric, ginger, pepper, banana, bitter gourd, okra, coconut, cardamom, jack-fruit sugarcane, bamboo, taro, indigo, sun hemp, amaranthus, gooseberries, etc. The gene bank of National Bureau of Plant Genetic Resources (NBPGR) has a collection of over 1,44,000 varieties.

Genetic Diversity

Genetic diversity is defined as variations in the genetic composition of individual within or among species. India being one of the 12 megabiodiversity countries possesses rich genetic diversity. Studies in genetic diversity are not as yet very widespread and sharply focused. However, studies in genetic diversity of wild crop relatives and domesticated animals have been carried out. There are several hundred species of wild crop relatives distributed all over the country. A major centre for wild relatives is North-Eastern hills and Tamil Nadu hills are rich in wild relatives of millets, wild relatives of wheat and barely have been located in the Western and North Eastern Himalayas so far.

Endangered Plant and Animal Species

About 427 endangered plant species have been listed by the Botanical Survey of India in its publications on the floristic synthesis enumerated for Red Data book. This contributes to about 20 per cent of India's total floristic wealth of higher plants.

The available evidences, however, indicate that human activities are eroding the prevalent biological resources and greatly reducing the biodiversity of the planet. Estimating the precise rate of loss, or even the current status of the species, is stupendous task because no monitoring system, whatsoever, systematic, is likely to match the huge diversity in its existing form. Thus, much of the basic information remains lacking, especially on the species rich tropics and the hotspots of agro-biodiversity.

18

Conservation of Biodiversity in Indian Scenario

Gujarat

Gujarat, located in the western region of India, has a variety of climatic and edaphic conditions. Inspite of continuous onslaught on the biological resources, the overall range of biodiversity of Gujarat is still unmatched. Gujarat has many kinds of natural ecosystem due to varied environmental, climatic and edaphic conditions. Some important ecosystems in this state are deserts, grasslands, scrublands, agri-fields, wetlands, forests, marine ecosystem etc. This state has got a large population of tribals dependent on the natural biological resources for their livelihood. The Southern and Eastern Gujarat are primarily tribal areas whereas in Saurashtra and Kachchh, there is a large population of nomadic pastoralist community called Maldharis, whose migratory lifestyle effect the floral and faunal diversity of the state.

Gujarat with an area of 1,96,020 sq. km. and about 1,600 km long coast. The eastern hilly region is formed by southern, western and northern extension of Arawalli, Satpura and Sahydri (Western Ghat) Mountain ranges, respectively, and vast extent of saline deserts form a part of this state. The broad spectrum of the eco-climate is defined by rainfall varying from 250 mm in North -West to 2,000 mm in south-east. Seasonal variations of tropical monsoon climate are distinct, on the coast, there are two important sheltered gulfs viz. Gulf of Kachchh and Gulf of Khambhat.

Of the ten **bio-geographic zones** of India, Gujarat has parts of four zones, namely, zone 3,4,5 and 10.

Gujarat Ecology Commission initiated a research programme of preparing a *Status Report of Biodiversity of Gujarat* (based on scientifically recorded species and their distribution) with the following objectives:

1. To discover and describe species diversity (i.e. describing biodiversity).
2. To analyze and synthesize the information derived from the research programme into a predictive classification system that reflects the history of species (i.e., understanding biodiversity).

This preliminary work of compiling secondary information on biodiversity of Gujarat would provide a basis for future studies and plans. India has over 1,26,188 species of bacteria, fungi, plants and animals already described, of which, nearly 7,850 species are recorded from Gujarat. These species occur on land and in freshwater and marine habitats or as symbionts in mutualistic or parasitic state with other organisms.

There are about 1,933 species of Algae reported from Gujarat, out of which, 210 species are marine. Marine algae are known since long as a source of food, fodder and fertilizer. Barring a few, all species and all genera found on the Gujarat Coast are cosmopolitan. Out of the 23,000 species of fungi reported from the Indian region, only 165 species are documented from Gujarat. Similarly, very scanty information is available on the groups like Bryophytes, Pteridophytes and Gymnosperms.

The angiosperms flora of Gujarat is mostly varied in extent and composition. There are 2,198 species of higher plants belonging to 902 genera and 155 families which represent 12.91% of the flora of the country. Majority of them are distributed in the semi-arid regions of Kachchh.

Twenty-seven species of mangroves are listed from Gujarat. Earlier reports show that there was a luxuriant growth of mangroves at Sikka, Salaya, Jodiya, Jakhau, Modhwa and elsewhere on the Gujarat coast. However, drastic reduction has been noticed in the recent years due to extensive cutting for timber, fuel and fodder. The ethnobotanical studies carried out by several investigators have identified nearly 1,200 plants species used by tribals for meeting their multifarious requirements.

The present survey of **fauna** has revealed that though a good deal of work has been carried out in floristic explorations, no comprehensive efforts have been made to explore the faunal wealth of Gujarat, barring the vain and mammalian species. Altogether, references of 3,462 species of animals belonging to different taxa recorded from Gujarat have been collected. The fauna recorded from Gujarat is 4.54% of the Indian faunal population.

Euglenoids and Amoeba are the protozoan groups studied systematically in Gujarat while the remaining phyla namely, Ciliophora, Apicomplexa, Microspora and Myxozoa remain untouched. Our knowledge on the Calcareous and Hexactinellid sponges of the Indian waters is still confined to the classical monographs that appeared in the pre-independence period.

Among the colorful marine Cnidarians only the corals are studied systematically along the Gulf of Kutch while other two classes of Cnidarians (Hydrozoa and Scyphozoa) have yet to gain attention. All the 37 species of corals in the Marine National Park are reported to be in the endangered/threatened category.

Platyhelminthes constitute one of the major phyla comprising about 17,500 known species out of which, 1,622 species have been recorded from India. One species of Planaria is included in the vulnerable category by the Marine National Park authorities. Records indicate that among Aschelminthes, only Rotifers have been systematically studied during the sixties in Gujarat.

The Molluscan fauna of Gulf of Kutch is fairly well known. About 350 species have been identified from this coast. However, the information available is mainly focussed on the edible marine molluscs as it is one of the important marine foods in Gujarat. According to the Marine National Park authorities, many Molluscan species are threatened due to over exploitation in the Gulf of Kutch. In contrast to marine Molluscs, little information is available on the freshwater molluscs in Gujarat.

Annelids constitute one of the largest phylum with 1,093 species reported from India of which, only 69 species are reported from Gujarat. The studies were mainly on marine Polychaete worms along the Gulf of Kutch.

The other largest phylum, Arthropoda, is represented by more than 50,000 species in India. However, only 743 species are reported from Gujarat. Some isolated studies on scorpions, pseudoscorpions, crop pests and mites have been carried out. Most of the forest areas in Gujarat are unexplored from the entomological point of view.

The Bryozoans of Gujarat (42 spp.) comprise a good proportion of the total 170 species recorded from India. On the other hand, of the total 765 species of Echinoderms reported from India, only 15 species are known in Gujarat. Hemichordata is a group which is not studied adequately from Indian waters, and there is an urgent need for taxonomic studies, particularly on the families Protobalanidae, Cephalodiscidae and Planctospaeridae.

In comparison to the invertebrates, the vertebrate fauna has been studied fairly well in Gujarat. The fish fauna is represented by 364 species, of which, 19 species are in the endangered/threatened category. So far, the non-edible fishes and freshwater fishes have not received adequate attention. 78 species of reptiles are reported from Gujarat and 8 of them are placed in the Red Data Book of ZSI. Many reptilian species are facing severe threat due to habitat destruction.

The avain fauna of Gujarat is unique and is represented by 454 resident/migratory birds. 8 species are placed under the endangered and 5 species in the vulnerable category in Red Data Book of ZSI. In addition to this, 11 species of birds are listed ad endangered/threatened by the Marine National Park authorities. The Mammalian fauna of Gujarat comprises of 69 species, which is 18.49 per cent of the total Indian Mammals. Four species are featured in the Red Data Book as endangered while ten species are in the vulnerable category.

Analysis of the livestock population of Gujarat has shown that over a period of ten years (1982-1992) the livestock population growth rate is 7.16 per cent. District-wise population growth of cattle, buffaloes, sheep and goat has also been worked out. The data indicates that measures are required to study the carrying capacity of the districts as well as the feeding practices in some districts. Altogether, the number of indigenous breeds of Gujarat is far higher than those of the exotic varieties.

Based on the available data, an attempt has also been made to categorize the existing forest areas into Unique, Rich, and Unique an Rich areas of biodiversity.

Management Suggestions

It is essential for a Wildlife Manager to study biodiversity and its habitat before attempting any addition/alterations. The following suggestions are made for conserving the biodiversity of Gujarat :

Captive breeding and reintroduction programme should be taken up for rare and endangered animals and birds of Gujarat. Sakkarbaug Zoo of Junagadh is already doing commendable job for breeding Asiatic lions, Indian wild ass and other important animals and birds.

Conservation and protection of existing natural forests of Gujarat, particularly that of South Gujarat, Panchmahals, North Gujarat, Gir and Girnar.

Environmental awareness and nature education programmes have been a great success in Gujarat. It should be extended to other areas. Nature conservation societies and clubs can be of great help in conserving biodiversity.

Extension of lion habitat can be attempted on the lines of GGE concept by taking up suitable habitat management and proper and sufficient water and development of prey base.

Intensive research should be carried out in key biodiversity areas to study the habitat, ecology and food requirements of wild animals, biotic pressures etc. Marine national park area in Gulf of Kachchh needs more immediate attention.

Key biodiversity areas should be protected from cattle grazing. In such areas, ecodevelopment programme (as being taken up in Gir) would be of immense help. In semi-arid regions, integrated watershed development works should be taken up on large scale.

Natural bamboo breaks/bamboo forests of Rajpipla form excellent habitat for several wild animals and therefore, need to be conserved.

Preservation plots should be protected and conserved with vigour and continuous study should be carried out to monitor the changes in the biodiversity status of these areas.

Preserving large sized trees for flying squirrels, grey hornbill, birds of prey and Owls. Study and conservation of sloth bear habitat in hilly-forested patches of Jessore, Ratanmahal, and Rajpipla forests. For the enlargement of Sloth bear habitat, Girnar and Jambughoda can be taken up on experimental basis as these areas are suitable to sloth bear population.

Protected Areas in Gujarat

1. Gir Lion National Park and Sanctuary

Location : Junagadh and Amereli district

Area :	Sanctuary	1153.42 sq.kms.
	National Park	258.71 sq.kms.
Total :		1412.13 sq.kms.

The last abode of Asiatic Lions (*Panthera leo persica*) in the entire world. The habitat represents the best example of the dry deciduous forests in the country. It has got the highest concentration of the two largest predators in the country i.e. Asiatic lions (304) and panthers (230). Other wild animals include cheetal, sambar, chowsingha, chinkara, wildboard, nilgai, ratel, rusty-spotted cat, pangolin, crocodile, etc. It is one of the best managed National Park and Sanctuary in the country.

2. Marine National Park and Sanctuary

Location : Sea coast in Jamnagar District

Area :	Sanctuary	458 sq.kms.
	National Park	163 sq.kms.

National Park area is included in Marine Sanctuary area

It is the first Marine National Park and Sanctuary in the country with rich marine floral and faunal diversity. There are more than 40 species of Corals and 70 species of Sponges. Bonnelia, which is recorded only at three places in the world, is found here. Three of the world's six marine turtles (leather back, green and Oliver Ridley's turtle) are found here. Besides, large marine mammals like whales, dolphins and dugongs are also seen in the Gulf of Kachchh. Other marine life includes sea horse, sea anemone, star fish, crabs, coral fishes, sea hare, octopus, squid, sea shells etc. According to late Dr. Salim Ali, the country's best aquatic and marine bird concentration and one of the best mangrove forests are found here.

3. Velavadar Blackbuck National Park

   Location : Bhavnagar district

   Area : 34.08 sq.km.

A grassland of very high ecological value in the saline Bhal region has almost flat terrain. It falls in the low rainfall region of the semi-arid Saurashtra. It has the largest concentration of blackbucks along with the only predators of this area, i.e., wolf. With effective improvement of grassland status in the park, the area now has perhaps the largest concentration of lesser floricans (above 42) during breeding season.

4. Vansda National Park

   Location : Valsad district

   Area : 23.99 sq.kms.

This contains moist deciduous forest which falls in the western ghats, and contains wildlife such as panther, cheetal, sambar, wildboar, barking deer, racket tailed drongo and orioles etc. Although, the habitat is well suited for wild animals, the animals population is very low.

5. Nal Sarova Bird Sanctuary

   Location : Surendranagar District

   Area : 12. 82sq.kms.

The largest natural wetland habitat in the state which attracts large number of winter migrants such as ducks, geese, storks, spoonbills, flamingoes, plovers, coots, cormorants and pelicans etc. A heaven for ornithologists, naturalists and bird watchers.

6. Khijadia Bird Sanctuary

   Location : Jamanagar District

   area : 6.05 sq.kms.

Being situated close to the sea coast, this sanctuary attracts marine as well as fresh water birds. Although small in area, it is probably richer in bird diversity than Bharatpur. The unique location and typical terrain of the area ensures the availability of both fresh as well as seawater in the same area. This sanctuary attracts migratory birds such as flamingoes, pelicans, cranes, plovers, spoonbills, arctic terns, gulls, skimmers, grebes etc.

7. Thol Bird Sanctuary

   Location : Mehsana District

   Area : 6.99 sq.kms.

This bird sanctuary attracts a large population of ducks, such as shovellers, brahminy duck, pintail, teals, widgeon, cranes and other migratory birds.

8. Porbandar Bird Sanctuary

   Location : Junagadh district

   Area : 0.09 sq.kms.

Located in the middle of the coastal city of Porbandar which also happens to be the birthplace to Mahatma Gandhi. The most peculiar feature of this city which falls in the semiarid drought prone Saurashtra is the terrain of the whole city, which is saucer, shaped. Due to this, even during low rainfall years, sufficient water is available for birds. This small lake attracts large concentrations of flamingoes, cranes, egrets, stilt, etc.

9. Gaga Great Indian Bustard Sanctuary

   Location : Jamnagar District

   Area : 3.33 sq.km.

A typical grassland habitat, which has great Indian bustard besides other birds like pipits, larks, coursers etc.

10. Kachchh Bustard Sanctuary

    Location : Kachchh district

    Area : 2.03 sq.km.

In this Desert thorn scrub and grassland habitat, good concentration of great Indian bustard is reported.

11. Wild Ass Sanctuary

    Location : Surendranagar and Kachchh District

    Area : 4,953.70 sq.km.

One of the largest sanctuaries in the country, represents true desert habitat. It is inhabited by Asiatic wild asses (*Equas hemionus khur*), one of the seven sub-species found in the world. More than, 2,000 wild asses roam over this vast arid habitat alongwith nilgai, blackbuck etc. This sanctuary also has some rare birds like great Indian bustard, houbara bustard, lesser floricans, cranes, pelicans, spoonbills etc. It is an unique ecosystem, spread over a large tract of saline land.

12. Ratanmahal Sloth Bear Sanctuary

    Location : Panchmahal district

    Area : 55.65 sq.km.

It is famous for sloth bears, having typical hilly, deciduous vegetation rich in floral biodiversity. The sanctuary falls in the tribal area of the state and it has richest biological diversity in the whole district.

13. Jessore Sloth Bear Sanctuary

    Location : Banaskantha District

    Area : 180.66 sq.km.

This is also a sloth bear sanctuary with dry deciduous scrub type of forests. Besides sloth bear, nilgai, wildboar and panther are other major species found here.

14. Jamughoda Wildlife Sanctuary

    Location : Panchmals District

    Area : 130.38 sq.km.

This area is known for its floral diversity and has panthers, wildboars etc. It is believed that till few decades back, the whole district had the same type of floral and faunal diversity.

15. Purna Wildlife Sanctuary

    Location : Dangs District

    Area : 160.84 sq.km.

This sanctuary has a typical tiger habitat with moist deciduous forests and is rich in floral and avifaunal diversity. This sanctuary is characterised by typical western ghat vegetation and it also represents the best forests of western ghats which falls in the Gujarat state. It is one of the most important gene pool of floral resources and a unique ecosystem of the state.

16. Narayan Sarovar Chinkara Sanctuary

    Location : Kachchh District

    Area : 765.79 sq.km.

This sanctuary has desert thorn scrub and grassland habitat with excellent concentration of Chinkaras and occasional Bustards. Recently, the area of this sanctuary has been reduced.

17. Shoolpaneshwar Wildlife Sanctuary

    Location : Bharuch District

    Area : 607.70 sq.km.

Falling in the catchment area of Karjan and Narmada rivers, Shoolpaneshwar has excellent floral and faunal diversity. Animals include sloth bear, sambhar, wildboar, rusty spotted cat, toddy cat, leopard cat, panther, giant flying squirrels, barking deer, chowsingha and a variety of birds like sunbirds, parakeets, birds of prey, grey hornbill, woodpeckers, etc,.

18. Baroda Sanctuary

    Location : Junagadh and Jamnagar Districts

    Area : 192.31 sq.km.

The habitat is hilly and dry deciduous, located in arid regions of Saurashtra. The rainfall is very low, scanty and erratic, often leading to droughts. The biotic pressure on this sanctuary is very high. The forest is degraded but during the past few years, a concerted efforts has been made to conserve and improve the ecosystem. This are has been envisaged as an alternative home of Asiatic lions. The moisture conservation efforts and improved habitat protection has given excellent results. This scrubs forest is very rich in floral diversity. Animals reported are sambhar, nilgai, wildboar, panthers etc. This area falls in the Greater Gir Ecosystem along with Girnar, coastal forests and grasslands.

19. Rampara Wildlife Sanctuary

    Location : Rajkot District

    Area : 15.01 sq.km.

A small sanctuary with grassland habitat, is mainly inhabited by nigai besides wildboar, wolf, chinkara, hyaena etc. The terrain is Saucer shaped and scarcity of water is quite common.

20. Kachchh Desert Wildlife Sanctuary

    Location : Kachchh District

    Area : 7506.63 sq.km.

Typical desert habitat with desert flora and fauna such as bustards, nilgai, caracal, chinkara etc.

21. Paniya Sanctuary

    Location : Amreli District

    Area : 39.63 sq.km.

It is a part of Gir forests with the same kind of habitat and wildlife as in Gir. In this sanctuary blackbuck, nilgai, asiatic lions, panthers etc. are reported.

22. Balaram-Ambaji Wildlife Sanctuary

    Location : Banaskantha District

    Area : 542.08 sq.km.

Typical dry deciduous habitat with animals such as nilgai, chowsingha, panther, sloth ber etc. are seen in the area. The area is rich in floral diversity.

23. Hingolgadh Nature Education Sanctuary

    Location : Rajkot District

    Area : 6.54 sq.km.

This Sanctuary is famous as a Nature Education Centre with excellent open deciduous type of vegetation. It is a major centre for imparting nature education. The area is characterized by the presence of fauna such as chinkara, reptiles and birds etc in semi-arid, drought prone region.

Madhya Pradesh

The State has 11 National Parks, and 33 Sanctuaries. This area of 16,976 sq km encompasses about 5,100 sq km of *sal* forests, 3,390 sq km of teak forests 7,518 sq km of mixed forests and 1,300 sq km of grasslands. The protected areas amount to 3.8% of the State's geographical area (4,42,841 sq km). This is 12.5% of the state's total forest area (1,35,396 sq. km).

The Wildlife (Protection) Act, 1972 was enforced in the State since 25.1.73. A total ban on hunting was also imposed. After the launching of Project Tiger in the early seventies, Kanha National Park was one amongst the first nine tiger reserves constituted in the country. This park has a glorious conservation history and is internationally renowned for its floral and faunal richness. It also has the unique distinction of harbouring the only known world population of the endangered Branderi Swamp Deer (*Cervus duvauceli branderi*). This species was virtually saved from the brink of extinction and its number has now increased to almost 400 from a precarious 66, two decades ago. Subsequently, four more tiger reserves have also been added, namely, Indravati (Bastar Distt.), Pench (Seoni Distt.), Bhandhavragh (Shahdol Distt.) and Panna (Panna Distt).

Many endangered species are conserved in the protected areas of the state, like tiger, branderi barasingha, wild bufffalo, great Indian bustard, lesser florican, gharial, mugger etc.

Research

Considerable advancements have been made in the field of wildlife research and several research papers have been published under the auspices of Project Tiger in Kanha on the following lines. Monitoring changes in vegetation and abundance of prey-predator population, use of remote sensing techniques (satellite imageries and aerial photographs) for wildlife habitat evaluation and utilization, grassland management for wild ungulates, translocation of barasingha and the like. Apart from these, species specific research on barasingha, Hanuman langur, tiger and gaur has been done for doctoral research. Separate management plans have been prepared for most of the protected areas in the State.

Special emphasis has been given to ecodevelopment of the villages near the protected areas so as to achieve the twin objectives of socio-economic upliftment of the local people to elicit their support for wildlife conservation, particularly in Kanha National Park and Nauradehi sanctuary. The following works have been envisaged under ecodevelopment :

- Improvement of agricultural land
- Irrigated pasture development
- Crop protection from wild animals
- Water sources for people & cattle
- Small irrigation schemes
- Well construction
- Use of non-conventional sources of energy

Some important managerial issues that need to be addressed are :

Population pressure vis-a-vis wildlife survival

- Cattle grazing
- Forest fire
- Illicit collection of wood and non-wood products
- Poaching.

1. Strengthening wildlife protection and management in Reserve Forests/Protected Forests and non-forested areas.
2. Control of damage to man and his property by wild animals like crop raiding, cattle lifting, man eating, etc.

3. Control of wildlife diseases.
4. Wee problems.
5. The need of trained wildlife managers and researchers and institutional improvements.
6. The need for larger zoological parks in near natural setting for conservation and breeding of endangered species.
7. The need of wildlife interpretation and extension services.
8. Wildlife based tourism.
9. The broad objectives for future planning could be as follows:
10. Development and protection of forest corridors connecting the wildlife protected areas.
11. Preparing wildlife data base and inventories.
12. Ensuring support of local public for wildlife.
13. Management plans for all the protected areas.

The above objectives would be achieved to a large extent under the Biodiversity component of the World Bank Forestry Project.

North-East India

India's north-eastern region is a unique biogeographical entity. This is a region of physical, ecological, biological, social, cultural and linguistic diversity. Over the centuries, the culture, tradition, belief and life style of the people of the region have had close links with nature; many of the communities in the region continue to live very close to nature and even worship it, as manifested in their socio-religious practices and life style.

Shifting cultivators in N.E. India plan their lives on the basis of the cycle of clearing, burning, tilling, planting, weeding, protecting, harvesting and regenerating vegetation in the fallow fields to recover nutrients over the subsequent years before the cycle begins a new. A wide range of crops are grown traditionally in the *jhum* field, thereby transforming the natural forest into a harvestable agricultural land, which does not necessarily loose diversity on a landscape scale, as such fields may contain 8-15 major crops, plus dozens of other minor crops totaling to at least 35 crops (Ramakrishnan, 1992). The North-East India is one of the mega biodiversity centres in the Indian sub-continent. It is also one of the 10 distinct biogeographic zones of the country (SAARC, 1992). Its ecosystems vary from tropical wet evergreen, moist decidous, sub-alpine, alpine forests, and grasslands to numerous freshwater lakes, rivers, swamps and marshy wetlands. These diverse ecosystems are the home of equally diverse range of flora and fauna.

One of the best manifestations of biodiversity conservation of the region is perhaps in the form of *sacred groves,* as seen among the different communities of the region, particularly among the Khasis of Meghalaya. Recently, Tripathi *et al.* (1995) has documented the existence of 71 sacred groves in Meghalaya, varying in areas from 0.01 ha to 900 ha, and of course, at different degrees of 'sacredness'. Some of these groves are still well preserved even to this day, acting as islands of relic forests and biodiversity.

Table 18.1 : Status of forests in north-east India (in sq km)

States	Geographical Forest Area	Total Forests Cover FSI 1993	Recorded Forests	Govt Forest	Community Used/ Controlled Forests
Arunachal Pradesh	83743	68661	51540	18591	32949
Assam	78438	24508	30708	21520	9188
Manipur	22327	17621	15154	5634	9520
Meghalaya	22429	15769	9496	903	8593
Mizoram	21081	18697	15935	7077	8858
Nagaland	16579	14348	8625	1163	7462
Tripura	10486	5538	6292	6292	–
	255083	165142	137750	61180	76570

Source : Ministry of Environment & Forests, Regional Office, Shillong

The history of the faunal diversity of the region is intimately connected with geomorphological evolution of the area. It is worthwhile mentioning here that the obliteration of pre-tertiary Tethy's sea began in North-East India producing in its wake a land bridge between Indian peninsula and main land Asia. This region served as a faunal gateway and therefore, a complex assemblage of faunal elements are observed in the region (Biswas & Ghosh, 1976).

Located in the north-east of the Indian sub-continent, criss-crossed by Himalayan ranges flanking the Brahmaputa and Surma valleys, the constituent units of the region, i.e. Arunachal Pradesh, Assam, Manipur, Meghalaya, Mizoram, Nagaland and Tripura offer a complex-variety of habitats and ecosystems. The rich forests have greatly suffered due to biotic interference and land exploitation such as shifting cultivation on the hills and rice cultivation in the valleys. This has drastically affected the distribution and abundance of the faunal wealth of the region which, however, still retains numerous unique and rare/endangered species of great scientific and economic value (Ghosh and Tiwari, 1984).

One way of arresting this trend of depletion was through the creation of managed forests and protected areas. In the context of North East India, such managed forests and protected areas constitute about one fourth of the total geographical area of the region.

Fauna

It may undoubtedly be said that the North East India is one of the single richest biogeographic zones in the Indian sub-continent, presenting such an array of richness and unique characteristics of faunal wealth. The faunal resources of North-East India has been described earlier by Ghosh and Tiwari (1984). Their description mainly includes vertebrate Phyla Mammalia, Aves, Reptilia, Amphibia, Pisces and some groups of invertebrates, notably butterflies and beetles. Since then, many more species were described from North-East India, particularly the invertebrate groups belonging to Rotifera, Annelida, Crustacea, Cladocera, Myriapoda, Odonata, Orthoptera, Phasmida, Dermaptera, Dictyoptera, Isoptera, Mallophaga, Hemiptera, Diptera, Trematoda, Nematoda and soil arthropods, notably Acari and Collembola. Many of the species recorded from the region were new to science. Some of the recently described new species from North-East India, whose para-types are in the holdings of Eastern Regional Station, Zoological Survey of India, Shillong include at least 94 new species belonging to different groups.

It is not that these animals are confined to the managed forests and protected areas of the region alone. In fact, as can be seen from the legal status of the forests in N.E. India (Table 18.1), most of the forest areas, except in Tripura, are under the ownership/control of the community in the form of village councils, district councils or private ownership. Almost all the areas of **Unclassed State Forest** are under this category and the local tribals enjoy excessive concession in the exploitation of the products from these forests. Faunal endemism is also very high in North-East India. Some of the fauna restricted to North-East India by Darlong *et al.*, 1989; and Alfred, 1990.

The soil fauna is perhaps one of the least known biotic frontiers from this region. But from preliminary studies on the soil fauna of this region, it is observed that the groups promises to constitute a huge reservoir of biodiversity, particularly the two dominant groups Acari and Collembola, whose species richness is also very high.

The soil fauna of some of the managed and protected forests studied from this region by Alfred *al al.*, 1989; Darlong and Alfred, 1993. They show that the soils of these systems are a reservoir of soil faunal diversity. On the contrary, the forest soil fauna could be much richer and denser if sampling is beyond 40 cm (the present studies were limited to 40 cm), a range of extraction methods employed, rather than confining to only Berlese-Tullgren funnel extractions as was the case for the present studies, and careful sorting and identification of all groups and stages could be done. The effect of disturbance, especially fires resulting from shifting or swidden cultivation, has a profound effect on all animals and especially soil animals. They have reported a decrease in faunal density to the tune of approximately 50% or more in most groups of soil animals.

Askot Wildlife Sanctuary of Kumaun (Himalaya)

With an area of approximately 600 sq. kms., ranging between 600-6700 m, Askot Forest Range forms the sanctuary area. An international boundary with Tibet forms its northern limit. The river Kali in the east separates it from Nepal. The boundaries of East Almora Forest Range and Pithoragarh Forest Range constitute the western and southern limits of the sanctuary respectively. The area also includes three major watersheds (Kali, Dhauli and lower part of Gori) of the Kumaun region.

The Askot Wildlife Sanctuary (AWLS) covers most of the area in Askot Forest Range (28943.2 ha). The spatial distribution of various forest types existing within the sanctuary area under administrative control of Forest Department. *Pinus roxburghii* (14%) and *Quercus leucotrichophora* (14.2%) are the major forest types at lower elevations, whereas *Quercus semecarpifolia* (16%) and *Tsuga dumosa* (14.9%) are represented at higher elevations. Samant et al. (1993), have reported the detailed general distribution pattern of flora and fauna along the vertical gradient.

19

Conservation of Biodiversity in Future Strategies for India

If you plan for one year, plant rice,
If you plan for ten years, plant trees.
But, if you plan for one hundred years,
educate the people

Development, both individual and societal, and environmental are two sides of the same coin. Due recognition is needed for global environmental risks while designing any environmental strategy, Environmentally and socially disruptive economic behaviour must be changed if we intend to provide discarding environmentally sensitive forms of development can only bring about decent livelihood to all citizens of the present and future generations and this change. Rather, the stress should be on qualitative development based on efficient use of energy and natural resources, recycling of end products and minimizing the use of nono-biodegradable products.

In a country with a burgeoning human and livestock population, concrete actions and sound strategies are easy to formulate but difficult to implement. Human beings have a tendency to prefer present goods to future goods and in the process, overlook the side effects. To sustain 1.5 billion people in India by the year 2025 heavy financial investment and state-of-the-art technology will be needed in agriculture, water resources, infrastructural development and so forth. Ecological balance will definitely be adversely affected to create liquidity despite our best efforts not to do so. Primary and secondary forest industries will use more and more wood, cork, bamboo, cane, gum, resins, essential oils, seeds, medicines etc. for home consumption as well as export in a liberalized economy.

Despite all adversities one has to adopt viable strategies for the conservation of species as this will only guarantee our future and lead us towards the goal of sustainable society, sustainable economy, sustainable use and sustainable development. The strategy should be aimed at policy makers, all Government and Non-Governmental organisations. This will help in achieving the objectives of biodiversity conservation and sustainable development. Any conservation measure has to be carefully planned so that :

1. Environmentally and socially disruptive behaviour is altered to provide a decent living for all
2. Production pattern is socially equitable
3. A human centered, knowledge intensive civilization is built
4. A balance is maintained between all forms of capital i.e. human, natural, physical and financial.

Some of the measures suggested as a part of long term planning are as follows :

Carrying Capacity of the System

Carrying capacity varies from region to region depending upon population pressure, their dependence on food, water, energy, raw material requirements, waste production and import and export. In order to live within the carrying capacity the society will have to change the life style and consumption pattern in such a manner that causes least damage to the ecosystem. Over population in a particular region has to be avoided and excessive pressure on land must be reduced. It calls for changes in the prevent land use policy, overhauling of land tenurial system and population control. Forest land will have to be used in such a way that maintains their biodiversity, productivity, regeneration capacity, vitality and their potential to fulfil, now and in future, relevant ecological, economic and social functions.

Enhancement of the Protected Areas

One of the best means of protecting the natural forest is through the establishment of protected areas, an effort that has taken more than a century to evolve from the concept of reserve forest to national parks and biosphere reserves. These protected areas serve three basic functions viz., benefit, education and recreation of the people.

Although, India has made steady progress in the management and expansion of the protected areas, yet, the problems are being compounded with time. More establishment of protected areas does not ensure socio-economic development and species conservation. The management has to take timely action for the speedy disposal for the settlement problems. Local people who consider themselves as the primary stakeholders must be involved in the use and management of the resources. Socio-economic well being of the people should be one of the top priorities for minimizing the resources dependency on the core of these protected areas. Biological resources in the protected areas can be assigned consumptive and non-consumptive values so that the conservation of biological diversity can be linked to sustainable socio-economic development.

At the same time, there is a need for change in the attitude of the field staff towards the interests of the local people. This conflictual relationship has to be transformed into cooperation and partnership. Field staff will have to appreciate that with inadequate work force prohibitive management strategy is impossible to implement. Therefore, in the long run a conciliatory, interactive and motivational approach seems to be the best.

Enhancement of protected areas will need effective management with clear objectives for each area. This will require increased investment in human resources, education and research.

Stress on Integrated Development

Of late it has been realized that a direct stress on conservation measures and neglecting the economic development of the people in and around the protected areas leads to impoverishment, and initiates the vicious cycle of subsistence, emigration, land fragmentation, degradation and deforestation. According to the new concept, people will have to change their consumption pattern and find alternatives on their own, once their purchasing power increases. The only fear in this integrated effort is that short-term conflicts may arise in the absence of appropriate policy measures causing hardships to the people. It is therefore essential that conservation, rural development and compensatory incentive programmes should be organised simultaneously.

Research Impetus

There are five main areas, which need urgent attention :

Forest management

Forest management has undergone rapid changes in the decades of eighties and the nineties. Geographical Information System and Remote Sensing techniques have the potential to provide a large amount of information on forest resources, growth and biodiversity.

Restoration of degraded lands

Research efforts should concentrate on enhancing the soil micro fauna and fertility status so that sustainable production can be achieved by adopting agroforestry systems that can support the economic development of the region. Efforts should be made for genetic improvement and dissemination of multipurpose tree species that can grow in degraded areas.

Research in harvesting, processing and marketing

There is a need to develop low impact harvesting systems, drying and conversion processes for non-commercial abundant species, especially the NTFP's Marketing management has been one of the weakest sectors of forestry and needs new directions and support.

Policy, institutional and socio economic research

Socio economic objectives are best pursued if the policies regarding land use, rehabilitation of degraded lands, benefit sharing, land and tree tenure, participatory planning and resource management are appropriate to the sustainable management and biodiversity conservation.

Research in biotechnology

Sustainable development and biodiversity conservation in the 21st century will depend upon proper exploitation of economic potential of genetic diversity using biotechnology for the production of goods and services required by the increasing population. Foresters will have a dual role in future; one as *in situ* conservationists and the other in *ex situ* maintenance of biodiversity for commercial use.

Last but not least the concept of conservation and development should permeate our thinking and perception. We are all stakeholders, participants and decision makers in this struggle and can achieve the objectives by our concrete actions, strength and determination with the active support and participation of the people.

Strategies, Opportunities and Options

In India, creation of Sanctuaries, national Parks and Biosphere Reserves has been adopted as the major task of achieving the objectives of conservation. Restricting human interference to conserve biological diversity is an act opposed by the rural tribals already facing complex problems, including small land holdings, meager employment opportunities and, over and above, natural hazards, reduced productivity of *jhum* fields and environmental uncertainties. Conservation strategies must address to the needs of posterity but, at the same time, should not neglect the present needs of the people.

Biodiversity conservation strategies and goals, should redefine biodiversity in the context of local needs, usages and values as priority of conservation. The current experience in the region is that there is lack of objectives of biodiversity conservation strategy, which need to be explicit, clear and achievable. It should not neglect the anthropocentric objective addressing the local peoples' priorities to provide, if not enhance, sustainable and economic benefits.

Current policies and practices of *exclusive protectionism* of conservation areas need a re-evaluation. After all, people and their culture have historically played significant role in conservation and management of what was there in the past and what is remaining now. There is the need to survey peoples' needs and attitudes around our conservation areas. Possibilities of controlled harvesting or sustainable use of at least certain wildlife resources and minor forest produces (MFPs) by the communities neighbouring protected areas should be considered. Activities that may be taken up as pilot projects may be identified. Such objectives may also ensure long term survival of wildlife and biodiversity conservation, since the people themselves would want to protect the protected areas. On the other hand, damage to property and human life by wildlife, and the amount and system of compensation payments are some of the issues that are not yet well documented.

Many of the protected areas in India are indeed forests for wildlife shelter in the midst of intense human pressure and activities. Such islands of shelter cannot always afford to accommodate many of the wild animals who possess migratory instinct. There is, thus, the urgent need of identifying animal corridors and bringing such corridors within *buffer zones* or some kind of management practices in order to reduce man-animal conflicts or wildlife depredation.

In India, it is a fact that there exists rich biodiversity even outside the PAs, such as in sacred groves, traditional village/community/individual forests, in cultural *jhum* fields and fallow lands. Such Lands/forests are under the ownership/control of the Village Councils, District Councils, traditional chiefs or private individuals. These traditional socio-political institutions must be taken into confidence while formulating strategies for biodiversity conservation. Management strategies and policies should ideally encompass the areas under the control of these institutions, which can be achieved through appropriate dialogue and negotiations without infringing on their land rights/ ownership.

There is ample scope for amalgamation of traditional practices and modern scientific knowledge for solving the problems of PAs and biodiveristy conservation in the region. Not that all traditional societies are always wise stewards of biodiversity or their practices are scientific, conservation oriented or sustainable. Such practices should also be identified and discouraged and alternatives evolved.

One of the weaknesses of current conservation strategy in the region is its heavy emphasis on conservation of wilderness areas of forest lands. Conservation of cultivated crops, ethnobiological species including medicinal and wild edible plants is a neglected frontier. This weakness must be rectified immediately as the region has a variety of food and medicinal plants, many of which are not much known outside the region and may have potential of providing alternatives for food and medicines in the region and elsewhere. Such crops have also been reported to be pest resistant and stand long storage.

In various silvicultural practices and compensatory plantations in the region, there is often a tendency to patronize one or more better performing species such as pine in the higher altitude and teak in the lower elevations. Such practices should be reviewed and minimized, as monocultures reduce the spectrum of biodiversity, both of plants and animals. Ideally, there should be an assessment of areas that may be covered under monocultures and native mixed species should be planted to maintain biological diversity.

The development projects in India are not sensitive to biodiversity issues. Many of them are harmful to biodiversity to a greater or lesser extent. Therefore, there is a necessity of improvement in project appraisal methodologies and implementation of programmes with a view to making them more sensitive to biodiversity, while at the same time, ensuring sustainable development, better valuation of renewable resources, improved forests/wetlands/wildlands conservation and increased agricultural productivity. There is also the need for sensitizing the project authorities towards biodiversities awareness and initiatives.

In India, the Environmental Impact Assessments (EIA) in any region tends to concentrate on local physical, biological and ecological impacts, their effect on the human population and local environment. However, their effect on the biodiversity is invariably neglected. One of the main practical constraints to improve EIA is lack of data and the difficulty of ascribing values to all but direct use-values of biodiversity.

Inter department approach

There is also a lack of inter departmental approach and vision in relation to biodiversity conservation in the region. For example, while the state Environment & Forests/Wildlife Department may be enthusiastic to conserve wetlands, the Fishery Department may like to promote commercial fish farming in such wetlands, or the Agriculture Department may like to reclaim the swamp areas for wet rich cultivation. Moreover, there is often random introduction of exotic species or hybrid varieties for short term commercial gain endangering patronage of local species or varieties. Such practices also dilute the indigenous knowledge systems or erode germplasm of the native species.

Urgent Need

There is an urgent need for an increase in research activities, information processing and dissemination system, either to improve priority setting, or to guide management of biodiversity conservation, both within and outside the protected areas and even for almost all habitat types. Funds and resources may be allocated to the best possible effect. There is also the need for chronicling traditional natural resource management systems, which should be evaluated, strengthened and encouraged. Such knowledge should be exchanged or transferred between tribes for higher ecological and economic efficiencies, particularly conservation and management of biodiversity under their control.

Environmental Education

Environmental education, awareness building and motivation initiatives on biodiversity issues using local arts, cultures and languages should also be part of the strategy. Ideal traditional plant and wildlife knowledge and skills should be identified, documented and used as educational tools. There is also the need to infuse traditional values, cultures, skills and knowledge into the young generation, many of whom are now exposed to the mainstream life and culture, for better appreciation of their own native wisdom and resources. Environmental education, particularly awareness in relation to implication of biodiversity trade, conventions, regulations, removal and extraction of endangered species, etc. should also be extended to the law enforcing agencies.

Future Strategy for the Conservation of Biological Diversity

The rate of growth of human population must be reduced. Greater emphasis should be given to better management of available biological resources rather than to just consider conservation of biological diversity in isolation. The reason is that increased production of fuel wood, fodder, building material etc. in areas poor in biodiversity in the country is essentially required to offset the pressure on biodiversity rich areas. A natural resource policy may be formulated to achieve four basis goals:

1. Maintenances of adequate resources
2. Conservation of resources through reduction in demand and achievement of greater end use
3. Maximum use of renewable resources
4. Reduction in dependency of non-renewable resources.

20
Management of Wildland Biodiversity

A tropical nation or large multinational region has three basic kinds of land-use : urban, ever more intensively managed agroscape, and ever dwindling wildlands. The latter are generally patches of comparatively biodiverse habitats on socially or physically inaccessible sites or on "poor" agriculture soils.

The urban habitat is viewed as productive even if restive. The agroscape is productive, with largely pacific and homogenized biodiversity. The wildlands largely are viewed as removable, conservable, or conserved, i.e. they have been set aside by someone "else" for strip-mining of their natural products or for social fossilization, outside of the national economy. They are like cash in a shoebox neither the bed, neither earning interest nor circulating, but of value to someone.

This perception of tropical wildlands is unfortunate, and fortunately it is waning in popularity. There are encouraging nuclei of voices dotted across the tropical and extra-tropical landscape arguing that conservable and conserved tropical wildlands are a category of highly productive land-use.

Such as a shift in social and economic attitudes demands that a conserved wildland be blessed with the level of planning, knowledge, investment, oversight, budget, technology, and political attention that long has been characteristic of the more productive sectors of the agroscape, and also of a nation's institutions - highway systems, hospitals, education, and communication. Traditional tropical conserved wildland management - "fence it and put a guard on it"- is to such a blessing as a guard at the bank's front door is to the stock market, Federal Reserve, free market economy, taxes, and trade barriers all rolled into one.

We may anticipate a new edition of "potential land-use" maps for tropical countries. This is really what the "thou shalt inventory thy biodiversity" component of the Biodiversity Treaty is all about. No longer will there be a soil and contour map marked "apt for agriculture," "apt for forestry", mid "apt for conservation," with conservation meaning "useless" and therefore to be assigned to the national park service or its equivalent. Rather, these maps will show what has been explicitly designated as agroscape and wildlands conserved formation biodiversity and its nondestructive use, with awareness that any hectare of a nation can be developed as either, depending on society and history rather than on soil type, rainfall, slope, and distance from a road or border war. The overall goal will be to render both types of land-use to be sustainably productive, high quality, and much valued by a nation and a region.

Finally, as the knowledge base of humanity increases in bull and interconnectivity, the intrinsic potential for multiple use of a unit of biodiversity increases. All of these increases are proportional to our interest in them.

The locations of this remnant will be the serendipitous outcome of a multitude of social and economic forces acting largely irrespective of biodiversity's traits. One example of this process is that there are no unambiguously conserved large tropical wildlands on "good" agricultural soils. This "happy accident" strategy for conservation of tropical biodiversity will continue unabated if there is no major shift in social attitudes and economic process. This strategy is a quite comfortable majority of individual, national, and institutional agenda in the contemporary tropics.

The major shifts in social attitude and economic forces that are required by the "use it or lose it" strategy that tropical conserved wildlands are conserved for non-damaging use by all sectors of society rather than because they are wastelands, for our grandchildren, for the sake of conservation, crown jewels, biodiversity-prospecting pits, observation posts for bar-coded horses, or to fill the agenda of any other single social sector. Each of these seven sacred and reasonable cows, and a whole lot more, become by products and ingredients, rather than the goal. This needs to be true even if each today is of major importance somewhere in the tropics. This attitude is somewhat akin to recognizing that the value of good agricultural soil or quality roadworks is not in the specific or the specific truck, but rather in being a platform on which society carries out a multitude of diverse activities.

Our globe is the playground of domesticates, we consign humanity to the doldrums of just those things that human can imagine, invent, and control. We thoroughly deprive ourselves as if we excise our colour vision, our sense of smell but, for ferrying chicken, our taste but for salt and sugar, our hearing but for high, low, and middle. The brain is a computer with tens of thousands of applications invented to deal with non-human nature. By the removal of tropical wildland biodiversity, we are permanently relegating it to word processing. But the other side of the coin is that our appreciation for superlative architecture does not demand that we have only those buildings that will win international prizes. There is a place. Oil the landscape for a healthy agroecosystem as well as the wildland crop.

Because the emphasis throughout this chapter is on "use it or lose it" there are several caveats, all of which are traditional in other social sectors but have been slow to be applied to conservation. They boil down to several equivalent expressions. The frontier is gone. You are always in someone's living room. Tropical biodiversity must escape the Tragedy of the Commons. There is no free lunch. The only sure things are death and taxes. Applying these age-old concepts to the case at hand:

- The more we know about wild biodiversity, the more we can use it without destroying it.
- Not all persons can use wild biodiversity as much as they would like.
- The use of wild biodiversity must be scheduled and monitored.
- There are all sorts of users.
- Users pay in all sorts of currency.

Tropical wildland biodiversity needs detailed, knowledgeable and dedicated management as much as does any other social sector. Ironically, since it is so underdeveloped, the returns on an additional unit of investment in fact are likely to be often substantially greater than is the case with many well developed sectors.

Wild Tropical Biodiversity

In order to begin to use wildland biodiversity, we must come to know :

Identification and taxonomy : We need to know what biodiversity is that we can :

- know its parts when we see them,
- communicate about it today,
- pool and massage our information about it, and
- ink what we find out with what others have found out.

The latter is of particular importance with tropical biodiversity, much of which has geographic distributions spreading across many countries.

Microgeography : We need to know where biodiversity resides, or at least where a portion of its resides so that we can get to its location "on call". A catalogue of the books in a Library of Congress is of severely reduced use if there is no knowledge as to where the books are to be found, even if they can be recognized and read once in hand.

Trappers, tricks and husbandry : We need to know how to get biodiversity to hand or eye so as to get, the information that we seek or can use from it. All hunter-gatherers and their field-biologist counterparts have experienced the circumstance where a species is known and appears to be absent, yet-with the appropriate collection method-the species appears in droves. We also need to know how to get biodiversity to hand so that we can care for it and multiply it, so that with this husbandry we can introduce it to the agroscape rural or urban.

Basic natural history : We need to know what biodiversity does - its natural history in the broadest sense-so as to:

- give us clues as to what it offers by itself and through its interactions;
- suggest how to farm it elsewhere;
- allow us to know the impact of our presence, studies, and sampling; and
- allow us to know when it is in trouble and what to do about it.

These four activities can be, and will be, carried out by a diversity of persons for a diversity of agendas in a diversity of wildland sites. However, a tropical nation with species-rich conserved wildlands may well be fortunate enough to have>100,000 hectare blocks containing 100,000 to a million species and all their interactions. In such cases, a major strategy for biodiversity management and development is to select-in the context of a nation's full gambit of users and managers-a site, and rapidly inventory all of its species. This is an All Taxa Biodiversity Inventory (ATBI), The function of an ATBI is to set up a major block of a nation's wildland biodiversity for all users. It :

- projects a massive block of diverse raw materials onto society's table,
- enjoys substantial economics of scale,
- forments mutualistic gains among executors as well as among users, and elevates biodiversity inventory far beyond being a taxonomist's tool or a conservationist's listing.

When ATBIs occur in various countries and are firmly networked, a global network of these four advantages of an ATBI can and should be achievable. For example, this has been envisioned through the DIVERSITAS network as visualized by the United Nations Educational, Scientific, and Cultural Organization (UNESCO), and will be realized in the biodiversity clearing houses of the Biodiversity Convention. An ATBI is a major advance over the diffuse and dilute approach currently in play, an approach clearly rooted in the time-honoured traditions of curiosity-driven and highly individualistic field biology as performed by taxonomists and ecologists.

Second, the goal of tropical wildland biodiversity management is to imbed it in society-all sectors of society and not just those with access to scientific journals and preprints. Third, the greater part of tropical terrestrial biodiversity is international; biodiversity represents a global efforts even if a nation is the primary custodian. What we know, and will come to know, of the wild silkmoth *Rothschildia lebeau* is based on the aggregation of information from studies in Texas, Mexico, Costa Rica, Venezuela, and Colombia, among others, and conducted for a multitude of reasons-schoolyard exercises, pharmaceutical prospecting, ecotourism guiding, silk research, insect disease transmission, and religious symbolism.

The other side of public domain in the responsibility to actually conduct ATBIs and other kinds of inventories and the subsequent management and development of wildland biodiversity. Much of what commonly has been the social responsibility of the traditional academic/museum community on the one hand, and "park guards" or distant government offices in the capital city on the other hand, can be passed most profitably to parataxonomists, paraecologists, ecologists, educators, ecotourism guides, administrators of biodiversity, and other forms of site-based paraprofessionals. There is huge potential in training residents that neighbour conserved wildland or live in it. They can accept a major portion of the responsibility to carry out non-damaging managements and user-processes. This transfer of power and decentralization is essential for moving beyond what is today largely management of tropical wildlands by absentee-landlords.

This transfer however, does meet with two major classes of social resistance. First, the scientific community understandably is reluctant to invest the energy and modification of tradition that will bring this about without compensation by senior administrators of science and by society at large. There are widespread benefits that could result from such a levelling of the playing field, and some of these need to feed back to its contributors. Second, such a transfer of political and economic power to rural areas-in modern parlance, decentralization and management irrationality-is theoretically attractive but very difficult to bring about in the face of contemporary vertically organized society. A large and properly managed conserved wildland dances dangerously close to secession from the federal state in virtually all tropical countries.

The comments run in direct conflict with the human behaviour of attempting to mould the newly emerging activity of conservation of biodiversity so that its energy feeds one's entrenched agenda rather than targets the goal that elicited the activity-conservation of biodiversity.

We do not need to know how many species there are in the world, in a country, in a large conserved wildland. We already know that there are hundreds of thousands to millions of species, and most are unknown in most respects. That is enough information to get on with knowing biodiversity and setting it up for non-destructive use. It is not shameful that "science" does not know whether there are 10 million, 30 million or 10 million species of organisms, and it is a waste of precious time and human resources to focus on refining this estimate. What does need our attention is our ignorance of biodiversity in organisms and processes. This ignorance does not imply that the numbers of species per se are of importance, though numerical relationships do play their usual role in sampling community properties.

We do not need to know the world-level or even national-level detailed geographic distributions of all butterflies, birds, or trees. The world is simply not a sandbox offered to scientists to reorganize as they wish so as to save their favourite higher taxon.

The function of detailed biodiversity inventory is not to choose sites for conservation. One invests inventory on in area that already has been seriously designated for conservation status, with the goal of ensuring that status through understanding.

The bulk of the significant blocks of conserved or conservable biodiversity in the Earth's terrestrial tropics are already known and largely delimited. Where this is not the case, there already exist knowledgeable field biologists and conservationists-national and international-who can quickly set the majority of those limits through rapid ecological assessments and other protocols.

We do not need to know about one more set of data on traditional wildlife management about this or that turtle, macaw, tiger, or deer. Yes, there are some large conspicuous tropical organisms that need a close look-vis-a-vis the real biological and social context of where they live. But, in general, it is the other 98 per cent of biodiversity that needs much of our attention in natural history. The attention should be in the context of the society that surrounds and infuses the biodiversity of the focal sites and not in the context of the time-honoured initiation rituals of academic titles and institutions. Let us stop making conservation science, the science of trying to figure out how to get more money for biological research by piggy backing on the biodiversity crisis.

Our tropical resources, always in short supply, should be directed towards saving the bulk of biodiversity in a few large and well-distributed blocks, hopefully robust in the face of climatic change through their elevational and multihabitat diversity. This comment is not meant to denigrate the small patches and fragments of widespread tropical species and the occasional isolated endemic-at times of high value to their immediate owners and neighbours irrespective of their eventual demise or impoverishment of biodiversity. Rather, it is meant to suggest that we coldly practice a biologically realistic triad so as to bring conserved wildland biodiversity into peace with tropical society at large and, more specifically, the agroscape.

All conserved tropical wildlands are islands, and an ATBI is the beginning of providing a ground zero for asking how its biodiversity reacts to such things as global climatic change, pesticide contamination, use and insularization. Tropical wildlands, fractured into habitat islands with distinctive conditions, and perhaps subject to severe global change, will be the sites of a global speciation event that will dwarf anything generated by, for example, Pleistocene drying. An ATBI sets up a gigantic canary in the mine for these reactions.

The known universe of each ATBI site will serve as the foundation in which to construct an extremely diverse array of question-driven ecological behavioural, demographic, and ecosystem studies. The possibilities are mind-boggling with respect to what sorts of ecological work can be done with complex habitat where virtually every species can be identified and categorized ecologically.

An ATBI site is, in effect, a well-organized wild zoo, greenhouse, and culture facility in which to prospect for wildland organisms, genes, and their products. An ATBI will remove the single largest obstacle to serious examination of large arrays of wild biodiversity for human use.

An ATBI site is a national museum, national library, art gallery, concert hall, national zoo, national botanical garden, and national university for ecotourists and other forms of students ranging from grade-school children to senior citizens.

An ATBI site can be a major provider of ecosystem services especially if the site is chosen with that as an additional criterion, and simultaneously can be known well enough to study the internal mechanics of ecosystem services. Carbon sequestration can be a boring monoculture, or it can have complex value-added properties.

Limiting Resources

Today we are in the throes of determining how to record, manage, and transmit information about wildland biodiversity - the structure of databases, networks, distribution of databases, image transmission, authority files, authorship attribution, clearing houses, bar-coding, retroactive data-capture, and computer-capture of literature-but tomorrow these technologies and protocols will have been resolved for the most part. Then, and for centuries thereafter, the resource in short supply will be the information about biodiversity itself.

Computerization and the transfer of knowledge

The all-invasive wave of computerization represents a quantum and qualitative change in the acquisition, massage, distribution, and archiving of information about biodiversity. It will change humanity's relationship with biodiversity more than has the printing press, the camera, or the chainsaw. Computerization is a great part of what allows the realization of all the prognosis mentioned or alluded to here and elsewhere in biodiversity management.

For the first time, it is possible for an individual and a site to acquire, massage, distribute, and archive the unimaginably large quantity of highly particulate information-images, specimen descriptors, species descriptors, habitat descriptors, circumstances, previous knowledge that is pertinent management and use of a conserved wildland that contains hundreds of thousands of species and has been or is being studied by tens to thousands of observers over the years or even at one time. The real art is how to massage information and, put in a multitude of formats for a multitude of users. The real question then becomes whether the developed world is willing to accept the levelling of the global playing field that all this implies. Most evident of all, the last thing biodiversity management needs is new hard-copy journals, more hard-copy books, and continuation of the stultifying hard-copy traditions of biodiversity information management from the past several centuries. What tropical wildland biodiversity management needs is for all the holders of information about biodiversity to get that information as fast as possible into the Internet, rather than waiting decades to see it frozen onto thin sheets of wood.

Taxonomy and the taxasphere

Taxonomy is the basic philosophical and technological infrastructure for wildland biodiversity management. Without taxonomy, there is no inventory, no collation and distribution of information about biodiversity in space and time, no inference among species. But taxonomy, like conservation, ecology and other specialities in science, has evolved to its own drumbeat. Most encouragingly, taxonomy currently is re-examining its mission through efforts such as Systematics Agenda 2000 and a multitude of international symposia. Government agencies of the United States are beginning to undertake and support a global responsibility in taxonomy, and taxonomy once again is coming to be support as a form of national development.

Some things are evident in the changes that biodiversity management asks of taxonomy. Expert systems, picture keys, coming together for taxonomic and specimen data, data models, and user-friendliness in computerization. We all needs to be headed towards identification guides that allow us to flip through an electronic picture book, with centralized or networked processors where an image or discussion of a doubtful organism can be sent for taxonomic confirmation. Close on our heels is the magic box into which a piece of a bug is dropped, sequenced, sequences compared with a library, and a name spit out if it matches. Then, with the name in hand, one calls up what the greater global network already knows about the biology and biodiversity of that species or population, a global field guide in a pocket. Once again, it is in the information about genetics and biodiversity from the field that becomes the resource in short supply.

Where is most of that information pool today? It is in the heads of retiring taxonomists. The new PEET initiative of the National Science Foundation is a major step in this direction. This information-capture might well be done in conjunction with the retroactive data-capture that is possible in the world's large museums, but if not, the highly perishable should be given priority over the pinned and dried.

Taxonomy is really a taxasphere with nodes of specialists, collections, knowledge-bases, and hard-copy data - all strong together on an Internet lattice and variably plugged into the world's biodiverse sites. Teh nodes interconnect as much for taxonomy's own work as for all the other users of biodiversity. There is a major question as to whether, and to what degree, it is worthwhile to retroactively capture the information museums. Ironically, museums were, on the one hand, the expressions of interest in species as manifest through specimens. Information contained in museums therefore often was not gathered in a manner conducive to its maximum use in biodiversity management.

At the very time when extant museums are rethinking the value of their collections, they are the logical recipients of the new and enormous responsibility of curating the mass of voucher specimens that will appear from inventories and other kinds of biodiversity management. While many of these specimens are perhaps of lesser direct taxonomic interest, they are of huge importance underpinning the current mass of information about biodiversity base which much more information will be built. We find ourselves in the ticklish position of explaining to the tropical world at large that the specimen is of little or no value per se, and thus should not be the focus of nationalistic possessiveness, while at the same time it may be a voucher specimen or source of genetic information that merits long-term maintenance costs. The more the biotechnologists tell us, the closer that specimen comes to being a legible cookbook for many of the things that it did in nature.

The taxasphere has long run on the engine of personal interest in organism by taxonomists and other kinds of field biologists, rather than on a true economic and social recognition of the critical nature of the taxonomic underpinning of the use of biodiversity. To the degree that society neglectfully accepts that taxonomy is run by such a volunteer work force, we are confronted with the advantages and disadvantages of trying to run an army or national park staffed with unsalaried volunteers, even very competent ones. While the taxasphere needs to reach out with joy for the finances and responsibility that should come with a reversal of this trend, this same taxasphere then is confronted with an increased accountability to the funder, a kind of accountability not usually associated with those who operate in a free-spirited and artistic social sector. It will be most helpful if the taxasphere can manifest some self-directed willingness to spread responsibility to those taxa and technologies previously unconsidered, as a response to society's willingness to put resources behind this action.

Targeting the small stuff

The bulk of biodiversity comprises very small organisms - easily 80 per cent of a conserved wildland's biodiversity weighs less than a few grams even as an adult. However, many of the traditions of information management, field ecology, species-use, conservation, wildland education, and evaluation of biodiversity have been little affected by the biology of the small stuff. On the other hand, the enormous biodiversity for small species constitutes much of the potential for the use for biodiversity and offers a huge part of the complexity of managing this biodiversity. This means that finding out which aspects of biodiversity reside in a site and getting it in order for society will involve a very large number of field taxonomists and ecologists spending their time getting their big organisms into situations where they can be poked and searched by the people who work with viruses, bacteria, fungi, mites, small insects, protozoans, parasites, algae, and other little things. This means that the quality of one-site laboratory facilities will need to take a megastep upward to complement the old tend and machete. This means that the conserved wildlands will be brought yet closer to society.

The fact is, we will lose 10-20 per cent of them over the coming century, important or not. So let's get busy delimiting the areas that will be conserved tropical wildlands, largely forget about those things that live outside, and get on with making high-quality conservation areas out of the 10-15 per cent of the Earth's surface that contains the remaining 80-90 per cent of the species. And let's make very high-quality agroscapes and urbanized habitats in which these areas are imbedded. All the mental energy and all the funds put into anguishing over the losses could be spent far better on quality survival of the survivors.

It might be useful to note that the current extinction differs from the cretaceous extinction in the following ways:

- We are not about to give the terrestrial world back to biodiversity to revolve after this is all over.
- We will have reduced terrestrial biodiversity to tens of thousands of analogs of the Galapagos islands and New Guinea, where speciation and higher taxon evolution will work apace.
- The surviving subset of species will be defined by area and habitat, rather than being ecological groupings, such as those small vertebrates that could aestivate or stay warm and survive on a diet of plant and animal carrion, seeds and other dormant organisms, and the insects and fungi that feed on the same, during a winter induced by meteor impact.

So the first self-denial that must become characteristic of biodiversity and practitioners of biodiversity management is the temptation to have diffeent government agencies and nongovernment organisations (NGOs) harvest their own information about biodiversity and wield it to their end. The Oaxaca Declartion, the U.S. National Biological Service, much inter museum collabortion in the development of databases, the inforamtion, clearinghouse of the Biodiversity Convention, and databases moving onto the Internet are recent examples.

Terrestrial conserved wildlands are habitat islands and will become more so. They are habitat islands joined only by a selected few aerially mobile organisms, islands positioned in an ocean of intensely managed domesticates. This insularity means that no matter how large and how well planned and inventoried, each conserved wildland will have a different and heterogeneous ceilling for intensity of impact by on-site users. We even have the irony, that established conserved wildlands can render the concept of "endangered species" an anachronism. If species are in truly conserved wildlands, they survive in those habitats at their naturally achievable densitties or they go extinct. Outside of the conserved wildland, they are basically forgotten. Let us use our energy now to make the wildlands into better islands.

Second, recognize the extreme contradiction inherent in biodiversity. In mainland habitats of broad extent, virtually every organisms in the natural community evolved as a little population somewhere else. It is an immigrant in most of where it is found today, and wildland biodiversity programmes on mainlands are put together mostly through on-site evolutionary fine-tuning. The survival of the horse in the Old World, while we extinguished it in the New World and then reintroduced it. All wildlands are strongly impacted by humans already-extinction of the Pleistocene megafauna, extinction of contemporary vertebrates, global warming and other change, hunting, roadside secondary succession, and introduction of bacteria, fungi, algae, mites, and herbs. There is no "pristine" nature, free of "introduced species" andhuman influence, to conserve.

Once designated as conserved wildland for the non-damaging use of its biodiversity, this land-use categorization needs to be inviolate. In this respect, wildlands conserved for their biodiversity are qualitatively different from other kinds of land-use, and are not easily interchangeable with other kinds of land-use. The agroscape can easily move from peanuts to sorghum to cows to peanuts over the years, but moving a given hectare from rice to forest requires considerably more cost and long-range structure, and often is not biologically possible. Restortion is a limited tool, not a panacea. However, restoration does offer enormous potential in the siting of conserved wildlands throughut the tropics.

Ironically, the direct use of a conserved site by people is perhaps one of the easiest of all facets of biodiversity - use to manage. Society at large, and specific individuals, are very good at using or visiting a conserved wildland area to the level of intensity allocated-if they are clearly informed of the limits and if the method of explanation is clear and cast in a socially perceptible format. This communication requires more or less direct human presence and interpretation, depending on the society and circumstance. The more specific harvesters-researchers, staff, biodiversity-prospectors, inventories, ecosystem services personnel-likewise are proved to be socially responsible in conserving tropical wildlands if they find themselves cast in a responsibly managed and forward-directed interaction between society and biodiversity. But we can never forget that the finest farm or ranch can easily be destroyed through overgrazing of pastures, improper irrigation, failure to rotate crops, poor selection of varieties, or sloppy agrochemical application. Wildland biodiversity is another kind of farm or ranch.

Throughout the tropics, lured by the ecotoursim dollar there has been a very strong tendency to use the dollar as the primary currency in valuating conserved wildlands. While this has its valid points, what seems to be forgotten largely through the inconvenience of levelling the national social playing field - is that the "poor" national user of a conserved wildland pays in votes and inemotional attachment to the conserved wildland.

Equally revealing, and long-term, is the biodiversity-prospecting loop. When a conserved wildland or its facilitators bring home the first contract for biodiversity prospecting, the returns seem very large when set against the background of tropical conserved wildlands, viewed as all cost and no visible income other than piddling ecotourism entrance fees. However, a ministry of natural resources will take notice when th first actual royalties from drug discovery flow into the national budget for conserved wildlands or, better yet, into the endowment fund of the conserved wildland from which the raw materials were collected. But even then, a ministry of the economy will take notice. That will occur when the pharmaceutical company decides to move some substantial portion of the development process-more than $200 million per successful drung-into the source country. Once again the levelling of the playing field reappears, with all its advantages and impediments. The North American Free Trade Agreement (NAFTA) and the General Agreement on Tariffs and Trade (GATT) relate directly to conservation of biodiversity. The art of valuation of nondestructive use of biodiversity rests heavily on being able to work in many currencies, to recognize the market value of information of many different sectors. A field guide to the birds of a tropical country is not jut "a bird book." It is essential technology in the ecotourism industry. It is fertilizer for the ecotourism crop. Yet also, without a conserved wildland in which to observe the birds and the things they do, it becomes just a bird book. The value of information about biodiversity is extremely dependent on context. A country may "have" the most marvellous set of endemic species or bizarre habitats, but if the information they contain and display is not put into various social currencies, those species and habitats will contribute little or nothing to their very survival in a world dominated by humans.

All of this can do nothing but reemphasize the critical need for institutions and processes that accept the responsibility and challenge of the specific task of gathering, collating, massaging, and distributing information from and about a nation's conserved wildlands. This essential process must occur at the level of each specific wildland and at the level of the national synthesis, at the least. Your society can hardly be expected to value that which is invisible to your. Ironically, the very salvation of biodiversity–its valuation by society—is a multiedged sword.

First, if the area is conserved for its value on just one or a few axes, then it is in the same risk zone as the country that depends on a monocultural agrocape—coffee bananas, and Costa Rica are close to mind. Fortunately, wildland biodiversity is in fact farmore diverse than is the agroscape and, as such, diversification of crops as well as diversification of markets is very feasible.

Second, information differs from agricultural produce that one consumes today and needs more tomorrow. On the contrary, once consumed, information is public domain and continually widely available, and is even more so in the electronic and ecomputerized age. Therefore, a given piece of new information is unlikely to have nearly the same value in next year's market as in this year's market. There is a high premium on rapid development of products, almost as one encounters in the newspaper business. However, as in the news business, naive consumers of information about biodiversity continually do appear through human biologic processes, and the amount of absolutely new information about biodiversity to be gathered and developed certainly limitless for many decades to come. Information abo biodiversity can also become "new" through the appearance of a new use.

Finally, it is no secret that a nation's conserve wildlands are its package of local varieties. All that a nation does to both share and profit from the varieties in its agroscape is pertinent by alanlog to the treatment of the breeding stock and genus from its conserved wildlands. Just like petroleum, which occurs in a multitude of countries, the value of any one of these species depends on what that country constructs on top of its national supply of this basic raw material.

Any conserved wildland will need to struggle with the queston of physical use and the impact of sampling observing, studying, experimenting and visiting. Given the all conserved wildlands are in fact impacted already by humanity, and always will be the question is basically what level of use falls within the "natural ups and downs and expansions and contractions of behaviour, demography, and interactions. Any user does leave a footprint or a beer can in ones knows enough biology to see it.

At present, perhaps the largest single near-sighted use of detailed tropical biodiversity is the international academic and museum community. In what currency will they pay for and value their use? Long we have cast our graduate students in our own image, and now and then we have done the same to a student from a tropical country. But it is not at all clear that this is the kind of payment we would make if we really were to think out what a tropical resident needs for example, to be part of the managerial cadre or biodiversity. Even more basic is whether we should be expending so much energy on yet more graduate students in a steady-state system, or expending that energy in collaborating with the tropics as it comes up to speed. We are letting the lifestyles in our developed world define the way that we examine and study tropical biodiversity."

21
Biodiversity Issues Impact on Diversity

Biodiversity is located in small areas of the planet. As much as 20 per cent of plant species and a still higher proportion of animal species are confined to 0.5 per cent of Earth's land surface. These species are endemic to their areas, so if the local habitats are eliminated, these species will suffer extinction. The areas in question are indeed threatened with imminent habitat destruction. It is the two attributes together than cause the areas to be designated "hotspots". The concept has been much advanced in recent years. In some localities, it applies well, in others less so. From the start, the hotspots concept was explicitly not intended to apply much outside the tropics and subtropics, since species in temperate and boreal zones to features far less localized endemism that those from lower latitudes.

The hotspots approach has been limited largely so far to tropical forests and Mediterranean-type zones. This is not to say that other biomass doe not qualify; they simply have not been subjected to much analysis for hotspots. For example, coral reefs are likely to feature hotspots. Something similar applies to certain tropical lakes and wetlands, which are briefly reviewed here.

Tropical Lakes

A number of lake and river ecosystems are unusually rich in biodiversity. Worldwide lakes and rivers contain at least 8,400 fish species, or roughly 40 per cent of Earth's species of fish that have been identified to date. In turn, this means that these freshwater ecosystem support almost one-quarter of the planet's known biodiversity in less than 0.01 per cent of the planet's water.

By virtue of their isolation, lakes often form "ecological islands." In turn, this situation can lead to a high degree of speciation and endemism, producing exceptional biodiversity in terms of fish faunas, especially in the tropics. Many observers would urge conservation for them on grounds of their remarkably speciose fish faunas alone with all that is implies for evolutionary radiation. That is to say, and as we shall see below, there is a qualitative rationale for enhanced conservation beyond sheer numbers of species.

Lakes Victoria, Tanganyika, and Malawi feature at least 1,167 identified fish species, with a true total of perhaps 1,450 species. Around 885 (76 per cent) of the known species are endemic. These "fish swarms" occur in approximately 121,500 km^2, which means that 11 per cent of Earth's species of freshwater fish are confined to 0.08 per cent of Earth's land surface. Another two smaller lakes, Edward and George, contain at least 60 endemic species. The evolution of these Rift Valley lakes has fish that produced a phenomenon of explosive speciation, generating exceedingly fish communities with far greater differentiation than in any other tropical lakes. It is even considered that certain of the cichlids can speciate in as little as 200 years. However, the basic biology of the leading lake in the chain, lake Malawi, has yet to be elucidated, and there is no scientific programme for long-term research of a substantive and systematized sort.

Lake Malawi's expanse 28,200 km^2 contains at leat 500 cichlid species, 495 of them endemic, plus roughly 40 other fish species, approximately 20 of them endemic. Each year a number of new species is discovered, which means that the true total could be rather higher than accepted figures. Indeed the latest estimate suggests a total of 800 cichlid species. Lake Malawi is suffering from sediment deposition due to soil erosion in its watersheds, pollution from industrial installations and hinterland agriculture, extensive overfishing, and proposed introduction of alien species.

In Lake Victoria's 65,000 km^2, there are-or rather there used to be-more than 3000 cichlid species, around 130 of them endemic, plus almost 40 other species, some 20 of them endemic. On top of the speciocity of the lake's fish flock, the evolutionary youth and its ecological diversity make the community an unrivalled subject for comparative biology and evolutionary study of vertebrates. Yet introduced predators, among other problems, have already reduced the cichlid stock by 200 species and are likely to reduce the endemics by 90 per cent within another decade at most. This must rank as the greatest extinction spasm of vertebrates in recent time.

As for Lake Tanganyika with its 28,399 km^2, there are species and 115 other species of fish, 220 of them (76 per cent) endemic. The least rich lake in the chain, Lake Tanganyika nonetheless features almost one-quarter more fish species than Europe's total of 192 species. Fortunately, Lake Tanganyika is subject to little pervasive degradation as yet, though there is much sediment deposition from rivers draining eroded catchments, and there are plans for various forms of disruptive development in the lake and its environs, notably oid extraction.

Indeed, both Lake Malawi and Lake Tanganyika could rapidly follow the impoverishing experience of Lake Victoria, unless there are precautionary conservation measures put in place ahead of time. To this extent, and, on the basis of their spectacular cichlid faunas alone, there is much scope for anticipatory conservation. It is likely that conservation of these ecosystems would safeguard large numbers of species of invertebrates as well, though the scale of this spin-off benefit is impossible to determine at this stage. Three lakes probably feature some 1,450 fish species, or 17 per cent of the world's 8,400 species of freshwater fish is 0.08 per cent of Earth's land surface. Almost 900 known species are endemic. Apart from these three, few other lakes could qualify even as subsidiary hotspot areas. Lake Atitlan in Guatemala and Lake Lanao in the Philipines, possess only a few dozen endemic fish species each.

Tropical Wetlands

Wetlands are areas that contain, for part of the year at least, enough water to foster the development of specialized communities of plants and animals adapted to waterlogged conditions. They include marshes, fens, bogs, wales, wet heaths and moorlands, peatlands, floodplains, deltas, and estuaries, some of these types of ecosystems are as different from one another as are forests and savannahs. Many dense human communities live in or near wetlands, and they traditionally sought to drain or otherwise modify wetlands in order to make them better suited to human needs. As a result, we have lost half the world's wetlands during the course of this century. In the year 1800, there were 80,000 km^2 of wetlands in the "lower 48" United States, but by 1975 there remained less than half that much today, an additional 1,800 km^2 are drained and filled each year. In Asia, more than 5,000 km^2 of wetlands are lost each year to agriculture, irrigation, dam construction, and the like.

Regrettably, we do not know much about endemism-a key factor in hotspots-in wetlands. Because wetlands areas within one country are often interconnected by virtue of their aquatic continuum, there may not be much endemism. The global situation in this respect remains almost undocumented.

An area with prime potential to rank as a wetlands hotspot is the Pantanal Swamp. It is the largest wetland in South America and perhaps in the world with almost 130,000 km^2. Four-fifths of its expanse in Brazil, other sectors being in Paraguay and Bolivia. Consisting mostly of the low-lying floodplain of the River Paraguai and several tributary rivers, and interspersed with various types of mesic to dry forest, savannah and grassland, the area is the most important in South America for waterfowl. Well over 500 plant species are reported from the Pantanal Swamp. Little is documented concerning vertebrates, let alone invertebrates, and there is no such indication of endemism.

The area has been disturbed since the early 1970s by agricultural encroachment plus a good deal of deforestation, dam construction, and mining and industrial installations. The water stocks have been contaminated by chemicals from agriculture, mining, and industry. There is also a growing problem with fires, commercial overfishing, and poaching for wildlife products and the live-animal trade. In short, a largescale biota is being progressively depleted. The Brazilian government has placed less than 1,500 km^2, hardly 1 per cent, under protected status, and this seems tobe theoretical protection for the most part.

Another notable wetland area is the Sudd Swamp on the River Nile in southern Sudan. In fact, it comprises a series of large swamps in several substantial lakes, making up a present expanse of 16,500 km^2 of permanently flooded lands and an additional 15,000 km^2 of seasonally flooded lands. There are at least 500 plant species, of which only 100 or so occur in the central sector that is permanently flooded and features 7,000 km^2 of a monoculture of papyrus, but only one such species is known to be endemic. Over 500 bird species have been recorded, and just under 100 fish species. The area also supports 400,000 people and 800,000 cattle.

Much of the biota is threatened by the Jonglei Canal, a 360 km conduit that would cause the Nile water to bypass the Sudd, reducing the permanent swamps by 21-25 per cent and the floodplains by 15-17 per cent compared to the level characteristic of the early 1950s, whereupon there would be an 80 per cent reduction in the expanse of the swamp and a 58 per cent loss of floodplains, hence entraining all manner of adverse repercussion for the biota. As it happens the construction of the canal has been suspended since 1983, due to political, technical, and financial factors-mainly the civil war that persists in southern Sudan. The three East African lakes, shows that, while certain areas may contian unusual concentrations of biodiversity facing unusual threat, they cannot qualify as other than second-order hotspots because their stocks of species hardly compare with those of hotspot areas in tropical forests, coral reefs, and Mediterranean-type zones, especially as concerns endemism.

The poorest tropical forest hotspots are the southwestern Ivory Coast, with 2,770 plant species, 200 of them (7 per cent) endemic, in an area of 4,000 km^2 of primary forest and south western Sri Lanka, with some 1,000 plant species 500 of them endemic, in 700 km^2 of primary forest.

Categories of Biodiversity

Biodiversity occurs not only at the level of species, but embraces both species subunits and ecosystems. Some populations will be better equipped than others to adapt to the swiftly changing environmental conditions of the foreseeable future.

We may well find that, if the depletion of biodiversity continues unabated, we shall lose, say, 50 per cent of all species, and many of the surviving species will have lost, say, 90 per cent of their populations and perhaps 70 per cent of their genetic variability. This latter outcome is all the more probable insofar as human-induced attrition of populations' habitats tends to occur around the fringes of populations' ranges- precisely the sectors that often feature the greatest genetic variability insofar as it is here that populations encounter the greatest variety of environmental conditions.

To illustrate the point of population depletion, consider the case of wheat. In 1994, the species flourished across an expanse of 23 million hectares, featuring a rough average of 2 million stalks per hectare. This means that the total number of individuals in 1994 exceeded 450 trillion, probably a record. As a species, then, wheat is the opposite of endangered. But because of a protracted breeding trend towards genetic uniformity, wheat has lost the great bulk of its popualtions and most of its genetic variability. In extensive sectors of the original range of wheat, wild strains have all but disappeared, there is a virtual "wipeout" of endemic genetic diversity. In Greece, native wheats have become extinct; and across Turkey, wild progenitors of wheat find sanctuary from grazing animals only in graveyards and castle ruins. Collections of wheat germplasm are viewed as quite inadequate.

Are Earth's Biotas Stressed?

Let us now consider a further dimension of the biotic cirsis that receives next to no attention and that could markedly affect the eventual outcome. Certain biotic communities may be "stressed" following events of the late Pleistocene, leaving them unduly prone to extinction. For want of a better term, they may lack "ecological resilience" or "survival capacity" of a long-term sort.

The evidence for this, such as it is, lies in the fact that a mini-mass extinction started to overtake Earth's biotas during the late Pleistocene. After very long periods of steadily growing biodiversity, there was a marked decline starting some 30,000 years ago and continuing until about 1,000 years ago. Whether through human overhunting or climatic change or both, the large mammalian fauna of several entire regions - notably North and South America, Oceania, and Madagascar-lost more than 100 geanra. Following this extinction spasm, there has been continuing elimination of verteberate species, albeit not on such a spectacular scale as the late Pleistocene episode, until the onset of the present mass extinction, which started roguhly in the middle of this century. These recent eliminations were due almost entirely to human activities. We have no idea how many associated species-especially inveretebrate species-likewise have been lost, but they must have been quite numerous.

So today's biodiversity was already somewhat of a departure before the arrival of the unprecedentedly severe human impact from around 1950 onwards. Many surviving biotas have surely been affected adversely along the way, at least through depletion of their subspecies and population. Large numbers of species must have lost much of these genetic variability, hence their ecological adaptability-leaving them the more vulnerable to summary extinction. The present mass extinction may have had some of its origins in events long past. It would be difficult indeed to determine how far today's biotas are not only impoverished but have been subject to stress in the manner postulated.

Consequences for Evolution

The loss of large numbers of species, let alone species subunits, will be far from the only outcome of the present biotic debacle, supposing it proceeds unchecked. There is likely to be a significant disruption of certain basic processes of evolution. The forces of ntural selection and speciation can work only with the "resource base" of species and subunits available and, as we have seen, this crucial base is being grossly reduced. To cite the graphic phrasing of Michael Soule ad Bruce Wilcox, "Death is one thing; an end to birth is something else". Given what we can discern from the geologic record, the recovery period, i.e. the interval unit speciation capacities generate a stock of species to match today's abundance and variety, will be portracted. After the late Cretceous crash, between 5 to 10 million years elapsed before there were bats in the skies and whales in the seas. Following the mass extinction of the late Permian, when marine invertebrates lost roughly half of their families, as many as 20 million years were needed before the survivors could establish even half as many families as they had lost.

The evolutionary outcome this time around could prove yet more drastic. The critical factor lies with the likely loss of key environment. Not only do we appear set to lose most if not virtually all tropical forests, but there is a progressive depletion of tropical coral reefs, wetlands, estuaries, and other biotas with exceptional abundance and diversity of species and with unusual complexity of ecological workings. These environmens have served in the past as prominent "powerhouses" of evolution, meaning they have thrown up more species than other environments. Virtually every major group of vertebrates and many other large categories of animals have originated in spacious zones with warm, equable climates, notably the Old World tropics and especially in tropical forests. The rate of evolutionary diversification - whether through proliferation of species of through emergence of major adaptations - has been greate in the tropics, especially in tropical forests. Tropical species, notably those in tropical forests, appear to persist only in brief periods of geologic time, which implies a high rate of evolution.

All this will carry severe implications for human societies throughout the recovery period, which is estimated to be a minimum of 5 million years longer. Just five million years would be 20 times longer than kind itself has been a species. The present generation is in effect a decision on the unconsulted behalf of at least 200,000 follow-on. This must rank as the most far-reaching decision ever taken on behalf of such large numbers of peole during the whole course of human history that Earth's population maintains an average of 2.5 billion people during the next 5 million years, and that the gene time remains 25 years. The total number affected will amount to 500 trillion people.

We are reaching a stage when there is less and less scope for protection of biodiversity through the traditional strategy of parks and reserves. These protected areas total more than 8,000 units covering 7 million km^2, or 5 per cent of Earth's land surface. True, there is great need for many more such areas. Ecologists, and biogeographers consider that, in tropical forests alone, we should at least triple the expanse protected, while recognizing that one-third for existing parks and reserves already are subject to agricultural encroachment and other forms of human disruption. Additional forms of degradation are likely to stem in the future from pollution : as much as 1 million km^2 or almost 15 per cent of remaining tropical forests soon could become subject to acid precipitation. There eventually could be further and more widespread degradation from enhanced UV-B radiation and global warming.

The time has arrived when, as a bottom-line conclusion, we must recognize that we can ultimately safeguard biodiversity only by safeguarding the biosphere as well-that this entails for agriculture, industry, energy, and a host of other sectors, especially the growth of both population and consumption.

While formulating our responses to the mass extinction crisis, we need to bear in mind the length of time still available to us. The critical criterion for our efforts is whether we are doing far more than before.

Of course, not all habitats ae going to be destroyed outright within the immediate future. But that is hardly the point. What looks set to eliminate many if not most species in the long run will be the "fragmentation effect" i.e. the break up of extensive habitats into small isolated patches that are too small to maintain their stocks of species into the indefinite future. This phenomenon has been widely analyzed through the theory of island biogeography, and appears to be strongly supportd through abundant empirical evidence, albeit with a numbr of variations on the general theme. True the process of ecological equilibration, with its delayed fallout effects, will be extended periods to exert its full depletive impact; in some instances, it will be decades and even centuries before species eventually disappear. But the ultimate upshot, which is what we should be primarily concerned with, will be the same.

Consider the environmental degradation that had already occurred. Through dynamic inertia, it will continue to exert an increasingly adverse effect for a substantial way into the future, no matter how vigorously we try to resist the process: much potential damage is already "in the pipeline." Ozone-destroying CFCs now in the atmosphere will continue their work for a whole century even if we were to cease releasing them forthwith. There is enough global warming in store through past emission of greenhouse gases to cause significant climatic change no matter how much we seek to slow it, let alone halt it.

The potential biodiversity depletion that humankind has already engendered, it is realistic to prognose that there will be a large number of extinctions in a post-2000 world, even if it were relieved of humankind's continuing disruptions.

22
Systematics and Biodiversity

Agricultural development is increasingly dependent on the interactions and knowledge base of systematics and biological diversity. Biodiversity itself is the grist for the agricultural mill - the germplasm. Systematics is the explorer, describer, organizer, and predictor of biological diversity. Agriculture is a primary user of the products of the interactions between biodiversity and systematics.

History and the Future

Before the advent of pesticides, herbicides, and inorganc fertilizers, the need to understand the biology and biological requirements of agricultural organisms was of considerable importance. Control strategies for insect pests, for example, depended on knowledge of life histories so that agricultural practices could be used to circumvent high pest population levels. With the development of chemical pesticides and fertilizers came the complacent attitude that these products would solve the world's agricultural problems, and there was no longer a need to know the identity and interactions of the orgnisms that they comprised.

The most important organisms in agricultural systems are the most poorly understood in the context of biodiversity and systematics, including fungi, fee-living and plant-parasitic nematodes other microorganisms, endoparasitic helminths, insects, mites, and other arthropods. Even for many important species of pests, superficial understanding of the group is a major impediment to solving important problems in agriculture.

Agricultural Dilemma

The agricultural community now finds itself in a difficult situation. It increasingly espouses environmentally positive approaches such as biological control, sustainable agriculture, and pest management, but lacks a comprehensive understanding of many of even the most common organisms that form the foundation for these approaches. Further, destruction of native ecosystems for new agricultural lands eliminates the organisms upon which agriculture must depend for its future. Ehrlich points out that it is the "less cuddly" organisms that are the most important for the future of human existence in the following sections.

Biological Control

Biological control programmes mandate detailed knowledge of both pests to be controlled and the natural enemies that prey on them. The first step is to seek out systematic information known about the target pest and its natural enemies, and based on this information, to survey the natural enemies in the field.

An example of insufficient systematic knowledge that caused the delay of a biological control programme concerns a mealybug that was decimating the coffee plantations of Kenya. The mealybug was misidentified for many years as *Planococcus citri* (Risso) and later P. *lilacinus* (Cockerell). Discovery, evaluation, and introduction of the natural enemies of these species had no effect on the pest and was a waste of time and resources. Ultimately, a systematist was asked to study the mealybug, who described it as a new species that was formerly restricted to Uganda. Natural enemies were located in Uganda, introduced to Kenya, and within a few years the pest became rare.

In the future, biological control will have an increasing importance in pest control. With the reality of bioengineered organisms, it now becomes possible to alter a pest organism so that it is beneficial rather than harmful. Development of pesticide-resistant natural enemies also adds an innovative new dimension to biological control. Such strategies will be much more effective if they can draw on the genes of the millions of natural enemies that currently exist in the pool of biological diversity, but are unknown to science.

Integrated Pest Management

Integrated pest-management strategies depend heavily on knowing the identity of the organisms that occur in the crop system, understanding the complex interactions among these organisms, introducing biological control agents that are compatible with management schemes, understanding the ecological dynamics of the system, and developing minimally disruptive methods for managing the pests without diminishing the value of the agricultural product. In recent years, considerable research has been conducted on exploiting the biodiversity of entomophagous nematodes as biological control agents in agroecosystems. One of the first major successes was with *Steinernema carpocapsae* in controlling tree borers. As part of urban pest-management strategies, these nematodes are sprayed on infested trees whenever damage is detected, and they control tree borers in about 80 per cent of the cases. The discovery and use of this nematode has reduced pesticide usage and enhanced the likelihood of control at many times of the year. One can only wonder what other benefits might be available if the biodiversity and classification nematodes were more completely understood.

Sustainable agriculture

Sustainable agriculture depends on detailed systematic knowledge of the organisms that occur in agroecosystems, and attempts to maximize the use of the native biota and natural resources and to minimize the use of non-indigenous organisms and external influences such as inorganic chemicals. Even in areas such as the Beltsville Agricultural Research Centre, where agricultural research has been underway for more than 50 years, the native biota is only partially known.

After exposure to the indigenous wasp fauna for 3-5 days, the eggs were brought into the laboratory and parasites were allowed to emerge. Fifteen different species of Trichogramma were reared from the moth eggs, inclduing nine that were new to science. The identity and role of these wasps and other organisms in the agroecosystem are crucial bits of informatin for succesful implementation of sustainable agriculture programs.

Pest Control and Quarantine

The difficult task of the Animal and Plant Health Inspection Service (APHIS) of the U.S. Department of Agriculture is to exclude these organisms. Unfortunately, the knowledge base and tools to assess biodiversity that this organization needs to make better-informed decisions are not always available, APHIS generally does not allow plant-associated organisms identified as pest risks into the country if they currently do not occur here.

APHIS also has the responsibilityof determining if organisms unwanted by other countries occur in areas in the United States that export agricultural products to that country.

Some examples of the impact of systematic knowledge on APHIS decisions are as follows. A smut associated with wheat imported into Canada from the United States was initially detemined to be *Tilletia indica,* a high-risk exotic pathogen. A ban on the import of wheat from the United States to Canada was suggested. A systematist was asked to study the material and correctly identified the smut as *T. barclayana.* This species occurs on rice and apparently became a contaminant on the wheat in question when it was stored in a warehouse that previously contained rice. This identification was confirmed with isozyme analysis and a potentially costly international incident was averted.

Due to the efforts of systematics in Maryland, populations of this nematode were detected soon after first introduction into North America. Surveillance programmes, measures of control, and extensive epidemiological studies were initiated to limit the potential impact of this newly introduced nematode. Aside from this highly visible parasitic helminth, there has been a history of introductions of nematode parasites with exotic bovid hosts from Africa, camelids from South America, and cervids from Europe. Many of these parasites represent potential threats to wild and domestic ruminants in North America, further highlighting the importance of systematics within the context of cosmopolitan groups and the necessity to develop accurate inventories of the world fauna.

Plant germplasm

In the area of plant germplasm, botanists have come a long way in expanding science's understanding of species of crops and the world's vascular plant flora, but, even in this relatively well-known group of organisms, much remains to be done. Millions of dollars have been added to the tomato industry through increased levels of soluble solids derived from an inconspicuous species of Andean tomato, and disease resistance has been added to the genome of cultivated corn from a nearly exitinct species discovered in Mexico. With recent developments in genetic engineering, nearly any species of plant has the potential to contribute genes of importance in enhancing agricultural crops. Ethnobotanical data suggest that of the approximately 250,000 species of plants, more than 7,000 species have been used for human benefit as food. Understanding the diversity and systematics' these plants will add significantly to the goal of developing a much broader spectrum of agricultural products and will enhance sustainable approaches with more natural environments.

Brushfire Approach to Agricultural Systematics

Under current conditions of small budgets and reduced personnel, resources devoted to systematics and biodiversity are woefully small and stretched to the limit. The impact is that each of the small number of agricultural systematics must serve many roles, including those of researchers, identifiers, curators, database managers, illustrators, and technicians. The best strategy to support the biodiversity and systematic needs of agriculture is to direct resources towards the groups of organisms that are most likely to cause future agricultural problems or towards the beneficial groups that show promise of enhancing agricultural systems, and undertake monographic research on these groups. Unfortunately, there is rarely any time or resources to take a coordinated approach, and this leaves agriculture with a systematic support systems that is patched together with chewing gum and rusted wire. In most instances, agricultural problems are solved using the brushfire approach of reacting when an emergency arises, undertaking a narrow research programme on the pest causing the problem, hoping to find a solution to the immediate problem, and going on to the next agricultural brush fire. This approach does not solve problems of the future and does not add significantly to the development of a predictive classification system. In fact, in many situations, systematic analysis of a single species removed from the context of phylogeny and biogeography and not carefully integrated into the classification system detracts rather than adds useful information.

Predictor of Biodiversity

Classification systems serve as the knowledge base from which predictions about organisms are made. Too frequently this predictability is taken for granted as an innate capacity of the human mind, but in fact it is part of the classification process. A simple example involves a person who sees a hornet sitting on someone's arm. A prediction could be made in any part of the world where hornets occur. As long as the hornet is a female, the prediction will have a high level of predictive accuracy. Similar, but more sophisticated predictions can be made from other classification systems. For example, recent research on the *Trichoderma*-like fungi used in biological control has demonstrated that a species previously placed in *Gliocladium* (*G. virens*) shares many characters with the genus *Trichoderma* and should be placed in the latter genus. This is important because many species of *Trichoderma* have mycotoxins that are useful for the biological control of plant-pathogenic ascomycetes, rather than basidiomycets, as is the case for true species of *Gliocladium.* The sexual state of *T. virens* is *Hypocrea gelatinosa* or a closely related species and it is predicted that other closely related sexual states of *Hypocrea* will prove to be valuable agents of biological control similar to *T. virens.* Scientists have already found high levels of mycotoxins in isolates of *Hypocrea.* Specialists in biological control often ask systematists where to look for natural enemies and which species are most closely related to the pest.

APHIS would have information on all of the organisms that occur in the United States and could make better-informed quarantine decisions at ports of entry. Species of Trichogramma would be classified, and sustainable farming would benefit from systematic knowledge of all organisms that interact to maintain and stabilize the system. Crop plants would be enhanced by genes from plants that are currently unknown or are poorly known. High-risk adventive species would be well known and their distribution would be monitored so that some of the billions of dollars that might be spent on these organisms in the future could be saved.

The significant progress that systematics has made in the recent past is phenomenal. Some larger-sized organisms are well known, and even groups of smaller-sized organisms have a knowledge base many times larger than that which existed at the turn of the century. Through computer technology, imaging systems, morphometric analysis, molecular techniques, and recent theoretical innovations, the systematics community is poised and capable of carrying out the Systematics Agenda 2000 initiative. Armed with comprehensive knowledge of the components of Earth's biological resources, we will be better able to maintain high levels of agricultual productivity and reduce environmental impact so that future generations can continue to enjoy the mysteries and aesthetics of complex biological systems.

Wider perspectives are essential in order to assess the potential and limitations of major choices related to the maintenance of agricultural diversity. Maintenance of agricultural diversity whose promotion has environmental, economic, social and nutritional implications whose combined consequences and requirements will decide the long-term success or failure of diversification. An excellent insight into these challenges is offered by a closer look at legumes. If maximization of cropping diversity is seen as a desirable goal not only as a preferable agronomic practice but as a significant contribution to socio-economic well-being, then the promotion of leguminous species should follow. These crops have many well-known advantages, and no intensive traditional agroecosystem could work without relying on them. Why is it then that legumes have been in a virtually global retreat? the home of soyabean and beancurd, is now producing less than that two generations ago, indeed less than the peak output during the 1930s? Has this largely been a matter of avoidable neglect or a consequence of unavoidable developmental trends? Answering these questions should help in appraising realistic limits to increased agricultural biodiversity through leguminous cultivation, and to design effective approaches to the promotion of this goal.

Both the agroecosystematic and nutritional advantages of leguminous crops have been appreciated for centuries in convinving qualitative terms.

The principal agronomic advantage of legumes is, of course, the ability of symbiotic *Rhizobium* bacteria to fix substantial amounts of nitrogen, and thus, either almost entirely eliminate, or greatly reduce, the need for any supplemental supply of that critical macronutrient. In addition, legumes are generally more drought tolerant than cereals, require less irrigation in environments, and are also fairly pest-resistant. And leguminous cover crops alfalfa, clovers and vetches being the most important by far provide outstanding anti-erosion protection, fix much more nitrogen than edible species, and they are also the richest source of new organic soil matter in agroecosystems when ploughed under as green manures.

The nutritional appeal of legumes rests on their unusually high protein content, and on the fact that their amino acid make-up complements that of staple cereals. While protein levels in wheat, rice and corn range between 7 per cent and 14 per cent, most legumes contian between 20 per cent and 25 per cent, and soyabeans around 40 per cent. These proteins are deficient in sulfur-containing amino acids, but have abundant lysine, the pattern exactly opposite to the amino acid makeup of staple cereals. The nutritional quality of legumes is further enhanced by their relatively high levels of riboflavin, thiamine, calcium and iron, and by the provision of excellent dietary roughage by mature seeds : diets containing cooked pulses require no fanciful bran supplements. And, of course, several leguminous seeds, above all soyabeans and peanuts, are rich sources of plant oils high in polyunsaturated fatty acids.

Finally, legumes offer impressive culinary flexibility. They can be eaten as vegetables, either as immature pods or seeds, most commonly those of peas, or as sprouted seeds. Uncooked mature dry or roasted seeds make good snacks. Cooked dry seeds, including vitually all varieties of pulses, are universal staples, and there is an array of ferment foodstuffs, ranging from Indian *Idli* to Indonesian *tempheh* and to now almost global soy sauce, and of milled and coagulated products, most notably the Chinese *doufu* and slightly different Japanese *tofu*.

Not surprisingly, every traditional society planted food legumes in often complex rotations with staple cereals, root, oil and sugar species. Dominant food species differed from soyabeans and beans in East Asia, lentils and chickpeas in the Indian subcontinent and throughout the Middle East, peas and beans in Europe, groundnuts in West Africa, beans in the Americans- but the basic pattern of rotations or intercropping was universal. And as their productivity of agroecosystems increased, both Asian and Western farming also began to rely more on legumes as green manures.

Global Retreat

The clearest nutritional evidence for this trend comes from basic accounts of food availability. Food balance sheets, be they prepared by national statistical offices or calculated in the FAO's Rome headquarters, do not tell us how much, on average, people actually eat. There is a gap, often suprisingly large, between aggregates of energy and nutrient availability derived from food balance sheets and actual intakes measured by much more expensive, and hence much not consumption surveys are conducted. But these accounts are good enough to assess long-term changes in nutritional patterns and to appraise approximate contributions of individual foodstuffs to overall energy, carbohydrate, protein, lipid and micronutrient supply.

The other key ingredients are declining intakes of whole grain cereals and tubers, and higher consumption of animal foods, refined sugar and fresh fruits. Affluence tends to bring a more expensive diet lower in carbohydrates and higher in lipids and animal protein shares.

Alfalfa grown mostly in Europe and in the Americas is still the world's largest forace crop but its worldwide plantings have recently been shrinking by about one million hectares a year. Global plantings of clover and vetches have been also decreasing. This retreat has paralleled the steady increase in applications of inexpensive synthetic nitrogenous fertilizers and the expanding production of high protein soyabean-based feeds.

China's Experience

The best available information on typical rates of traditional Chinese food consumption comes from extensive surveys conducted during the 1920s, and from figures gathered by the Department of Agricultural economics of the National Agricultural Research Bureau during the 1930s. These data show legume intakes which were widespread and important. Nationwide, one-fifth of all families regularly consumed broad beans, one-quarter green beans, one-third field peas and two-fifths soyabeans, with regional rates rising to as high as 70 per cent for soyabeans in the Southwest, and 75 per cent for peas in the area of double-cropped rice. Average per capita legume consumption during the year 1931-1937 reached just over 25 kg per year, equal to India's mean and hence just about the highest recorded. Soybeans, beans and peas supplied about 7 per cent of all food energy and 16 per cent of all protein during the 1930s, with the latter share as high as 25 per cent in the winter-wheat region.

The current low status of other food legumes in China is perhaps best illustrated by the fact that unlike for soyabeans and for cereal, oil and sugarcrops the State Statistical Bureau offers no nationwide data on on their planted area and annual harvests, and that on the provincial levels such figures were available in the early 1990s only for Yunnan, Sichuan and Anhui.

Recent changes in the cultivation of green manures show even faster decline trends. Complex rotations, including a variety of green manures most commonly the Chinese vetch. Common vetch and luceme-used to be one of the characteristic features of traditional Chinese farming.

The reason for abandoning green manure cultivation is clearly the higher pressure to produce more food on limited land. To return to the peak plantings of the 1970s would mean to forgo grain harvest on some 6 million hectares, or roughly 20 million tonnes of food cereals, equivalent to about 5 per cent of China's total output and sufficient to feed some 75 million people. These grains and their residues, will incorporate about 600,000 tonnes of nitrogen derived largely the contributed amount so highly variable that it may often account for less than one-fifth of all nitrogen requirements, but there may be no net gain at all. Totals of nitrogen removed by leguminous crops cannot be estimated with precision by using simple averages for nitrogen tissue contents and crop residue multipliers. Nitrogen content of both seeds and crop residues can commonly vary by up to 25 per cent, and the amount of stalks and vines can differ by as much as two-fold.

The first convincing quantitative evidence of the often limited contributions made by leguminous crops to soil nitrogen levels came from a series of experiments in the US Midwest which began in the late 1970s and used labelled nitrogen. Rotating non-leguminous crops had no smaller benefits on crop yields than rotating legumes and non-legumes; wheat and corn did as soybeans and corn.

Detailed nitrogen budgets prepared for soyabeans, now by far the most important grain legume both in the US and in China, showed that when the crop yielding about 2,400 kg per hectare derives 40 per cent of its nitrogen from symbiosis, then it causes, even with complete residue recycling, net nitrogen loss amounting to more than 80 kg per hectare. Only when soybeans can derive more than about 80 per cent of their nitrogen need from symbiosis, or when they are planted as green manures, do they bring a major nutritional benefit for the subsequent crop : they can add up to 90 kg of nitrogen per hectare, the enrichment surpassed only by ploughing in good stands of clover or alfalfa.

Prorating these findings to the typical Chinese crop would mean that 1,500 kg of soybeans harvested per hectare gointrogens, cyanogns estrogens, antivitamin factors and toxic amino acids are less widespread, but prominent in some species.

Species-specific antinutritive factors incldue vicine and covicine in faba beans, which may cause hemolytic anemia and hematuria in genetially sensitive individual, and monoamine oxidase inhibitors, also infaba beans, engendering headaches and palpitations. Common, and not infrequently perilous, allergies to peanuts are wll documented, as is the risk from aflatoxins not only in unprocessed whole seeds but also in peanut butter.

Of course, many traditional cultures learned to live with the poorer digestibiity of legume and they also discovered how to destroy, or at least substantially eliminate, many antinutritive factors by prolonged cooking and by specific processing. Consequently; it is hardly suprising that the foods not requiring such elaborate preparation, yet providing equal or better nutritional value, will be preferable. As soon as modern societies get richer and increased purchasing power, this enables people to buy more convenient and less risky foodstuffs. They follow one of the most notable universal nutritional shifts and begin reducing their consumption of legumes. Although the initial stages of incipient affluence may be accompanied by a slight increase in demand for pulses but later stages bring a sustained decline in the eating of traditional pulses.

Legumes as the Sources of Nitrogen

While the nutritional drawbacks of pulses have been appreciated for a long time, it is only during the last two decades when a better understanding of symbiotic nitrogen fixation, the first detailed studies of nitrogen balances in agroecosystems, and a greater appreciation of the complex dynamics of the bio-spheric nitrogen cycle revealed some surprisingly limited contributions of common leguminous crops to the supply of the critical macronutrient.

These findings challenge the traditional agronomic notion that nitrogen-fixing legumes can be relied on to contribute substantial quantities of nitrogen to be used by subsequent crops. The reality is far more complex : not only is from chemical fertilizers, a mass equivalent to less than 5 per cent of China's recent application of synthetic nitrogenous compounds.

Certainly the most obvious disadvantage of pulses in modern agriculture are their relatively low yields, inherently lower than those of staple cereal crops. While the recent global means for wheat and rice are, respectively, abut 2,500 and 3,600 kg per hectare, those for pulses rose from around 500 kg in the early 1960s to less than 900 kg per hectare in the early 1990s. More importantly, typical yields of many traditional varieties remain so low that there is no incentive to expand their cultivation in areas of inadequate or marginal food supply. Even a rather poor cereal crop in the Sahelian zone of Africa will yield 1.5 tonnes per hectare, while traditional legumes may bring less than 500 kg. But low yields of legumes are just one, and not even the most important, factor. The lowered desirability of legumes in changing diets and their less needed contribution to nitrogen balances in intensive agroecosystems, whose nutrient supply is dominated by synthetic fertilizers, matter more.

Nutritional Drawbacks

Some of these detractions, above all complications in cooking and digesting many legumes, have been known since antiquity. Minimum cooking times for may mature seeds are much longer than those for cereal seeds and flours, on the order of two hours if the seeds were not presoaked. Although the food energy content of legumes is nearly identical to that of cereal grains their different composition makes it outright impossible to process them into such staples as bread and noodles.

Modern bio-chemistry has added along a list of other objectionable, and often unique, ingredients. All food legumes contain at least one, and most of them several, antinutritive factors. Enzyme inhibitors, mostly those interfering with the functioning of proteases, are especially common. Ironically, protease inhibitors make it more difficult for our bodies to use the proteins present in food legumes which may be eaten specifically for their high amino acid content! Concentrations of lectins are relatively high in virtually all legumes, while would remove about 105 kg of nitrogen in seeds and 27 kg of nitogen in pods, leaves, stems, roots and nodules. Even if two-thirds of all nitrogen were derived from symbiosis and all residues were recycled, the crop would still cause a nitrogen deficit of almost 20 kg per hectare.

Obviously, the possibility of such outcomes greatly reduces the appeal of legumes as providers of nitrogen for subsequent crops. Instead, non-leguminous crops fertilized with synthetic compounds may be chosen for rotations with cereal staples or, indeed, rotations of two staples may be more appealing, a choiice which would further strengthen the already high reliance on staple cereals.

And although green manures may supply substantial amounts of rather rapidly available nitrogen, their declining cultivation is easily explained by the quest for overall maximization of food harvests. As their population densities increase, countries with a limited amount of arable land cannot afford to use that space for a green manure leguminous fixer whose grooth would preempt the cultivation of a food crop and must turn to synthetic nitrogen fertilizers.

Putting intensive agricultures on a more sustainable basis is unimaginable without reversing the worldwide trend towards monocultures. Rational cropping must include rotations : they have been the mainstay of all sustainable traditional agroecosystems and their apid decline and sometimes outright abandonment. The past two generations should be seen as an undesirable aberration. The environmental advantages of rotations have been well documented by long-term field experiments. They include, above all, reduced soil erosion and water runoff, nutritional benefit of symbiotic nitrogen production by legumes, better sol tilth and interruption of weed, insect and crop disease cycles. Will food, feed and forage legumes continue to have an essential place in future rotations or will they mater only in a few areas specialized in the cultivtion of a single species? Legumes will be a widely sought part of rotationss only when they will provide a number of clearly demonstrable benefits. Only a multipronged strategy can achieve this goal.

To begin with, much more attention should be paid to the productivity of established edible varieties. An outstanding recent example of this approach is the release of an improved strin of mung bean by the Asian Vegetable Research and Development Centre. Between 1985 and 1990 this cultivar was planted on over 360,000 hectares in China outyielding local varieties by about 50 per cent and bringing a major economic benefit.

An enormous global potential of similar improvements is illustrated by the fact that the differences between average global yields and the lowest regional performances are much larger for pulses than for cereals. Taking peas as an example, the global mean is now 1,500 kg per hectare, but the rates are commonly below 500 kg per hectare in many Latin American and African countries.

Substantial gains can also be made in maximizing the protein content of commonly cultivted legumes. While doubling of the overall protein level is not out of the question for some species for example, protein content of pea cultivars ranges from just 15 per cent to well over 30 per cent, a more realistic goal would be to increase the shares by 25 to 35 per cent.

Increasing the nutritional appeal of legumes throughout the rich world does not appear to be very promising. There has been no shortage of innovation, especially soyabean-based processed foods, and many of them are now regularly available in Western supermarkets. In addition, a growing appeal of non-European cuisines has created new openings for many old foods stuffs. However, these changes have led at best to marginal increases in Western demand for food legumes.

Perhaps the most realistic expectation would be to maintain a modest per capita level of pulse consumption which would guarantee their continuing importance in all intensive agroecosystems. The growing demand for meat in industrialized countries should mean good prospects for planting of soyabeans needed to contribute protein to mixed feeds. Again, China exemplifies this universal dietary trend and the need for better feeds. In order to meet this demand the country's mixed feed industry has grown rapidly, but it still provides only a small share of total nutrients.

While it is unlikely that there will be yet another explosion of soyabean cultivation akin to the Brazilian experience, soyabean cultivars can be grown in a variety of soil types, and they also have a wide range of photoperiodic responses. Consequently, it is almost inevitable that in the long run the strong trend towards higher meat consumption in industrialised countries will be tied to more frequent soyabean planting.

23
Biodiversity for Tropical Region

Life on earth takes many forms and comes in all sizes, from microscopic one-celled plants to blue whales and human beings. Together these organisms and their interactions constitute our planet's biodiversity. Among this profusion of life are the beetles and their insect and arachnomorph relatives, which, taken together, constitute most of Earth's biodiversity. There are 1.4 million species of insects described in the scientific literature, which is about 80 per cent of all life currently recorded on Earth. Taxonomists, those who name and classify species, have been describing species of insects at about 4,400 per year for more than 235 years, and in the last 25 years, have described about 8,680 per year (±363). This written record is at best perhaps only 3.4 per cent of the species actually living on the planet. Recent estimates of insect species, mostly in tropical forests, indicate that the descriptive process is woefully behind. These estimates indicate there may be as many as 30-50 million species of insects, making this pervasive terrestrial arthropod group 97 per cent of global biodiversity.

Insects and their relatives are the most dominant and important group of terrestrial organisms, besides humans, that affect life on Earth. They affect human life in a multitude of ways, both for good and bad. Profound ignorance about insect life permeates most of human society, even among the highly educated. Insects and their relatives, in fact, are little credited for their beneficial environmental services but overblamed for their destructive activities. Despite lack of general human interest in insects E.O. Wilson wrote that they are "the little things that run the world".

Insects and their relatives live on all continents and occupy microhabitats from deep in the soil and underground aquifers to the top of trees and mountains, among the feathers of penguins on Antartica, and even deep into caves and in our eyebrows. Many lineages have evolved adaptations for living on and under ice fields, others at the margins of hot springs, and still others on the open ocean. land arthropods, by virtue of their pervasiveness, are incredibly important to the balance of life within ecosystems. Insects and their relatives eat virtually everything and compete even for the rocks under which they hide, mate, and rear their young. For one thing, most broadleaf trees and shrubs would not be pollinated, and there would be no fruits and seeds. For another, instead of penetrating dead matter, decomposers such as bacteria and fungi would live only on the surface, taking years or perhaps millennia to break the substrate down into recyclable nutrients for plants, and thus soils would be much less fertile. Many fish and birds, and even some mammals, would have no food and would cease to exist.

Among the insects, the bettles are the most speciose, the most pervasive, and the most widespread across the face of the globe. During dry seasons in tropical forests, they are the third most numerous individuals, after ants and termites, making up 12 per cent of the total insect community after ants and termites.

Beetles are found everywhere on our planet except in the deep sea. However, they do occur commonly in the sea's intertidal zone and estuarine salt flats. Beetles even occurred on Antarctica not long ago. Most families of beetles, about 140 of them, are worldwide in distribution, and their species provide equivalent ecological services wherever they occur. The"play" is generally the same everywhere, only the "actors" themselves change from place to place.

The histroy shows that two major faunal changes took place, the first in the mid-Jruassic Period when primitive lineages of beetles lost their dominance, and the second in the mid-Cretaceous Period, at which time modern forms acquired dominance over all other terrestrial arthropods. In terms of species and number of guilds, they still have this dominance in nearly every biotope. By any broad measure, beetles are the most successful lineage of complex organisms ever to have evolved.

The described species of beetles, about 400,000+comprise about 25 per cent, of all described species on Earth. This dominance of beetle taxa in the literature has resulted in Coleoptera being perceived as Earth's most speciose taxon. Thus, it has garnered further taxonomic attention from young taxonomists which in turn has resulted in more species of beetles being described than in other groups. Beetles are relatively easty to collect, prepare, and describe, significantly adding to their popularity. Such unevenness in taxonomic efforts may or may not give us a false picture of true relative insect diversities. Nevertheless, the dominance of beetles has been used to arrive at an estimate of 30 million insects overall, and even to designate the group most endeared to God. While this dominance may be arguable either scientifically or philosophically, it is certainly interesting.

The reasons probably lie in the overwhelming numbers of species, individuals, and the ever-plodding course of traditional taxonomy. Potential users of data on beetles simply have to wait too long to get names. Taxonomists have to wait too long to receive money to visit museums in which name-bearing type specimens are held; monographers take too long to produce documents with which users might identify their specimens by themselves and specialists are reluctant to take on a large identification load for other scientists, such as ecologists and conservation biologists.

Neotropical beetles are second only to ants and flies in numbers of free-ranging individuals of arthropods in the canopies and subcanopies of neotropical trees; Psocoptera are a distant third. However, per species, beetles are not abundant. Beetles participate in virtually all aspects of ecosystem processes; they are predators, herbivores, folivores, detritivores, scavengers, fungivores, wood-eaters, and grazers, and they tunnel, mine, and chew nearly every substrate. Some are ectoparasites, others are nest parasites, some even live in the fur of vertebrates. Still others are subsocial, with adults participating in the raising of young. A global perspective based on beetles could provide a much more fine-grained view of biodiversity than the coarse-grained one we get from less speciose groups such as jaguars, birds and monkeys, which heretofore have garnered most of the attention.

Collection of Data

Because the interface between insects and their environment is at a small resolution, the information they provide may well be critical for ecological restoration. Management will depend on what we really know rather than what we surmise. Conservation cannot now deal with insect information, but will be compelled to do so in the not-too-distant future. We will need a system for data gathering that is just now becoming available.

One goal of Janzen's ATBI is to inventory all the taxa within a given area. A biotic inventory includes finding the area's species, classifying them, making voucher collections, and storing these data in a way that they are easily retrievable. Additional information about the species, either gathered during the process of inventorying or aded later from literature or follow-up studies, can be piled on top of the four basic elements in a growing database.

The first ATBI are is destined to be at the Guanacaste Conservation Area (GCA), Costa Rica, a site with dry forest in lower elevations ascending through cloud forest and containing intermediaries between these levels. Based on the experience in nearby Panama with a similar range of habitats, it is estimated that GCA should have about 50,000 species of beetles. Since this estimate can be only a first approximation, it is used for purposes of designing a sampling regime for the project; budget and time must be considered to be modifiable as the project narrows to better estimates.

Each species of tree, shrub, and herb/grass may have its own host-specific species of beetles. Riparian strands in various watersheds will have different types of substrates, water quality, vegetation, etc. contributing to their distinctive biodiversities. In addition, the GCA is distinctly seasonal, hence both dry and wet seasons need to be sampled for each microhabitat.

The sampling regime must consider the above in its attempt to record as many species as possible in the shortest amount of time. The guiding priciples are as follows:

Phase 1 : mass co-occurrence sampling; rapid processing with bulk cold storage; identification process using matching specimens; interim naming with alphanumerics; accumulation of data using linked spread sheets, including curves showing sampling progress and character filing with the Quick Taxonomic Assessment System (QTES).

Phase 2 : send target taxa and QTES data into the taxasphere for formal species names;

Phase 3 : replace EXCEL 4.0 spreadsheet and QTES interim names with formal one, transfer these data to the database at Instituto Nacional de Biodiversidad (INBio).

Phase 4 : generate illustrations and three-dimensional laser images; produce documents.

An Agenda for Sampling Beetles in an ATBI

Sampling

The following criteria must be met for acquiring samples of beetles that can provide a reasonable inventory and serve both immediate and future needs of research, as well as detemine to an order of magnitude the species present in the target area for us in subsequent samling projects:

1. Sampling assumes the use of a fogger and three per cent Resmethrin for all microhabitats from one in above ground through the canopy rim; this should capture one to seven species per inch of foiage and much more in compacted microhabitats such as suspended dry leaves, vine tangles, and complex canopies. Leaf, litter and soil layers are sampled by photoeclectros and sifting/Toplgren extractor techniques. Berlese banks can substitute for Tolgren, if electricity is available. The stratum of herbs and grasses is sweep-sampled by sweep-netting.

Methods of trapping by attraction and even passive traps that catch flying insects produce catch without biocontext, i.e. specimens that are not tied to any microhabitat, substrate, host plant, etc. Much time and effort goes into preparing, identifying, and storing such bulk lots, yet the quality of data is at the lowest level. These methods of collecting are simply not worth the efforts, unless one is interested solely in recording the presence of species in the general area or in building collections. However, these techniques can be used as a test of the methods that incorporate biocontext to determine if microhabitats exist that are not being sampled with the other techniques.

(2) A standard set of field data includes precise locality; type of forest; type of microhabitat and its volume or surface area; inforamtion on species of plants; date; and collector(s). Lot numbers are assigned to each individual fogging collection, sifting series, photoeclector sample, sweep series, etc. Thus, all specimens taken from the same microhabitat or plant or trap at the same time get the same lot number so that the set of species, including non-beetles, can be reassembled at a later date if desired. Non-beetle specimens will be directed to anther Taxonomic Working Group (TWIG), along with appropriate sets of data.

(3) Specimens of beetles are preserved in 70 per cent alcohol in the field. Alcohol must be changed the same day at the lab and subsequently each time the specimens undergo processing. Egg carton inserts are placed in funnels or on suspended sheets to catch dry specimens. After these are selected from the carton surface, the remaining specimens are dumped into the alcohol bottle.

(4) Sampling design involves taking replicate microhabitat samples in sets of 10 throughout each type of forest and open area. During preparation and data entry of the 10 samples, species accumulation curves and Chao's estimator track the progress of the inventory. A complete inventory for smaller families will require fewer replicates, but the leaf-beetles and weevils will require more than 10 replicates, based on data from over 5,000 species acquired at Pakitza, Peru, from 1988-1992. A decision needs to be made at the outset as to when to stop, because it will not be "humanly" possible to get the "last" species on the list in the larger families. however, 100 per cent likely will be reached in smaller families.

Preparation

All tropical forest samples are replete with beetles. The object of preparation should be to make the species and their whereabouts and abundance known in the shortest amount of time possible. Traditional preparation of all collected specimens, therefore, is not feasible. The following method leads to one prepared specimen per species per samples, with cold storage of the bulk lots that can easily be accessed lter by taxonomists and other workers who need series.

(1) Each sample lot is sorted to families using a 6.2 cm white ceramic dish with 70 per cent alcohol. Families are gathered in small plastic lids set inside the bottom of a petri dish, the top of which is ringed with vaseline to create a seal when the specimens are sitting unworked. Parataxonomists and beginning graduate studens can be trained to sort at this level quickly.

(2) Each family then is sorted in sequence to species, with one good specimen selected for pinning/pointing. The specimen is placed on damp filter paper inside another petri dish. On the filter paper, numbers one to 20 are written and the chosen specimen is placed next to a number according to its abundance in the sample. If over 20 specimens of a species

are in a sample, a small label is written with the number, and this lable is affixed to the pin to be removed later in the process. The rest of the counted but unprepared specimens are returned to the lot vial of 70 per cent alcohol, bar-coded, and sent to cold storage. Sorting to species across the Coleoptera can be done only by a highly trained taxonomist with long experience, and this person becomes the key to the entire project. Preparation and storage procedurs can be handled by a technically trained person.

(3) Each specimen from the filter paper is pinned or pointed with Elmer's glue to pins or pre-prepared points in the traditional manner and placed in a unit, tray with strips of numbers sequenced from 1 to 20. Each specimen is aligned next to one of these numbers according to its abundance in the sample.

(4) Preprinted lables are attached to each specimen as it is "identified-by-matching" using the synoptic collection. The name of the species is an alphanumeric in the form of "family code + number" and lot number. All families of Coleoptera have a standard code of four letters. The margin of the label is colour-coded with pencil for instant recognition of microhabitat, although the lot number references this too. Once a family is represented by more than 100 species, "identifying-by-matching" becomes less and less efficient. For very large families, such as weevils, staphylinids, and chrysomelids, use of QTES is recommended.

Interim identification

Each prepared specimen from a sample is compared with its corresponding family-level synoptic collection. Smaller families are easy to keep in one or a few units. Larger families may be subdivided by subfamily so that the amount of matching necessary for recognizing the specimen's status is kept to a minimum. As a specimen is identified or is kept to a minimum. As a specimen is identified or recognized as a species new to the synoptic collection, it is placed either in an interim unit tray awaiting entry of the data before going to the duplicate collection or added to the synoptic collection, from where its data are entered. All species of a family tht are sorted from the sample are labelled, then the data are entered in EXCEL before preparing the next family.

Data storage

EXCEL linked spreadsheet templates for families of beetles contain about 13 kilobytes of forms that are based on microhabitats. Entry of data from a sample involves simply number of specimens per species per lot. The programme automatically computes all basic inforamtion and accumulates the data on summary sheets for easy viewing. The programme is exceedingly user-friendly.

Building Collections

The resulting synoptic-unit trays of families of beetles are ready for special scrutiny at any time during the process if the specialists is on-site to make formal identifications. The duplicate collection built from second through *n* occurrences of a species across samples can be sent through the taxasphere regularly and results can be fed back into the EXCEL data system, making it easy to move the information to the INBio standard data files. As additional microhabitat replicates aresampled and specimens processed. Those species represented by uniques in the synoptic collection will be duplicated and can then be sent through the taxasphere.

Serious taxonomists who must do a revisionary study immediately can read the database to find lots with series and arrange to extract those from cold storage themselves. Common species that are found in many or most lots will have that many more prepared specimens ready for study in the duplicate collection.

24

Plant Species Richness and Global Warming

Fossils and sedimentary rocks form a record of changes in the Earth's climate and biotic responses over geological time. This record can be used in two main ways to develop a more complete understanding of global climate and of the long-term effects of climatic change on ecological systems. First, comparing climatic conditions indicated by fossils and sedimentary rocks with computer stimulations of global climate for the same period is the only way of testing the ability of climate models to simulate conditions other than those that exist today.

With the exception of the Holocene deglaciation, the Paleocene-Ecoene transition is probably the best documented example of a geologically rapid warming of global climate, although just how rapidly it occurred is still a matter of investigation.

Typically, events that are less than 10,000 years apart appear synchronous, and records from different parts of te world cannot be correlated with greater than 100,000 year resolution. Still, the Paleocene-Eocene warming event is a close match for extreme predictions of human-induced global warming in terms of the absolute magnitude of warming and final climatic conditions.

Reconstructing Climates of the Past

Climatic conditions for the distant past can be inferred in a variety of ways, but the two techniques most widely applied to cintinental paleoclimates are based on fossil plants. One method, called the nearest living relative (NLR) method, relies on the assumption that fossil species grew in climates similar to those preferred by their extant relatives. the validity of the NLR method depends on how closely related the fossil and extant species are, how much evolutionary change there has been in climatic preferences of the lineage, and the degree to which the range of the extant species is controlled by climate. Living taxa that have relictual distributions or few species are especially unreliable paleoclimatic indicators because they are far from occupying the full range of climatic conditions they can tolerate physiologically, NLR inferences about early Cenozoic climate are more reliable if they are based on diverse, nonrelictual taxa that have strong climatic limitations on their present-day distribution.

Palms are a good example of a speciose group with clear climatic limits on their distribution. Palms do not naturally occur in regions where means of cold months are less than about 6°C, or where forsts persist for more than two days. Experimental work on a wide variety of living species of palms shows that sensitivity to cold temperatures is typical of the whole family and results from absent or poorly developed physiological mechanisms of frost hardening. The combination of geographic and physiological information on palms makes a convincing case that extinct species in the family are unlikely to have had significantly more frost tolerance than their living relatives, so palms can be used with confidence to fix lower limits on paleotemperature estimates for early Cenozoic floras.

The second major method for inferring paleoclimate from fossil plants, leaf physiognomy, relies on correlations between the shape and size of leaves and climate that are observed in living floras. Geneally, floras growing under warmer climates will have a higher proportion of species that have leaves with entire margins. The correlation between mean annual temperature (MAT) and the proportion of species with entire margins was first observed over 80 years ago, but has been documented and quantified most extensively by Wolfe causing floras from East Asia. However, the high correlation exhibited in Wolfe's data set is due in part to the exclusion of living floras from seasonally dry climates. Although the relationship between te percentage of species with entire margins in a local flora and MAT has been used widely to obtain numerical estimates of MAT for Cenozoic fossil floras, it may give misleading results for fossil floras that grew under seasonally dry conditions.

Many characteristics of leaf shape and size are correlated with temperature and rainfall conditions and can be used to infer paleoclimatic conditions. The leaf characteristics were then compared with parameters of temperature and rainfall for the sites using ordination analysis. Relationships between physiognomic variables of leaves and climatic variables can then be used to infer climatic conditions for fosil leaf assemblages based on their physiognomy. Wolfe's Central and North American data set also has been analysed using multiple regression techniques. Mean annual temperature estimates based on the multiple regression approach have errors of 24°C; precipitation estimates are generally no more than ballpark figures.

The Equable Climate Paradox

Global climate through most of Earth's history has been much warmer than at present. The early Eocene has long been recognized as the warmest part of the Cenozoic and one of the warmest periods of global climate in the last 100 million years. During the early Eocene, polar icecaps were absent in both hemisphere and midlatitude continental interiors has much warmer winters than they do now.

The causes for such globally warm climate are not well understood. Higher sea levels, lack of polar icecaps, and more dispersed land masses have all been thought to play a role in maintaining a warmer world. In recent years, general circulation models (GCMs) have been applied to paleoclimates. To produce "predictions" for temperature and precipitation for a time in the past, GCMs use starting conditions such as ancient coastline positions, paleotopography, continental positions, sea-surface temperature gradients, and global ice distribution. Weather patterns are generated using the same equations that describe the dynamics of the atmosphere today, then the weather patterns are averaged to yield a paleoclimate. the early Eocene is arguably the most interesting case of conflict between proxy data and model output, because it is close enough to the present that the proxy data are extensive, and uniformitarian assumptions about the climatic tolerances of animal and plant lineages are probably valid. Additionally, there is little evidence for very high CO_2 levels in the early Eocene atmosphere, so this factor cannot be invoked in an unconstrained fashion to explain differences between proxy data and model output.

The sharpest discrepancy between model output and proxy data occurs in mid- to high latitude continental interiors because to low thermal capcity of land and the isolation of continental interiors from the moderating influence of the oceans result in model predictions of high seasonal variation in temperature in these areas hundreds of early Eocene plant and animal fossil localities scattered across much of the continental interior of North America. The distribution of frost-intolerant forms, as well as weak development of seasonal growth rings in some fossil wood, high diversity of small arboreal frugivores and insectivores requiring year-round food resources, and the physiognomy of fossil dicot foliage, all yield similar conclusions. In the Early Eocene, the interior of North America as far as 50°N experienced no significant winter freezing.

Mean temperatures in cold months were certainly above freezing, and probably higher than 50°C in most areas during summer. Mean annual temperature was 15-20°C. There are fewer proxy data available for the interiors of other continents, but southern Australia, which was 50-60°C in the early Eocene, has fossil faunas and floras indicating even warmer winter temperatures than similarly high latitudes in the interior of North America. The proxy data consistently show equable climates with low seasonality of temperature and warm winters. GCM simulation results consistently yield highly seasonal climates in continental interiors, with freezing temperatures far south of fossil sites containing frost intolerant froms. The are three basic kinds of explanation for warmth in the mid- to high latitudes during the early Eocene: (1) more heat was transported from equatorial to polar regions by ocean currents or winds, (2) more solar radiation was absorbed at high latitudes because albedo was lower, and (3) more heat was retained at high latitudes, and everywhere else, because of higher concentrations of "greenhouse" gases such as water vapour, methane, CO_2. The persistent discrepancy between GCM output and proxy data arises because mechanisms in the first two categories–such as increased heat transport by ocean currents or no polar ice–are not strong enough to explain the levels of warmth indicated by proxy data, especially in continental interiors. The last mechanism, a CO_2, methane, or water-vapour greenhouse, implies that tropical as well as high-latitude regions would have been substantially warmer than at present. The best proxy data for the Eocene tropics, however, indicate little difference from present temperatures. Furthermore, neither proxy data nor geochemical modelling are consistent with Eocene CO_2 levels more than about 900 parts per million (ppm), and it is possible than Eocene CO_2 levels were no different from those at present.

Thermal Paleocene Event

The global climate warmed considerably from the late Paleocene to the early Eocene, roughly 57-52 million years age. This warming trend has been quantified in the marine realm by measuring changes in oxygen isotope ratios in the tests of benthic and planktonic foraminifera recovered from deep sea cores and in North America by phyiognomic and floristic analyses of fossil plant assemblages. In the mid-latitudes of interior North America, MAT increased from approximately 10°C to nearly 20°C between the late Paleocene and the mid- to early Eocene, a period of about three million years. Although this rate of change averages to only a few thousandths of a degree per millennium over the whole period, there is growing evidence that the rate of climatic warming was not constant.

In cores that recover laminated ocean bottom sediments, it is possible to resolve time in tens of thousands of years, even in the early Cenozoic. Recent detailed stratigraphic studies of the Paleocene/Eocene boundary interval recovered in cores from the southern Pacific and southern Indian Oceans have shown an excursion in isotope values that occurred approximately 100,000 years before the Paleocene/Eocene boundary. The sudden increase in the light isotope of oxygen took place over less than 10,000 years and is thought to represent an interval when the temperature of bottom waters and surface waters in mid- to high latitude oceans increased by 5-8°C. This geologically short period of time, marked by isotopic shifts, is referred to here as the "Terminal Paleocene Event".

Oxygen isotope analyses of planktonic foraminifera tests indicate that surface waters in the equatorial Pacific mainained a temperature of about 200°C during the Terminal Paleocene Event, but analyses of the tests of tropical benthic species imply warming similar to that seen in the benthic and planktonic foraminifera of higher latitudes. The isotopic shift coincides with the most several extinction of benthic foraminifera from the deep ocean during the Cenozoic; the extinction is believed to have caused by rapid warming of deep ocean waters and a concomitant decrease their dissolved oxygen levels. Flowing in the negative isotopic excursion there was a rapid rebound in the positive direction, followed by a renewed decrease in 6180 value that continued to a Cenozoic minimum in the mid-early Eocene about 52 million years ago. This minimum indicates the time of maximum global warming during the Cenozoic.

The terminal Paleocene oxygen isotopic excursion is paralleled by a similar excursion in carbon isotope values. The shift from ^{13}C to ^{12}C is probably related to the source and rate of delivery of organic carbon the deep ocean, and possibly to decreased rates of oxidation of organic material in bottom waters. The rapid isotopic and biotic changes during the Terminal Paleocene Event may all relate to a change in ocean circulation in which the source for bottom waters shifted from cool high latitudes to warm low latitudes.

In this hypothesis, global warming during the later Paleocene increased the warmth, and therefore decreased the density, of high-latitude surface waters. At some point, warm, oxygen-poor saline waters generated by evaporation in the low latitudes exceeded the density of cooler, more oxygenated waters from the polar regions, and deep ocean cirulation began to move in a poleward direction. This is the opposite of present-day conditions in which dense bottom water is created in cold, moderately saline high-latitude oceans. The hypothesized "reversed" bottom-water flow at the end of the Paleocene transported heat to mid- and high latitudes very effectively, resulting in a rapid increase of bottom water temperatures, but also a sudden warming of mid- of high latitude surface waters and continental surfaces as the warm bottom water upwelled. Although the precise triggering mechanism that led to the sudden reversal of bottom water ciruclation is not understood, the effects of the Terminal Paleocene Event appear to have been global, because rapid changes are seen at this time in sediments deposited in continental and shallow marine environments as well as in the deep sea.

One effect of increased poleward heat transport by deep ocean currents would have been a decrease in latitudinal temperature gradients. There is direct evidence for decreased temperature gradients both in ocean surface water and on the continents. Decreased surface temperature gradients would be expected to lead to a reduced intensity of zonal atmospheric circulation, which is largely driven by equator-to-pole temperature constrasts. Reduced surface wind velocities were responsible for the sharp decline in the size of wind-blown dust grains in Paleocene/Eocene boundary sediments of the central Pacific; sluggish winds are not capable of carrying larger dust particles far out to sea. A secondary effect of reduced wind velocities may have been a decrease in coastal upwelling zones, which are powered by wind shear.

Greater warmth at mid- to high latitudes may also have resulted in increased precipitation and chemical weathering on land surfaces. Evidence for increased chemical weathering is seen in the sudden, widespread increase during the Terminal Paleocene Event of kaolinitic clays that are characteristic of leached soils formed under "tropical" weathering regimes.

Paleocene-Eocene Transition on Land

The carbon isotope excursion associated with the Terminal Paleocene Event is a geochemical marker that can be detected in continental carbon sources such as hydroxyapatite preserved in tooth enamel of fossil mammals and calcium carbonate preserved in fossil soil nodules. Because the largest short-term reservoir in the global carbon budget is the ocean, shifts in the $^{13}C/^{12}C$ ratio of the oceans, such as those observed during the Terminal Paleocene Event, should force similar isotopic shifts in the carbon reservoirs represented by the atmosphere and continental ecosystems. Investigations of Palecene/Eocene boundary sections in the Bighorn Basin of northern Wyoming have detected a carbon isotope excursion that corresponds to the Terminal Paleocene Event in direction, magnitude, and timing, thus establishing a datum that permits correlation of terrestrial and marine events.

The terrestrial carbon isotope excursion coincides with the Clarkforkian Wasatchian boundary, a major and rapid change in the composition of mammalian faunas which has long been thought to be approximately correlated with the Paleocene/Eocene boundary. It is at the beginning of the Wasatchian that North American faunas first included representatives of the perisso-clactyle, artiodactyla, and euprimates. Similar faunal changes are known to have taken place at about the same time in Europe and Asia and in Europe the faunal change is also associated with the carbon isotope excursion. The nearly simultaneous appearance of these forms on all three northern continents indicates that high latitude land corridors in Greenland and Beringia were open to mammalian migrants, but the place of origin for these groups is still not known. Although rates of mammalian taxonomic turnover were high near the Paleocene/Eocene boundary, there is no evidence for a substantial decline in the number of mammalian genera. In spite of maintaining high numbers of genera, there is evidence that, at the level of species, the latest Paleocene mammalian faunas from North America were characterised by high dominance of a few species, possibly an indication of stressful conditions.

The effect of the Terminal Paleocene Event on terrestrial floras is less well understood, although it has been known for some time that floral similarity between North America and Europe reached a maximum in the early Eocene. Two studies have shown a substantial reduction in the number of species of plants near the end of the Paleocene in North America, approximately coincident with the mammalian faunal exchange and the Terminal Paleocene Event. This rapid decrease in the number of species of plants at the end of the Paleocene was geologically short-lived; by the mid- to early Eocene, as global climate reached its maximum Cenozoic warmth, floral richness has recovered to levels considerably higher than those preceding the Terminal Paleocene Event. Stratigraphic resolution of the terminal Paleocene drop in flora richness is not yet sufficient to determine whether it coincides with increasing temperatures at the start of the Terminal Paleocene Event, decreasing temperatures at the end of the Terminal Paleocene Event, or is spread throughout the entire excursion.

Scenario for Paleocene/Eocene Plant Extinctions

The stratigraphic resolution and geographic scope of our knowledge of faunal and floral change across this time period is increasing rapidly, but it is not yet sufficient to privide a clear answer. A key to understanding the cause of the loss of species among plants is the difference between the curves of species richness for mammals and plants. In the present, the diversity of both groups of organisms generally increases towards the equator, but the migrational abilities are obviously quite different. The difference in rates of migration provides and explanation for greater loss of species among plants than mammals at the Paleocene/Eocene boundary.

The Paleocene flora of the northern continents has long been noted for its relatively low diversity and high homogeneity, with many lineages of deciduous plants distributed throughout the middle and high latitudes of North America, Europe, and Asia. Warming at middle and high latitudes during the Terminal Paleocene Event and the early Eocene is likely to have caused local extinction at the southern ends of the ranges of cool-adapted lineages with broad circumpolar distributions. Warming at middle and high latitudes would also have permitted the poleward migration of subtropical and tropical taxa in both mammals and plants. For floras in the northern Rocky Mountains and the Gulf Coastal Plain of North America, there was a significant delay, as much as several hundred thousand years, between the loss of species at the Terminal Paleocene Event and the arrival of subtropical elements in the early Eocene. It was this lag that created the rapid drop and rapid recovery of plant richness at middle latitudes. What in mammals shows up as a rapid intercontinental mixing of faunas manifests iteself among plants as a extinction event at mid- to high latitudes, followed perhaps 100,000 years later by a major immigration of species from the south.

The migrational lag explanation seems at first glance unlikely because plant migration can be very rapid geologically; the movement of forest taxa in the wake of the retreating Holocene glaciers totally modified the vegetation of vast areas of North America in 5,000-10,000 years. The Holocene example is not wholly applicable to the Eocene situation, however, because of the different continental configurations in the Eocene, and because Holocene plants were migrating into a recently denuded landscape that probably was not occupied by established forest vegetation.

Tropical taxa may have had limited access to Europe along the northern shore of the Tethys, but the straits of Turgai divided Euorpe from eastern Asia at about the present-day position of the Caspian Sea, so the appearance of subtropical forms in Europe might also have been delayed. By contrast, Asia had a broad land connection with tropical continental areas during the Eocene, so that the ranges of terrestrial plants could have shifted northwards without confronting an oceanic barrier.

The paleogeographic differences between Asia and North America can be used to help resolve whether a migrational lag was important in creating the pattern of change in the number of species across the interval of the Paleocene/Eocene boundary. Because of eastern Asia's connection to tropical land masses, the migration of lineages from warm climates into middle latitudes should have been much easier than in North America or Europe.

The cause of the loss of species during the Paleocene-Eocene transition can also be evaluated by comparing the pattern of floral immigration within North America. If the delayed immigration of subtropical forms to the northern Rocky Mountain region was largely a consequence of the water gap between North America and South America rather than slow migration across the North American continent, then forms from warm climates should appear at essentially the same time in the southern and northern part of the continent.

Two possible general patterns emerge from a consideration of events at the Paleocene/Eocene boundary. First, species richness of plants is much more dramatically affected by the climatic change than is taxonomic richness of mammals. This should not be surpirsing in light of the sensitivity of plants to climate, but it has been proposed that plants are generally more "extinction resistant" than other groups of organisms. This conclusion was based on the observation that most major lineages of plants survived mass extinctions at the Permian/Triassic and Cretaceous/Tertiary boundaries, wheras higher taxonomic groups of animals were decimated. Analysis of the Paleocene/Eocene extinctions of plants supports the idea that the cause of an extinction may be more important in understanding its effects than its "size," as measured by the percentage of taxa that are lost.

The Paleocene/Eocene extinction of plants also illustrates the principle that the rate of climatic change may be as or more important than its direction in causing extinction of plants. Rapid climatic shifts initially may decrease species richness of plants through accelerated local extinction regardless of whether the climate is warming or cooling.

25
Diversity in Community

When a number of species come together to form communities each fits in a different niche and plays a different role in the internal dynamics of the community. Dominance is the relative control exerted by organisms over the species composition of the community. Species exerting this important control are called dominants. Plants are more frequently dominant in terrestrial communities than are animals. In aquatic communities, animals are relatively more important in this role. Although dominance is often not developed. Dominance is most commonly expressed in the reaction of an organism on its habitat.

Dominants shoulder the full impact of teh climate or the environment but modify this effect for other organisms within the community by tempering light, moisture, space, and other conditions. Only those organisms that find these modified physical conditions tolerable can exists within the community. Furthermore, dominants are ordinarily the most prominent species in the community, make up its greatest mass of living material, and serve as the major source of food, substrate, and shelter for the animals that are present. In a forest community, trees are dominant. They decrease light intensity increase the relative humidity intercept precipitation, monopolize most of the moisture and nutrients in the soil, decrease wind velocity, and furnish shelter and food for animals.

Grasses play a similar, though less conspicuous, role in prairie communities; sedges, rushes, and cattails in marsh communities, sagebrush in the arid habitat of the Great basin, mussels and barnacles on a rocky seashore; and so forth.

Sometimes dominance is demonstrated in coactions, direct effects organisms on each other. In some fresh-water ponds, carp and suckers may consume much of the submerged vegetation. This coaction thus prevents the plants constituents from assuming their usual role in the community, and by so much prevents the occurrence of animal species that depend directly upon the plants. These fish also react upon the habitat by stirring up the bottom, from which they derive organic matter, thereby greatly increasing the turbidity of the water. Penetration of light into the water is reduced, greatly handicapping sunfish, bass, and other species which locate food visually.

In primeval days, bison our western great plains fed on the luxurient taller grasses more extensively than on the short grasses, with the consequence that, over extensive areas, short grass species replaced tall grasses almost entirely. Thus bison were coactant with and dominant over the composition and character of the community (Larson 1940).

Although animals are more common coactors than plants, plant pathogens may occasionally exert dominance in this way. As a notable example, chestnut tree blight (a fungus) virtually eliminated the chestnut tree from the deciduous forest of eastern North America during the first few decades of the twentieth century. This fungus infects the cambium, forms pustules under the bark, and causes the bark to fall off and the leaves to wilt. The consequent opening up of the canopy with the death of the trees has allowed the extensive invasion of new species of shrubs, herbs, and animal inhabitants.

Trees are the dominants in a forest community, but species in the lower stratum of shrubs modify the habitat still further and even the herbs exert some control over the physical conditions on the surface of the ground. A subdominant species must tolerate the conditions established for it by the dominants; but it in turn is a modifier of the community composition in a secondary manner.

Influence

By influence is meant effect upon the abundance, health, and activities of other organisms but not to the extent of directly excluding species from the community. Influence is conspicuously expressed through coactions, but it may be effected through reactions as well. Insects many partially or wholly defoliate a tree; a pack of wolves may diminish a population of deer over winter; squirrels may bury acorns and nuts and thereby aid germination of them; parasitic or poisonous plants may lower the vigour or destroy the life of some other plants or animals; animals may burrow into the soil and there by increase percolation of water and air, a benefit to plants; and all organisms, upon death, add organic matter to the habitat. These and other actions influence become extreme they do not determine whether or other species will occur in the community. Influence, then is of essentially the same nature as dominance but is less vigorous in the modifying role that it plays.

Evaluating and Classifying Animals Ecologically

One the most important, yet difficult, problems in ecology is determination of importance values of organisms to the structure of importance values of organisms to the structure of communities and of their roles in community dynamics. Various statistical criteria have been used, but we will be concerned only with some of the more general evaluations.

A basis for classifying species is exclusive when it occurs only in a single area, habitat, or community; characteristic (selective, preferential) when it is abundant in one area or community but also occurs in small numbers elsewhere; or ubiquitous (indifferent) when it is found more or less evenly distributed in a wide variety of communities. The terms given in parentheses are synonyms used by plant ecologists. Exclusive species are often rare and of little importance in the dynamics of a community, but when they are conspicuous they often make useful indicator species for identifying and recognizing community units.

The recognition of characteristic species presents special difficulties since one must decide how much more abundant a species needs to be in one community to be sure a definite preference over another is indicated.

The time and duration of occurrence of species in a community affect the amount of influence it exerts. Generally, the longer the yearly period during which a species is active, the more important its role becomes. Species may be classified on a temporal basis into perennials, those which are active in a community throughout the year, year after year seasonals, which are present or active only part of the year; and cyclics, which may be very important some years but of negligible importance other years, as evidenced by their wide fluctuations in numbers. Even though present, a species is usually considered inactive when it is hibernating or dormant or when it is represented only by eggs, spores, or encysted stages of its life cycle. The effect produced on the community by individuals and species may be modified by the way they form secondary groupings within the community. These minor aggregations of plants and animals are called societies and are of various sorts. Layer societies occupy different strata, such as the subterranean, ground, herbs, shrub, and tree societies in a forest; local societies are usually parts of layer societies but are more confined in area, as groups of animals occupying an ant hill, a rotting log or stump, or a restricted but distinctive area of ground; and seasonal societies include all the organisms at particular times of the year.

Other factors that affect the influences of a species in the community are the size of individuals, their metabolism, food habits, and general behaviour. A moose consumes more food than a mouse, and a warm-blooded mouse more than a cold-blooded salamander of the same size. A carnivore at the top of several food-chains affects the lives of more different species in the community web of life than a herbivore feeding on plants. Borrowing rodents react on the habitat more than do most birds. One factor may cancel another. An individual of a perennial species of carnivorous mammal certainly eats more than an individual cold-blooded herbivorous insect of small size and active only during the warm season. Yet there may be 1000 insects to one mammal, so that in the aggregate single species of inset may actually produce more disturbance than a single species of mammal. The difficulty of evaluating the relative effects of species is partly alleviated by calculating their respectiveth biomasses and amounts of energy used.

The biomass of a species is the average weight or volume of individual, multiplied by the total number of individuals present. The computation of the biomass of each species thus corrects for differences between species in size and number of individuals. Because of differences between species in body moisture and amount of inert substances such as endoskeleton, chitin, shell, and the like, biomass is expressed with greater accuracy in terms of dry weight than wet weight or volume, and is even more accurate if given in terms of carbon or nitrogen content, or calories.

Attempts have recently been made to compute more significant biomasses or bird populations, using physiological constants. A biomass composed of few but large individuals has a lower metabolic activity than an equal biomass composed of a large number of small individuals.

Productivity

A characteristic of communities that has become of considerable importance in modern ecological research is productivity. The number of individuals or biomass in a community at any one time is the standing crop at the beginning of the year or reproductive season the standing crop is usually small, but as reproduction and growth take place there is an increaein the amount of organic matter making up the biomass of the community. The production of organic matter per unit of time and area is productivity. Productivity is thus a rate and is commonly indicated on a yearly basis, but it is also possible to measure monthly, weekly, or daily production, small standing crops may have a high productivity and large standing crops a low productivity, hence biomass alone does not give a measure of productivity. The largest standing crop which a habitat can support without deterioration, or the maximum number or biomass of animals that can survive the least favourable yet tolerable environmental conditions during a stated period of time, is the carrying capacity. Carrying capacity is determined not just by the amount of food available, but also by shelter, social tolerance, and other factors. A variety of methods are being used to measure productivity of different kinds of organisms and of different habitats. It is desirable to indicate productivity as accurately as possible in descriptions or analyses of community dynamics.

Succession

Communities are in a more or less continual process of change. These changes result in part from the reactions and coactions of the organisms themselves and in part from such external forces as changing physiography, changing climate, and organic evolution. The habitat is usually affected as well as the community, and as the habitat changes, new species invade it and become established, and old species disappear. These changes are especially noticeable in dominant species, since these species exert a controlling role over the composition and structure of the community as a whole. The replacement of one community or ecosystem by another is succession, and succession continues until a climax or final stage is reached.

Succession is a process. The series of steps or communities comprising a successional sequence leading to the climax is the sere. Seres are sometimes classfied according to the predominant force that is bringing them about. These forces are biotic, climatic physiographic, and geologic and their resultant seres are commonly called bioseres, cliseres, eoseres, and geoseres.

Succession that begins on a sterile area where conditions of existence are not at first favourable-as, for example a newly exposed sand dune or a recent lava flow-is termed primary succession. The term secondary succession refers to community development on sites previously occupied by well developed communities or succession on sites where nutrients and conditions of existence are already favourable such as abandoned croplands, plowed grasslands, cutover forests or new ponds. As would be expected rate of change is much more rapid and time required for completion of the sere is much shorter in secondry succession. Finally it is important to distinguish between what may be called autotrophic succession and heterotrophic sucession. The former is widespread type in nature that begins in a predominantly inorganic environment and is characterized by early and continued dominance by autotrophic organisms. Heterotrophic succession characterized by early dominance by heterotrophs occurs in the special cae where the environment is predominantly organic as for example in a stream heavily polluted with sewage or, on a smaller scale, in a fallen log. Energy is maximum at the beginning and declines as succession occurs unless additional organic matter is imported or until an autotrophic type but is usually maintained or increased during succession.

Biotic Succession

Biotic succession is brought about by forces inherent within the community and in the activities of the plants and animals themselves. The most important of these activities are the organismal reactions and coactions that produce modifications in the habitat and interrelations between species. Important reactions involve filling in of ponds with plant and animal remains, the addition of organic nutrients to sterile soil, and the reduction in light intensity by increasing density of plant growth. With progressive improvement of the soil and changing light and moisture conditions, a series of new dominants come into the area. When invasion of new species occurs, intense competition develops; if the invaders are successful, the old species disappear as a new community replaces the old one.

Succession is considerably influenced by the kinds of propagules available in the vicinity. Seeds, spores, and the like are dispersed more or less readily, depending on form. Some kinds of animals roam more widely or spread more readily into new areas than do other kinds. The composition of the community that develops and the rapidity with which it becomes established depend, in large part on the rates at which different species invade.

If a forest is cut, and the land cleared to the soil, a succession of plants will invade and grow on the exposed surfce. The first plants which invade are typically those capable of seeding in quickly on disturbed land or capable or generating from viable roots left within the soil. This depends not only upon the former vegetation, but also upon seeds from surrounding plant communities, as well as the characteristics of the soil and climate. Typical invaders inthe United States are herbaceous weeds such as ragweed (Ambrosia sp.), lambs quarters (Chenopodium album) T dock (Rumex sp.), horse weed (Erigeron canadensis), plantain (Plantago sp.), crab grass (Digitaria sp. et al.) and woody vines such as honeysuckle (Lonicera sp.)and poison ivy (Rhus taxico-dendron). As soon as the first plant germinates and becomes established, the community increases vastly in physical complexity. A surface of bare earth now has shaded areas which differ in light, moisture and humidity. This enables other seeds to grow, and new species become established. Woody plants appear, again those capable of with standing disturbed environments and harsh conditions, such as black locust (Rohbinia sp.), sumac (Rhus sp.) and hawthorn (Crataegus sp.), Pioneer animals also arrive, including ants, beetles and flying insects. Birds begin to forage over the new community searching for seeds and insects. Small mammals may venture forth from adjacent forsts or grasslands. Lack of animal traversing the area adds organic nutrients. Within a year or two a surprisingly complex community has arisen, and is growing rapidly. As plants grow, they create more shade, and thus continuously modify the light and moisture conditions on the surface of the ground. These initial pioneers which thrived on total sun exposure are now less favoured and more shade-tolerant seeds can germinate more successfully. Woody shrubs emerge above the herbaceous layer and compete more successfully for the ambient light.

The ground cover and intermediate layers have now matured to include those shade-tolerant species capable of living in the forest under study. The insect and bird populations have changed from open field forms to forest species. By 70 to 80 years, the forest may be approaching relative maturity, with tall trees in the 60 to 80 foot range, deep shade and cool moist conditions on the forest floor capable of supporting the rich array of life. The climax first may be approached in 80 to 100 years, when it becomes a stable and self-perpetuating community. Processes of change still occur. Old forest trees die and crash to the ground, creating a microcosm of renewed sucessional development, but the main forest continues in equilibrium as a permanent community as long as environmental conditions remain favourable. If it is cut, the process of succession begins again. If it is defoliated, flooded, burned or dehydrated, it will change and attempt to seek a new form of equilibrium with its environment. The tree species that finally emerge as the dominant-members of the climax community vary according to the climate, topography and soil type.

The above discussion has mainly concerned a deciduous forest community, but a similar account could be given of coniferous forests, and, in fact, of all biotic communities. Thus, a prairie grasslands, a tropical savannah, a desert, and estuarine marsh, a cedarbog, a pond, a lake, a delta, a mangrove swamp, or a coral reef all have characteristic patterns of development and succession. In some communities, successional development may involve hundreds or even thousands of years before mature climax conditions are achieved. Primary tropical rain forests, for example, are the culmination of thousands of years of succession development. Once cut, they are not recreated in a few decades or a hundred years.

It should be emphasized that succession is also a naturally occurring process and is not confined to disturbed habitats. Theoretically, each biotic community tends toward maturation and stable climax conditions, but there is sufficient natural change in ecosystems so that succession is a continually recurring process in many situations. The grassland and forest are subjected to natural fire as an ecologic process, the desert to flash flooding, the marsh, bog, pond, lake, delta and swamp to filling, and the coral reef to the continual processes of construction and destruction. Human activities often increase and accelerate these processes far beyond the natural rates of occurrences. In doing so, man impedes the development of maturity and ecologic stability in biotic communities.

An example of natural succession is the filling of a small pond and its eventual conversion into a terrestrial community. As the pond ages, it becomes more filled with aquatic vegetation and silt. Emergent vegetation surrounds the edge creating a marsh community. Accumulations of organic material and sedimentary deposits fill in the water volume, soil is formed at the edge, and terrestrial grasses and hebs become established. Eventually, shrubs and trees complete the process. Man accelerates this process by increasing the siltation process and enriching the waters with nutrients so that more extensive plant growth occurs. The process can be retarded bydredging the pond or manipulating the water level to reduce aquatic vegetation. Lakes proceed through some of the same stages, but if sufficiently deep, a point is reached in lakes where the scouring action of currents slows or even balances the situation and sedimentation rate.

Another example is the volcanic island of Krakatoa in the East Indies which blew up in 1883 and virtually destroyed all plant and animal life on it on two adjacent islands. In 1886, it was apparent that the pioneer plant stage consisted of crust of blue-green algae covering the lava. A few mosses were also present, as were many ferns, and a scattering of some 15 species of flowering plants including 4 species of grass. The vegetation stage following the algae consisted predominantly of ferns, but the grasses had become dominant over most of the island. By 1906, woodland had appeared, which has since developed into an increasingly luxuriant mixed forest. Dispersal of seeds, spores, and other propagules was affected by wind, sea, animal, and man, inthat descending order of importance. The order in which the propagules reached the islands greatly influenced the succession that occurred; as vegetation developed it reacted on the soil and habitat, bringing about conditions amenable to the return of the tropical rain forest (Richards, 1952).

One year after the explosion, the only sign of animal life on the island was was a single spider, six years after the explosion flies, bugs, beetles, butterflies, and lizards were also reported. In 1908, 202 species were found on Krakatoa and 29 on a nearby island. There were no earthworms, snakes, or mammals present, but there were many spiders and centipedes, a number of insects, 2 species each of land snails and lizards, and 16 species of birds. Bats and earthworms were found in 1919. A more through survey in 1921 revealed 770 species of animals, including rats, apparently introduced in 1918. In 1933, 1100 species were found, but true forest mammals had not yet appeared and many families of forest birds were unrepresented. As with plants, the invasion by animals depended on wind, sea a man and other animals, in that descending order of improtance. Survival and establishment of the animal species were correlated with the stage of vegetation that was reached. It is of interest, however, that scavenger species appeared first, then omnivores, herbivores, and finally predators and parasites. Succession of animals depended in large part on the speed with which they reached the island and on their finding proper food and shelter.

The early stages or communities that make up the prisere depend largely on type of bare area on which the prisers originates. As succession proceeds, however, later stages in the various seres in any area having a relatively uniform and humid climate come to be more and more alike. The successional development from widely diversified communities in initially different habitats to closely similar or identical climax communities in habitats that have also become much alike is called convergence.

Biotic succession proceeds rapidly and appears to reach a final, permanent stage in just a few decades or centuries. The climax stage of teh biosere is undoubtedly more nearly stabilized, self-maintaining, and in steady state in its particular habitat than are the seral stages, yet it also is subject to gradual change over long periods of time. The cliamx, as well as the seral stages, changes with climate, physiographic forces, and evolutionary processes. However, the clisere usually requries a few thousands of years before changes in the community structure or composition become evident. The progress of an eosere is even slower, that of the geosere, slowest of all. The climax, therefore, is best defined as the last obvious stage in the biosere.

The climax may be recognized by the fact that in a uniform climatic area all seres tend to coverage into it, and by its steady state in respect to structure, species composition, and productivity. In the climax community, all species, including the dominant species, are continually able to reproduce successfully, and there is no evidence that new and different species are invading.

It is thus important for man to understand successional processes to that he can manage biotic communities for maximum productivity and aesthetic concerns. In agricultural lands and harvested forests, man has learned to maximize productivity for his own economic gain. This is done by artificially maintaining communities in early successional stages, where productivity is greatest, and by channelling all possible production into the species to be harvested. That is, man maintains a cornfield or a wheatfield in an early productive stage excluding all other pioneer species. He maintains a forest in an early successional stage and directs all productivity into the species of tree he wishes to harvest. This is all well and good provided the cost of such managementis not too high in terms of toxic chemicals that must be applied to eliminate other species, or in terms of chemical fertilizers that must be applied to support an artificial single-species community.

Quite another principle prevails in the management of parks and sanctuaries where man wishes to maintain balanced plant and animal communities for a esthetic and scientific values. Here it becomes essential to understand and promote natural succession so that mature and stable communities are achieved.

Hydrosere

Ponds, Marshes, Swamps and Bogs

Ponds is a popular term for lakes of the third order that are small, shallow, and when mature, have rooted vegetation over the bottom. There is no clear distinction between the ponds and lakes of the first and second orders. The littoral zone of eutrophic lakes, for instance is pond like in habitat and organisms. Floating and emergent vegetation commonly occurs around the margins of ponds to form extensive tracts of marsh, the pond habitat may originate as shallow basin, as a large pool in a stream as the result of the filling in lake, or form a stream dammed by beaveas, man, or landslide. Slow-flowing rivers are essentially elongated ponds, and have a similar fauna.

Because of the slight water movement in ponds, the surface film becomes an important microhabitat for some species, pond water temperature is often unform at all depths, but during warm sunny weather, ponds well protected from the wind may show considerable stratification, not only in temperature, but also oxygen content and other characteristics (Wallen 1955). Daily and seasonal variations in temperature may be great because of the small volume of water. Ice forms earlier and lasts longer in ponds than in lakes, freezing shallow ponds to the bottom in severe winters. Light penetrates to all depths, encouraging growth of vegetation except in high turbidity.

Plant Sere

The plant hydrosere, or pond sere typically contains the following stages and characteristic species :

Submerged vegetation: Water lily, pond lily, pondweed, smartweed, duckweed, and water hyacinth. All except the last two are rooted in the mud, often at a depth of 2 to 3 meters and may have rhizomes from which long petioles extend to the leaves floating on the surface. Duckweeds and water hyacinths are unattached floaters, and cover the surface extensively in some localities.

Emergent vegetation (marsh) : the dominant species are : cattail, reed, wild rice, and sawgrass. They invade water of over 1 meter depth, but in shallower waters or in secondary succession; they are replaced by the sedge meadow composed of sedge rush and spike rush.

Swamp shrubs : Buttonbush, alder, dogwood, swamp rose and sometimes shrubby willow and cotton-wood.

Swamp forest : Red and silver maples, elm ash swamp white oak, pin oak.

Suceeding stages depend on the climate region. In arid regions the swamp forest may be poorly developed, and grassland or desert vegetation may come quickly. In the mesic climate of the eastern stages, and oak-history associes follows the swamp forest, replaced in turn by climax sugar maple-beech or mixed mesophitic forest. The hydrosere in the broad leaved evergreen climax area of south-eastern North American brings in cypress and a number of other unique species.

Animal Sere

Animals characteristic of marshes and ponds constitute a distinct pond-marsh biocies which extends into sluggish or base-leveled streams and the littoral zone of eutrophic lakes. Most animal species are restricted to single stage or community of the plant sere, but commonly occur in several stages in varying abundance and for activities. Fish for instance, feed in open water but spawn in shallow water among the emergent vegetation. Submerged, floating and emergent vegetation represent different levels or strain in a single biotic community and each stratum has about the same degree of distinctiveness as forest community strata.

With the invasion of swamp shrubs and the ground level well above the water table most of the year, ponds and marsh species largely disappear, replaced by many characteristic new species. This animal community represents the swamps facies of the deciduous forest-edge biocies. The swamp forest is often quite open at its outer margin, and forest-edge species remains common. But as this forest develops a closed canopy and drier ground stratum, it is invaded by the swamp facies of the deciduous forest biociation. In the ponds sere there is a succession of animal adaptations from aquatic to amphibious to terrestrial.

Bogs

Characteristics

Bogs or moors typically develop in the hydroseres of cold northern regions; while marshes and swamps, which are markedly different from bogs, are characteristically southern in their location. Bog commonly develop into a coniferous forest climax; swamps succeed to deciduous forest or other southern climax types. Several thousands years ago, when glacial climates gripped the northern states, extensive bogs developed and have persisted as relic communities inspite of the warming of the climate. These bogs are slowly being replaced by pondmarsh species at equivalent seral stages by cliseral succession.

Bogs occurring in the Great Lakes region are ordinarily small in area and have little or no drainage. There may be oxygen present in the open water of the larger bogs, but it is characteristically in very low concentration at the most in small bogs or in the marginal zones. Bog water has distinct brown colour a low nitrogen content; a low temperature beneath the surface; a low pH at least in the marginal vegetated zones; and a low dissolved salt content.

A false bottom is characteristic of bogs. It consists of finely divided plant material of light brown colour, held suspended in the water at varying depths below the surface. This false bottom may be extended downward several meters before a true solid botom is reached. The material disperses on slight disturbance and may render all the open water turbid. Ordinarilly, the surface waters are quiet and clear. Dead vegetation does not completely decompose; as it accumulates, it becomes compressed to form peat.

Plant Bog Sere

In the early stages of development of the bog, organic detritus may accumulate mostly in the deepest portions. As time goes on, however, a definite concentric-circle zonation of vegetation is established around the margin. As peat accumulates, each zone encroaches on the next inner; the in most shriks until all open water disappears. The ara finally is covered with climax forest. In some bogs the first plant stage may be composed of floating vegetation (Nuphar, Naymphaea potamogeton, Sparaganium), but floating vegetation is often absent and the first stage is a sedge-mal composed of sedges cotton-grass, and buckbean. The rhizomes of the sedges grow out into the water and become so interlaced that they form floating mat.

The next plant stage is dominated by low shrubs, which encroach on the floating mat. The leatherleaf, bog rosemary, laurels, labrador tea, sweet gale, and cranberries are important species.

A high shrub stage commonly follows the low shrubs at such times as the mat becomes thicker or grounded. Common shrub species are holly, willow chokeberry, alders, and dwarf birch. The first tree of the bog forest to invade the shrubs is commonly the tamarack, but this species is now less common than formerly because of fire, logging, and depredations of the sawfly larvae

Lygaonematus erichsonii. Black spruce may either invade the shrubs directly or follow the tamarack. Later, the northern white cedar may become dominant and pesist for a very long time, but the ultimate fat of the bog, upon addition of upland soil or the water table, is to be covered with the climax forest of the region.

Animal Life

In bogs that have a large body of open water, or an inflow of water entraining oxygen, and in which the pH is not extreme, invertebrate life comparable to that found in ponds and marshes occurs. True bogs, however, have little oxygen and a low pH, and many pond species do not appear. Mollusks are characteristically absent; sphaeriids may persist but their shells become very thin. Bottom organisms in general are poorly represented because of the tenuous physical nature of the substratum.

Desmids predominate among the phytoplankton, although dinoflagellates, Chlorophyceae, and Myxophyceae are common. Rotifers and a variety of Protozoa are the principal Zooplankters.

Amphibians and reptiles are not characteristic of bogs, although the leopard frog is sometimes numerous on the sedge mat of bogs in Minnesota. Proceeding northwards boreal species of birds become progressively more prominent in the bog avifauna. The muskrat and beaver persist into northern Canada.

In general, the productivity and economic value of bog is very low compared with ponds and marshes. Liming experiments, calculated to improve productivity, are being made. Calcium combines with the humid colloids, which then flocculate and fall to the bottom. This clears the water, light penetrates deeper, pH is raised and greater algal, zooplankton, and fish growth is induced. Peat is a special product of bogs. It is of importance in northern Europe. It is cut out in blocks, dried, and used as fuel and for mixing in the soil of gardens and lawns.

Rock, Sand and Clay Sere

The origin of life was undoubtedly in the sea. physiological adjustments were necessary before organisms could occupy fresh water. Although some organisms may have become air-breathers and invaded land habitats directly from the sea, most evolution of terrestrial forms has doubtless come from fresh water. Relatively few major groups of animals have been successful in this invasion of land, the most notable being oligochaete worms; gastropod mollusks; many arthropods, especially the insects and spiders; reptiles and their evolutionary offspring, the birds and mammals.

Adjustments to the Terrestrial Habitat

Living on land presents many problems. Our present concern is to analyze the ways in which animals have met these problems and to trace the succession of communities in the extreme terrestrial habitats of rock, sand, and clay. A variety of techniques have been developed for measuring environmental factors.

Gravity

In water, organisms counteract gravity by means of various floating and swimming mechanisms. Fluid buoyancy permits water-dwellers to attain huge size; considere the whale. A land animal on the other hand, must support its entire weight. Some terrestrial animals gain a modicum of support by burrowing into the soil. Others drag their bodies over the ground surface. But the supportive advantage they thus gain is costly in other directions, for they are slow moving and relatively helpless before predators. The animals best adapted to terrestrial life have evolved appendages in the form of legs or wings that not only raise the body above the ground but are also the means of more or less rapid locomotion and adroit movements over the surface or through the air. Terrestrial adaptation has involved the development of a tough body covering to hold fluids and internal organs in place; a skeletal frame work to give permanent shape to the body and, as a system of levers, to furnish means of locomotion; and powerful muscles to lift and move the heavy body.

Moisture

In sharp contrast to aquatic forms, terrestrial animals are not constantly enveloped with a continuous watery medium, with the limited exceptions of protozoans, nematodes, and other small organisms living in moist soil. Land animals lacking constant contact with the water medium, are faced with the problems of obtaining water and preventing excessive water losses from the body.

Water becomes available to animals in varying amounts in the forms of rain, snow hail, frost, and fog. Whatever the form it arrives in the significant things are the amount of free, liquid water added to the sustratum, accessible to plants and animals and the humidity of the air. Considerable amounts of moisture are lost to organisms as runoff water flowing into streams, by evaporation back into the air, and as water bound in snow and ice.

Water is obtained by a land animal by various devices. There may be some direct *absorption* through body surfaces such as occurs in the toad in moist soil and in some beetle larvae in moist air, this device is important in only a few species. Large mammals frequently travel several miles each day to water holes to imbibe drinking water. Many, but not all, birds require *drinking* water; some species,for instances quail and partridge, get it as morning dew on vegetation. Butterflies may frequently be observed drinking water from small pools. An important source of water is the *free water in food,* particularly in succulen vegetation and in the blood and body fluids of animals. Desert animals depend almost entirely for water on that contained in their food and on *metabolic water,* liberated when fats and carbohydrates, and to a lesser extent, proteins, are oxidized in their bodies.

Water is lost from the body through the skin and lungs as insensible moisture perspiration. Rapid and largely uncontrolled loss of moisture through the skin of amphibians, snails, annelids, and insect larvae is a limiting factor confining these animals to moist habitats or to activity only in times of high humidity. Adult insects and other arthropods, reptiles, birds, and mammals have evolved body surfaces of chitin and waxes, scales, ro cornification of the surface layers of the skin that largely prevent uncontrolled loss of moisture. Moisture loss through the respiratory surfaces in these forms remains considerable however. Water is also lost with the faces, although in some species much water is reabsorbed by the large intestine before the feces, are ejected. The amounts of water removed from the body by excretory organs particularly the kidneys varies directly with the amount of water intake and immersely with the amount lost through other devices.

The kidneys are critical to maintenance of proper concentration of salts in the blood and body fluids. It is important that water intake balance water loss. Organisms are very sensitive to disturbances in bdoy water balances and this factor is significant in determining the type of niche which a species comes to occupy.

In order that animals could exist in terrestrial habitats, they had to acquire the ability to carry on reproductive activities in the absence of water. The chief reproductive adaptations involve the following (Pearse 1950): internal fertilization; a shell covering the egg to conserve moisture and salts; food provision to the embryo and young by yolk in the egg cell, placenta in the uterus of the mother or direct feedings by adults; reduction in number of young with more efficient parental care; reduction or elimination of free-swimming larval stages; and greater segregation of species into different niches to avoid interspecific disturbances.

Temperature

Aquatic animals are not ordinarily subject to temperature below freezing and are in a relatively stable temperature environment, but terrestrial species are exposed to highly variable temperature that may reach extremes of about 680°C and +550C. No single species is required to with stand such a wide range of temperatures, however. Optimum and tolerance limits vary from one species to another.

No aquatic species has evolved control over its body temperature. Aquatic warm-blooded mammals and birds are derived from terrestrial forms. A ability to maintain a constant body temperature has survival value in terrestrial habitats however, and consequently physiological mechanisms for homoithermism developed independently in birds and mammals. All other land organisms are poikilotherms; that is, the have no physiological mechanism for maintaining a constant body temperature. Some poikilotherms, such as bees have developed special behaviour patterns that enable them to maintain fairly constant conditions in the hive by cooperative efforts; some moths and other insects with hair-like body coverings maintain a degree of contancy hair-like body coverings maintain a degree of constancy in body temperature by shivering their wings; some lizards, snakes, and turtles are able to exert some control over their body temperatures by moving into and out of sunlit areas.

Rates of activity, food consumption, metabolism, growth and other physiological functions increase, to a certain limit with rise of body temperature. Homoitherms maintain a continuous high rates of functioning because of their constant high body temperature, but the rate at which poikilotherms function varies with the temperature of their habitats.

Oxygen

Dry air 760 mm Hg pressure contains approximately 21 per cent oxygen, 0.03 per cent carbon dioxide, 78 per cent nitrogen and traces of other gases. Oxygen is thus much more abundant, constant, and available at all times in air than it is in water. Oxygen availability seldom becomes a critical factor for land animals, with occasional exception of forms that live in the soil or invade high altitudes.

Although terrestrial organisms have evolved simple moist chambers, branched trachael systems, or complicated lungs to replace the gills found in many aquatic forms the fundamental requirement of moist membranes for the exchange or oxygen and cabon dioxide between body fluids tissues, and surrounding medium remains the same. The skin still serves this purpose in some terrestrial forms annelids and some amphibians but in most forms the moist membranes are within the body. Internal placement decreases the loss of water through evaporation. The evolution and any ability to take oxygen directly out of the air apparently preceded the actual invasion of land, and may have been induced invasion in the pond and marsh habitat when oxygen dissolved in the water becomes reduced or absent during summer stagnant periods. The evolution of internal airbreathing organs was probably concurrent with the evolution of mechanisms to prevent excessive water loss from the exposed surfaces of the body.

Rock Sere

Plant communities

Stages in the plant sere on bare rock include lichens, mosses, herbs and grasses, shrubs, and forest. The species composition of each stage varies with the chemical nature of the rock, the prevailing climate, and the locality.

In the first stage, various kinds of ichens compete for a foothold, but crustose types usually precede boliaceous types. Mosses such as foliaceous lichens as Cladonia follow foliaceous lichens; or may initiate the sere telescoping the earlier stages.

Lichens and mosses soak up moisture in wet weather. They derive mineral nutrients from the underlying rock. Carbon dioxide secreted from the rhizoids forms a weak acid with water and dissolves the binding material of the smallrock particles. Rhizoids may penetrate rock for several millimeters. These plants tap wind blown dust and obtain nitrogen from organic compounds in it. When the plants die, they become an addition to the accumulation of organic matter. Herbs, grasses, ferns invade to continue the crumbling of the rock and build up of the soil. Freezing and thawing of water may crack the rock, and in these cracks wind and water borne soil lodges and supports plants. Once shrub and tree roots get started in crevices, their growth exerts a powerful force, further splitting and crumbling the rock.

Animal Life

Animal life in the pioneer plant stages on rock is scanty. Ants and spiders roam over the bare rock, and insects of various sorts may stop there, temporarily. Spiders may construct webs and nests in rock crannies or among the lichens. Some tardigrades find preferred niches in lichens. Mosses offer a somewhat more substantial micro-habitat, but only those animals that can tolerate great extremes of flooding, dessication, heat and cold can survive. Such forms are found in the rhizopod protozoans, nematodes bedloidid rotifers, tardigrades, copepods, small insects and mites. They often have special means of attachment to keep them from being blown away by the wind, such as strong claws or cement glands on the feet, long bristle-like threads to entangle among the moss filaments; stickers or spines covering teh eggs. Since wet periods are often too short to permit complete development, all stages must be tolerant of dessication, at which time activities and growth are largely suspended.

Animal life in general and land snails in particular are usually more abundant in vegetation established on soils delivered from sandstone, granite or volcanic rock. Calcium carbonate is a mineral essential to the metabolism of most animals and for building such skeletal structures as bones and shells.

Clay Sere

Plant Communities

Erosion or calculated removal of overlying material in strip mining for coal may leave bare areas of clay exposed and impound small bodies of water, often with low pH values. These water bodies undergo a modified pond succession. In clay above pH 4.5 annual plants, of which smartweed is particularly important appear within a few weeks to 2 years; the higher the clay pH, the quicker the appearance of vegetation. Within 2 to 5 years there after sweet clover invades and develops nearly complete dominance over large areas. Sweet clover is a liennial and an exotic species unimportant in the sere in some parts of the country. Prior to its introduction, this stage in the sere on bare clay may have consisted of the perennial grasses still found in small scattered patches. The first trees begin to invade early in the sere, out they are scattered and slow or growth, and do not attain dominance for 25 to 30 years. Willows occur in wet spots herbs species of the first two plant stages disappear, for the most part, in the shrub stage. The Herb stratum now consists largely of wood nettle. Advance forest stages of oaks, hickories, basswood and sugar maple will likely invade in the future, as they occur now in adjacent areas.

Animal Life

The number of invertebrate species tends to increase as the sere advances, although not always regularly. In a study of a formerly stripmined area, 18 species were found to be important in the annual stage, 41 species in sweet clover, 40 species in shrubs, 32 species in the early forest stage, and 67 species in the unplan climax. More species would be found in advances stages because of the greater variety of niches available. Thus, in the initial bare area there is only the ground stratum, in the annuals and sweet tree strata. Since the early forest is on a flood plain, the ground is frequently swept by floods, and the shrubs stratum is poorly represented. The climax forest has all strata, richly developed, and possesses the greatest number of animal species. There is an increase in the abundance of individuals per m^2 with the progression of the sere: annuals 268; clover, 531; shrubs, 532; early forest, 748; climax 2445.

Beetles, spiders, ants and mites are the most abundant animals in the annuals stage, and along with aphids remain most abundant also in the sweet clover community. Grasshoppers are fewer in number but especially characteristics of the first three stages; they practically disappear in the forest. Earthworms are absent in the annuals and scarce in the sweet clover, as are the springtails; as the amount of soil humus increases with the development of the sere, both groups become more and more numerous. Snails first appear in the sweet clover stage and increase in importance in the forest stages.

Starting with bare ground and continuing through early and late shrub stages to bottom and forest, there is a progressive increase in bird species diversity/density and biomass. The first small mammal to invade the area is the prairi deer mouse followed by conttontail rabbit and fox-squirrel.

Floodplain Sere

A stream continuously deepens its channel, thus lowering the water table of the surrounding land. At times of heavy rain, the channel may not be deep enough to carry the discharge and the river overflows its banks. In mesic climate this commonly happens three or four times per year, usually during late winter and spring, and each flood may last up to several days. The more extensive floods, persisting for long periods, come at less frequent intervals. The low land area between the river and the bluffs on each side is called the *flood plain.* In the course of time, the river in its erosion meanders back and forth across the flood plain, cutting new channels and the surrounding upland. The higher the terrace or the farther area is from the channels and abandoning old ones, and frequently leaves a sequence of terraces between its present channel the less frequently and for shorter periods is it flooded, so that there is commonly a sequence of plant communities that constitute the plant sere.

Plant Communities

Gravel, sand, or silt is deposited on the inner side or river bends. Attached aquatic vegetation may occur in the water. On land, such herbs as smartweed, cocklebur, ragweed, beggar's ticks occur. At some bends small sand dunes may occur, displaying their characteristic plants and animals; usually this stage is narrow at most, and may be entirely absent. On sandly islands in the river or on sandly shores the sandbar willow often forms dense, shrubby thickets. The first tree stage is frequently black willow mixed with eastern cotton wood,and sometimes silver maple. On the floodplain on the Canadian River in Oklahoma, the sere proceeds next to an edaphic subclimax of either tall grass prairie or elmoak. The climax on the surrounding upland is mixed prairie. The normal sequence of stages in this region has become considerably modified by the extensive ecesis of the exotic tamarisk tree, introduced from Asia.

Elsewhere in the eastern United States, the cottonwood-willow stage gives way to a narrow zone of a sycamore. Two or three species of elm, white ash, and boxeleder follow: then a mixed forest that includes black walnut but ternut, blackmaple ohio buckeye red mulberry, American basswood tuliptree, and hackberry; next an oak-hickory stage; and finally the beech-sugar maple climax. The herb and shrubs strata are usually well developed in mature floodplain forests. Telescoping or skipping of stages is not uncommon in this sere, since variation in ground level or in height of terraces is considerable and the transition between heights is often abrupt. The later stages occur only on the very oldest terraces and may be hard to find at all.

Animal Life

In the bare areas, in the herbs, and among the invading trees occur such beetles as *heterocerus pallidus* and *Bembidion laevigatum* that feed on the algae and detritus present on te shore. They make their burrows insand. Frly larvae, a cocklebur weevil, a cocklebur mirid, and a cocklebur fly also occur. The tiger beetles *Cicindela hirticollis, C. cuprascens,* prey on the ground species and may even dig them out of their burrows. Spiders, ground beetles, and rove beetles invade from higher stages. In the herb stratum and in the shrubby growth of willows, adult midge flies and other flies are sometimes very abundant. Tarnished plant bugs, 12-spotted cucumber beetles, and other insects of open-area habitats are present, and there in invasion of various species from the forest itself.

The animal life of the flood plain forest in much the same as that of the deciduous forest in general and does not need to be discussed here except for its unique features. Annelid worms make their appearance in the ruderal stage, become very abundant in the moist soils of the elm-ash and mixed floodplain forests, then decrease in numbers in the direr soils of the late seral stages. They occur mostly in the first 5 to 10 cm below the surface in moist soil, but up to 30 cm or more in dry soil. During the winter they keep below the frostline, and in very dry weather they roll up in small knots. Ten species occur in the floodplain of the Snagamon River in central Illinois, and each species has its particular range of moisture requirements between the river's edge and the upland forest.

Snails and slugs and moisture-living animals and occur in large animals and great variety in floodplain forests; its is not hard to find 15 to 20 species with a little searching. *Mesodon thyroidus* in a commonsnail, and on a floodplain in central Illinois an average population of 6.3 individuals per m^2 was found during the autumn. This amounts to a biomass of living flesh (shell excluded) of 15.8g/m^2 (141lb/acre). Snail flesh is an important source of food for such small mammals as the short tailed shrew.

26

Bio-Resources Protection

A resource may be defined as any item that could be put to use by man with a view to improving his quality of life. It is clear that with such a definition, every single item that has come within the handling power of man is a resource; from wind, water to stratosphere to all the living and the non-living components of this earth, plants and animals and the by products of their metabolism and the minerals lying in the bowels of the earth and the various atmospheric gases.

There is the wide range of gadgets and substances created by man. It will be too ambitious a task to undertake even a casual survey of these sources, let alone their management. We will confine ourselves to a very small range to these resources and limit our discussion of their management to the extent to which such a management (or mismanagement) affects the physical status of environment and thereby the future of man, viz. Bio-resources.

Man is part of a biosphere, a term applicable to all the living organisms together with the environments in which they live. The seas with all the life-forms found in them; the atmosphere pervaded by a variety of spores, bacteria and to the life-forms; the top loose crust of the earth which we call soil with its vast array of living objects; and their dead remains in different stages of decay; the ponds, the rivers and such other fresh water situations. Add to these environments, the back waters, the icy patches of the polar regions and high altitudes, the crevices in the barks of trees, the interiors and exteriors of the bodies of animals and plants-dead and alive-the hot water springs the crannies within the tectonic plates, the dark caves, the hot deserts; the grass-lands, the scrub jungle, the deciduous forest of the cold and hot climates and the ever green tropical rain forest of the equatorial belt and what have you. All this is biosphered and each form of life inhabits it constitutes a resource of one kind or another to man.

Every one is familiar with human dependence on other forms of life; for this food, for his clothing and for his shelter. One can discern a criss-cross pattern of mutual support, sometime one species getting the better of another; but often times, the so-called dominant species getting worsted in the long run. Through this see-saw struggle, eventually a balance emerges. And that is what we call ecological balance. Fortunately or unfortunately, man plays a unique role in this matter of ecological balance. He tends to bend every other species to suit his own ends.

Often times, he exploits a resource to such an extent that sooner than later he finds this resource in danger of disappearance. Not only that in the course of such an exploitation, he had planted seeds of imbalance between species, whose (that of imbalance) working had been so insidious that he fails to notice it. When finally the disastrous consequences catch up with him he gets panicky and sometimes succeeds incombating the situation but many times he finds himself unequal to the job of restoring the balance that he so blindly destroyed. It is in a context like this, conservation and management techniques acquire significance.

The story of Agricultural development is replete with tragic comic inadequacies of human conduct. The slash-and burn method of agriculture of the tribals in India must have sent a phenomenal number of life-forms into oblivion. Even the conventional mode of Agriculture would contribute to this permanent loss for after all, no agriculture is possible without denudation of forests in the first place. Recent surveys showed that is the tropics only 1/6 of all species had been scientifically recorded.

It is thus quite clear that several hundreds of specie, which could well have served as a bioresource of great potential, have disappeared it to the limbo of time, thanks to the growth of 'human civilization'. The damage to natural potential in inestimable when we take into account the fact that 11 million hectares of forests are being lost every year. Even in countries where there is a realization of the damage being inflicted through deforestation, there is compensatory afforestation only up to the extent of 1/10th. In the so-called developing countries, the burgeoning increase in human population is exerting unbearable pressure on the bioresources of the world. In the meanwhile, a new concern to preserve whatever is left has led to the establishment of several lead-organizations charged with the responsibility of devising ways and means of achieving what at the moment looks like impossible. We will briefly outline some of these management programmes, of which the pride of place should be given to International Biological Programme (I.B.P.), Man and Biosphere programme (M.A.B.), and World Wildlife Fund.

International Biological Programmes (IBP)

The IBP is a world-wide plan of research concerned with the "biological basis of productivity and human welfare". The programme had been thought of as early as 1959. But it was only in 1964 that the first general assembly of IBP was held. This assembly drew up a time-bound programme for achieving its goals. The first 3 years through which Phase 1 ran were used for drawing up a broad outline of all the feasible studied Phase II began to operate in 1967. Though initially intended to run for 5 operate in 1967. Though initially intended to run for 5 years; Phase II actually got extended by another 2 years. Phase III which commenced in 1973-74 set for itself the tasks of synthesizing all the observations made and transferring the informations to the different groups of the programme. The Aims of IBP : The crux of this programme lies in two premises; (1) the ever increasing human population and (2) the inelastic natural resources. These two processes pull in opposite directions.

The main aim of the IBP is to reconcile the every increasing human population and the consequent increasing demand for food production and the limited nature of the natural resources. Necessarily the programme concentrates on the rational management of the resources. This calls for extensive education of farmers and other populations with a low level of attainment. Thus, one may say that the aims of IBP is to fill the gaps in knowledge by means of a coordinated comprehensive approach including sufficient standardization of methods and thus ensure a comparative study of the results.

The IBP has been divided into 7 section

(1) Productivity of Terrestrial community (PT) i.e., Primary productivity of the green plants, secondary productivity of the consumers dependent of plants and decomposition of organic matter.

(2) Production process : (PP) i.e., Utilization of radiant energy in photosynthesis and transpiration by plants and the N_2 cycle.

(3) Conservation of Terrestrial Communities (CT) i.e., To provide scientific conservation of areas and species ensuring present and future needs of natural habitats.

(4) Productivity of Freshwater Communities (PF) : Like PT, this section is concerned with primary and secondary productivity and decomposition in acquatic systems.

(5) Productivity of Marine Communities (PM) i.e., Studies of inshore waters and estuaries and conservation of marine habitats.

(6) Human Adaptability (HA) i.e., Genetics growth and physique tolerance to heat cold and high altitudes, working capacity and population dynamics of man.

(7) Use and Management of Biological Resources (UM) i.e., Establishment of plant gene stocks, biological control, growth and production of cereals and the development of possible biological resources for man.

An annual meeting of IBP is held by SCIBP. Meetings are also held at national levels. The progress of the programmes is published annually through handbooks and IBP news.

Man and Biosphere (MAB)

MAB is an interdisciplinary research. It emphasizes ecological approach to the study of inter-relationships between man and environment. It focuses its attention on the structure and function of biosphere along with it ecological divisions, systematic data about the changes bought about by man in the biosphere and its resources, overall effects of more changes in human life. It undertakes to provide information on these aspects to its member bodies. The MAB is a programme with in time limit and this is borne out a per the Resolution Number 2.313 of the UNESCO passed at its 60th session.

The MAB aims at rational use and Conservation of resources and biosphere. It also aims at improving the global relationship between man and environment. It endeavours to predict the consequences of today's activities on tomorrow's world. Thus, it makes for a conscious effort at managing national resource of biosphere more efficiently. The MAB ha a two-fold approach : (1) Analysis of the ecosystems, and (2) study of the impact of man on environment and of environment on man.

The MAB has so far identified 14 projects

Project-1: deals with ecological effects on increasing human activities on tropical and subtropical forest ecosystems. This project assessed the consequence of selective cutting of forest species, studies strip cultivation, analyses the consequences of extensive clearing of forests for agriculture, pasture and plantation studies, tropical and subtropical forests and lastly studies the flow of mineral nutrients from the upland forests to the lowland cleared areas.

Project-2 : makes a study of ecological effects of different land uses and management practices. In particular, this project studies the effects of the introduction of the fast growing exotic species, assesses water balance of forests, studies the impact of air pollution and composition and performance of forest ecosystems, studies the consequences of leisure and tourist activities.

Project-3 : studies the impact of human activities in relation to land use. This project studies the effects of grazing on an ecosystem, makes a comparative study of domestic and wild animals in order to find a more efficient source of protein, studies the effect of fire in savanna and related tropical forest lands, studies the herbs and shrubs that could be employed for grazing and other general purposes, studies the problems of wind crosion, studies the long-term-changes in N_2 content caused by ploughing and cultivation studies the consequences of cropping in relation to microorganisms and so on.

Project-4 : studies the impact of human activities on the dynamcis of arid and semi-arid zone ecosystems and also studies the migratory and nomadic movements of human and animal populations.

Project-5 : studies the ecological effects of human activities on the value and resources of lakes, marshes river, deltas, estuaries, and so on.

The other projects include study of mountain ecosystem, island ecosystems; steps to be taken for the conservation of natural areas and the genetic potential ecological management of fertilizers use and pest control, the impact of major engineering works on man and his environment and so on.

The Indian national committee on MAB has been constituted by the Government of India under the Department of Science and Technology.

Besides the above mentioned projects, many private agencies have taken up the job of conserving the environment. The best known movement in India is the Chipko Movement, led by Sunderlal Bahuguna, launched in Himalayan Tracts of Tehri-Garwal, U.P., Mr. Bhahuguna educated the local population on the disastrous consequences of indiscriminate tree-felling, woke up the people of the area to the dangers of floods, soil erosion and so on. As a result of this education, the people of the area prevented the forest contractors from felling the trees by hugging on to them. Hence, the name Chipko Movement. The Green Peace Movement is mostly devoted to protection of marine animals such as the whales, dolphins, etc. They also protest against atomic explosion in pacific atolls. The members of this movement prick the conscience of whale hunters and such other groups and hope that such expeditions would be given up a new lease of life to the endangered animal and human populations.

Conservation of an ecosystem, very often means its maintenance in its natural state. In the natural state, an ecosystem has a balanced populations of predators and prey, dominant and the subdominant species. Each species in such an environment will come to terms with species in all the other species of the area. Many of the animal and bird sanctuaries established in several parts of our country and throughout the world aim at maintaining the ecological balance. One must say that there is no better method of conservation that to let an ecosystem be.

World Wildlife Fund

The world-wide concern regarding the danger posed to wildlife everywhere led to the formation of International Union for Consevation of Nature and Natural Resource (IUCN). This body gave birth to Survival Service Commission, charged with the responsibility of providing information about every species of animal that faces the danger of extinction. The Commission's labour resulted in Red Data Book; it listed 277 species of mammals and 321 bird species as the endangered ones. This was in 1966. Waking up to the danger of losing these species for ever, the World Wildlife Fund had been created. This fund finances projects designed to offer protection and conserve endangered species. The best known of these in India is the Project Tiger, and thanks to this project, the tiger population had increased ten-fold through the last twenty years, a no mean achievement by any standard. The Indian Lion, the one-horned Rhinoceros, the Indian Wolf, the Snow Leopard, the Fishing cat are some of the mammals that got a new lease of life as a result of the activities of the Fund. Other beneficiaries include the Flying Squirrel, Black Buck, Chinkara, the Four-horned Antelope, the Himalayan Tahir, the Swamp deer, the Yak and the Wild Buffalo. The Blue Whale and the Gangetic Dolphin are among the aquatic mammals that escaped extinction, thanks to the Fund's activities. The Ghariyal and the Crocodile and a good many snakes also benefited by the activities.

Among the birds, the great Indian Bustard which almost disappeared staged a come back, thanks to the ban imposed on the shikaris from the rich gulf countries who used to come to India mainly for shooting the bird. It is not our aim here to list out all the species threatened with extinction, the emphasis is more on what even a modest endeavour at conservation would achieve. In the most immediate past, i.e. through the last five years, it is realized that the Indian elephant is fast becoming an endangered species. This is not so much due to poaching by greedy ivory hunters as because the area over which the elephants could freely roam about for obtaining their food requirements is just shrinking.

When therefore elephants trespass into agricultural fields where crops like sugarcan are cultivated, they are shot down mercilessly. So, a project elephant very much like project Tiger is receiving a active consideration. What do these sanctuaries mean? In plain language they only mean that man is willing to tolerate the presence of wild life so long as it does not pose any harm to his interests and secondly he is not certain if some animal or plant species he allows to perpetuate itself would not prove useful in some manner at a future date and this conservation might come in handy as and when usefulness is discovered. Overriding this utilitarian outlook the sense of oneness with nature which is at the bottom of his religious thought puts on man a moral obligation to let live the other forms of life. And his aesthetic sense also is a factor of some significance in this regard. But sometimes, man acts as though he kills for the sheer joy of it.

The massacre of whales that is going on unabated in spite of a strong conservationist movement is a case in point. Indeed, Nobel prize for Literature was awarded to Earnest Hemming way for his book, The old man and the sea for depicting the indomitable human spirit. If only the panel had looked at the picture from the other end of the telescope they might have been appalled by spotting a spirit that must prevail no matter what the cost is to the marine life. But all that is neither here nor there : though it brings us to talk of the bioresources of our aquatic environments marine and fresh water.

India has a long coast line of 6,100 km. and its exclusive economic zone extends for 200 nautical miles into the sea. This gives us an area of 2,59,000 sq. km. of sea from which a rich haul of fish, prawns, lobsters etc. could be harvested. If this source of food is fully exploited, we do not have to go hungry. But it is only in recent times, that the presence of this great potential had been realized. The numerous seafood Corporations that have come into being through the last 30 years is testimony to this realization. The Government of India and private agencies have embarked upon schemes of mechanized fishing and fisherman millionaires are no more uncommon. The inland waters, including the estuaries of major rivers, backwaters and brackish water lakes and the river systems have resources that are as yet untapped. Efforts in this direction are in their infancy and the emergence of University courses in acquaculture, apart form the traditional courses in Fisheries and Marine biology, stresses the point. An ICAR Report (1976) estimated that in 1976, the marine fish production was of the order of 13,52,855 tonnes. Surely, in the intervening years it must have doubled, prawns form a good percentage of this catch.

A different kind of management is called for in this connection : effluent treatment such that the quality of waters in which the fish thrives is not affected. There is yet another kind of managerial dimension, a practice adopted in the matter of inland fisheries. It has been found that growing carps (fish type) of different kinds in combination, Indian and exotic, (polyculture) yields better harvest than growing them pure. Intensive fertilization, supplementary feeding and higher rate of stocking doubled and sometimes trebled the yield. In the recent past, it has been found that rearing fish in the waters of paddy fields had benefited Rice crop.

The foregoing account clearly demonstrates that conservation/management techniques in respect of bioresources must keep pace with the changing face of the situation. It must be emphasized that there is no one-time solution for any problem. But the goal is clear exploitation without endangering the resource to the point of extinction.

The goal is more easily spelt out than achieved. This difficulty is best illustrated in the matter of food crops. Prof. M.S. Swaminathan, the doyen of Indian Agricultural Scientists, estimates that just about 150 plant species with about a quarter of a million local strains meet the calorie requirements of human populations. This restriction of human dependence, has in way simplified the problem of food resources for man but at the same time, it also poses very difficult problems in regard to the maintenance of the large genetic diversity, so essential for meeting the threat of crop diseases.

Agriculture is monoculture, i.e. growing a particular individual species populations to the exclusion of all others. Infact, weeding is done so as to eliminate other plant species competing for water, food and other resources with the crop species. With improvements that came to attend the agricultural pursuit, this monoculture has become even more specific in the sense that in a cultivated field, a particular variety of the crop species, with proven genetic traits alone is permitted to grow. Of course, now a days we hear of mixed cropping which is but a slight variation of the monoculture. Now, agriculture affects the ecosystem as a whole and that is obvious. But the stress is on rearing varieties of crop species with a particular manipulated genome. It has insidious effects and the worst of these is the loss of the genetic diversity that a wild population of the species would exhibit. For all practical purposes, it is as if 'clones' are grown, though the whole exercise goes in the name of pure strains. This sort of situation is true not only in respect of agriculture but also in the realm of Animal husbandry. It is difficult to realize the havoc caused by this practice of 'genetic improvement' in respect of crop species but it is easily perceivable in respect of animal breeds. All that we need to do is to recall the breeds of domesticated animals that were known to our fathers and grand fathers and their cries of agony over their disappearance through their lifetime.

No wonder, one of the important programmes is to develop 'gene banks' in respect of each crop species. Towards this end, an organization called 'International Board of Plant Genetic Resources' has been floated in 1974. This is entrusted with the job of identifying crop and geographical variants and suggest priorities in the matter of conservation. The UNESCO is seized of this question and organized an International Congress of Biosphere Reserves, the first of which was held in USSR in October 1983. 150 Scientists from about 50 countries attended that Congress had drafted an 'Action Plan' to protect the world's Wild genetic resources.' Its chief aim is to conserve for posterity the existing genetic variability.

The plan is three pronged and consists of : "(a) *Ex situ* protection, as the chief and generally the only means of conserving wild material. (b) *In situ* protection, as the chief means of conserving wild material and (c) develop a system of sound planning, allocation and management of land and water uses to achieve a sustainable combination of development on the one hand and of maintenance of the biological resources required for development on the other. Thus conservation of genetic resources must be a world effort."

The Scientific aspect of genetic conservation involves first, defining the categories of material worth preserving and then, working out the methodologies of conservation. One may say that the following are the categories of materials worth preserving : cultivars (Cultivared varieties) in use at present, obsolete cultivars (varieties that have been used for centuries in traditional agricultural systems), wild species and closely related to cultivated varieties, wild species of potential value to man (whose importance is yet to be recognized) and Special genetic Stock (including mutants, breeder lines with identified genes and gene combinations, chromosomal variant type such as polyploid, aneuploids, translocations and deficiencies). The methodology of conservation (of genetic resources) involves : (a) preservation of entire biomes, (b) *In situ* preservation of land races and wild relatives, sometimes incorporating their genes through hybridization and (c) *Ex situ* preservation of seed under appropriate conditions for long-term storage.

The practice of such a methodology requires "Considerable knowledge of genetic structure of populations, sampling techniques, methods of regeneration and maintenance of varietal gene pools, particularly in cross-pollinated plants." The data can be staggering. But thank God, we are in the computer age and it should not be difficult to evolve a system of information management.

Ex situ preservation involves exploration, collection, storage, evaluation and documentation not to say breeding for the effective use of desirable genes.

Additionally important crops like Sugarcane, Sugar beet etc. have to be taken care of. Over and above there are fruit crops, tuber crops, rhizatic species, drug yielding plants, fibre yielding plants, timber trees, trees that would help in the reclamation of waste lands and so on. If only we recall how Israel successfully converted desert land into flourishing orchards by employing the dew absorbing succulents, we would feed foolish in dubbing any plant species, for that matter any organism, ranging from microbes to mammals as one that deserves to go into extinction. After all, every species has its own relevance in the scheme of the living world. So, how do we go about conserving all this genetic diversity?

Obviously, the first task is to explore and identify the collected material. Indeed, what with the rate at which the species are getting endangered, due to the expansive activity of man, now a days the explorations have become rescue operations. In so far as the plant species that have to be collected and conserved, we may classify them into three broad categories : Seed crops, vegetatively propagated crops and fruits and timber trees. The collection of seeds must be so done that all the alleles in the population are collected. One possible way of achieving this goal is to collect at least fifty seeds from fifty different plants and treating them all as one whole sample. In the case of vegtetatively propagated crops, as many clones as possible from areas separated by not more than 50 kms should be collected. Collection of seeds from fruit trees pose problems as their storage qualities are not good. Having collected, the collected material should be stocked in 'Gene Banks.' Many seeds retain their viability if stored at about 5 per cent humidity and about 20°C. Periodic germination tests must be carried out in respect of these orthodox seeds, and when germination falls below the levels of 85-95%, the stock must be replenished by resorting to 'Isoclimatic regeneration, under conditions where every type of competition is totally excluded. Isoclimatic regeneration involves creation of climatic conditions closely similar to the one from where the original stock of seeds had been collected.

Elimination of any type of competition is recommended so that pheno-types do not suffer loss of genetic integrity, a possibility that is more than likely if there is competition. If In fat, more than 100 genetic resources centres working in different parts of the world on these lines and let us hope they will conserve all the genetic diversity for the posterity.'

27

Diversity in Ecosystem

There are in most ecosystems, some organisms that eat other carnivores like a snake eats a frog, or a bird eats all types fish including carnivores. These are called consumers of the third order. Then there are the top carnivores like lion, vulture etc. which eat other carnivores also but are not killed and eaten by other animals. The animals take plants or other organic materials as food, break them down in their digestion process and produce new type of organic materials in their body. Therefore, animals are also busy in the process of organic production. Based on this concept, the green are called primary producers and the heterotrophic organisms secondary producers.

Both the primary producers and the secondary producers have their life cycles, and new generation of populations develop while old ones die. What happens to the dead? If the materials so locked in the body of the organisms are not returned to soil and atmosphere, the cycling of materials will stop and the earth will be full of dead organic matter.

There is a continuous breaking up or decomposition of these organic materials staking place every where in all ecosystems. Some organisms with a specialized mode of nutrition and life, like fungi, bacteria, etc. constantly decompose the dead organic materials into simple inorganic substances and during the process derive from them their food and energy. These are called decomposers or reducers. In fact decomposers are also a form of consumer who derive their food from dead parts of the primary and secondary producers. The role of decomposers is very special and important in the ecosystem as without their activity the entire cycle of minerals is liable to get blocked. Decomposers are also called scavengers, Figure-1 illustrates the ecosystem structure and function in forest, crop-plant and pond ecosystems.

Some of consumers live on or in other organisms (hosts) and derive their food from the host's body. These are called parasites. Some organisms prey upon the other organisms like a frog preying upon a grasshopper and these are called predators.

There are certain other types of close biotic interrelationships among biotic components of the ecosystem and they are termed as commensalism, mutualism and symbolizes. Commensalism is mostly met with in animals where two different types of organisms associate themselves together and one derives benefit from the other without adversely or favourably affecting the benefactor. This term in ecology includes relationships between different type of organisms for their requirements other than food also. Mutualism and symbiosis are the terms applied to very similar types of associations where one derives nutrition from another and in turn benefits the food supplier in some other way. For example, in lichens, the algae provides part of food for consumption by the fungal component.

In turn the fungus affords protection to the algae against desication, etc. Many of these relationships and terminologies of biotic components are based on food habits. The quantity of biological materials in a given area or volume of space in an ecosystem is termed standing crop. if the quantity is referred in terms of weight it is called biomass. The functional aspects of a species (of course with reference to the place of occurrence) is referred to as its ecological niche. Odum has divided ecosystem into six compoenents based on six functions. The biotic components are placed under three sections viz (1) producers which include all types of autotrophic organisms; (2) macroconsumers or phagotrophs (phago = to eat) which include animals which eat or ingest other organisms; (3) microconsumers or saprotrophs (sapro = decompose) or osmotrophs (osmo = to pass through membrane) which incude, heterotrophic micro-organism like bacteria and fungi most of which decompose dead organic material.

Abiotic Components

The abiotic components of two type or the non-living environment are usually-materials and energy. Material are like light, heat, stored energy in chemical bonds, etc. The quantity of abiotic materials like minerals present at any given time in an ecosystem is termed the standing state similar to standing crop which refers to the quantity of biotic components. The material continually on cycling i.e., entering the living system and through death and decay returning to soil and atmosphere. This process is called mineral circulation or the biogeochemical cycle. While the materials keep on cycling, fresh energy is continuously trapped from the sun by the green plants on the one hand lost in space through respiration and loss of heat by all types of organisms on the other. Thus energy flows like one way traffic.

Odum (1971) distinguishes abiotic part into three components viz (1) inorganic like carbon, nitrogen, hydrogen, etc. which are involved in cycling, (2) organic compounds like proteins, carbohydrates, etc. which make the body of organisms and (3) the climate like temperature, light duration and intensity etc.

The structure and function of terrestiral and fresh water ecosystems have been illustrated in Figure-1. A landscape shows a crop field, forest and small pond. The solar energy is trapped in all these three ecosystems by chlorophyll in green plants; carbon dioxide is taken from the atmosphere by terrestrial plants and from its dissolved condition in water by submerged part of aquatic and with the photosynthetic process green plants produce energy rich organic compounds. Necessary minerals for metabolism are absorbed in dissoved condition from soil solution or water. A large number of microscopic phytoplanktons and other macrophytes do the primary job of energy fixation.

The (the primary producers) are eaten by herbivores. In turn the herbivores are shown to be eaten by carnivores or predators in different ecosystems. The series of organisms fixing energy, eating and being eaten is called a food chain. Food chain can be traced in any ecosystem such as grass, grasshoppers, frogs, snakes and vultures in a grassland. Thus there are four to five steps or links in a food chain where food energy is transferred to the next trophic level. At each step a large portion of energy is lost as heat and only small fraction goes to the eater. Therefore, the quantity of energy goes on decreasing at a rapid rate successively from producers to top carnivores. The ratio of energy intake and energy of the produced biomass i.e. of input is called ecological efficiency. This can be studied at any trophic level.

The number of primary producers is 1,500,000 is 1000 sq. metre and the number of herbivores, carnivores and top carnivores decrease to 200,000, 90,000 and 1 respectively in the same area. But in an English Forest the number of herbivores is 150,000 for a mere 200 primary producers (tree) in 0.1 hectare, thus making the pyramid of inverted shape pyramid of Biomass. In order to explain the inverted nature of a pyramid of number, the idea of pyramid of biomass is given where the weight or primary producers forms the base.

The pyramid of biomass, therefore, becomes upright in a case where it was inverted if numbers were considered. From an unpublished work of Golley an Child, Odum (1971) has stated that is a Panama forest for 15 grams for hetertrophs in a sq. metre are 40,000 grams of primary producers are available. But there can be instances where the pyramid of biomass also gets deformed or inverted. The biomass of diatoms and other phytoplanktons is quite negligible as compared to the small herbivore fish that feed on them. The biomass of large carnivore fish feeding on small fishes still higher. In English Channel the biomass of primary producers is only $4g/m^2$ whereas that of the consumers is 21 g/m^2. In fact this is the case in most aquatic bodies. This can explained if the time factor is also taken into account. The pyramid of energy explains this.

Pyramid of energy

As against the pyramids of number and biomass the shape of the pyramid of energy is always upright because in this the time factor is always taken into account. The pyramid of energy represents the total quantity of energy unutilised by different trophic level organisms of an ecosystem per unit area over a set period of time (usually of energy utilized by different trophic level organisms of an ecosystem per unit area over a set period of time (usually per square metre per year).

The quantity of energy trapped by green plants in an area over a period say one year is highest compared to that of organisms of other trophic levels and therefore, the base of pyramid is broad. In aquatic ecosystem also the population of phytoplanktons quickly complete their life cycle and sets of new generation or crops of phytoplanktons are formed every few hours of day. The cumulative energy content that these generations of phytoplanktons trap in course of a year is certainly much more than that of only a few generations of herbivores fishes in the corresponding time and space.

The energy content of top carnivores (utilized in one year) is the least. Therefore, the pyramid of energy can never be of any other shape except upright pyramidal. In the even of destruction of primary producers or organism of any trophic level, the organism of next higher trophic level, the organism of next higher trophic level will automatically die for want of food (or source of energy) and ultimately the upright pyramidal shape is maintained. In Silver Springs in the U.S.A. the energy contents of different trophic level organisms as collected by H.T. Odum (1957) over a period of one year make an upright pyramid. The energy contents in $kcal/m^2/yr$ from 20,810 in primary producers decrease to 3368 to 21 in consumers of the first, second and third orders. Thus a top carnivore for the organic production of 21 kcal needs a very broad base of green plants equivalent to 20,810 kcal of energy.

Productivity Concept

In the pyramid of energy a very important factor has been introduced and that is the rate. The rate of energy trapping by green plants governs the rate of production of organic material from simple inorganic substances (in a given area over a given period of time). This rate of energy conversion or increase in organic biomass produced is called primary productivity. The total production as a result of photo or chemosynthesis is called gross primary productivity.

Some of the primary product is all the time lost by way of break down or organs matter in respiration. Net primary productivity is the rate of production when loss due to respiration is deducted from the gross production rate. Primary productivity refers to total increase in weight in all the parts-roots, stems, leaves fruits, etc. as against agricultural productivity which refers to useful parts as grains or fodder parts. Ecosystem efficiency greatly depends upon the production rates of primary producers. Oceans from the largest ecosystem and its productivity differs in different climatic regions.

On sea-shores the productivity per square metre per day may be 2 to 3.5 grams and in deep seas only 0.5 grams. In highly productive lakes this value may be 5 to 10 grams/m^2/day and even up to 50 grams for short periods in exceptionally favourable situations. The net productivity of crop plants (whole plant) is in the range of 0.3 kg to 1 kg to 1 kg to 4 kg/m^2/yr or even higher. All productivity values are expressed on dry weight basis of the organic matter or on the basis of the energy content in unit time and space.

Systems Analysis and Ecomodelling

As already explained, a system is a unified whole, a structural and functional entity, made up of several interacting and interdependent component and the components of an ecosystem are different kinds of a tutotrophic plants, animals, decomposer organisms, the abiotic parts like soil, air, water, solar radiation, etc. by virtue of their interactions, there is a continuous flux or changes in the quantities of each of the components. The energy needed to bring about any change ultimately comes from sun but solar radiation by itself cannot be utilized in such interactions.

The ecosystems, function through the he process of flow of energy (starting with solar radiation) and cycling of materials. In ecosystem analyse, is basically we therefore need to know the biotic and abiotic components, their properties, the quantities of materials needed to build the them, the rates at which the built up or produced organic matter is disappearing or getting used up in respiratory process, the quantity passed on to next trophic level organisms, the driving forces of light, temperature, rainfall, soil structure, thickness of the profile, nutrient status etc. We also need to know the rate of litter far, death of organisms, the rate of decomposition, mineralization and recycling of inorganic salts into the biotic components. From such worked out values, we learn in a very broad way about the working of any ecosystem.

When we wish to learn in greater detail, we have to further analyse each category of broad components into subcomponents such as in the autotrophs, we record the names of different species, their density, frequency, biomass, etc. on a time scale, i.e. recording the pattern of changes and the driving forces in term of measurable quantities. A very specialized role is performed by each species and the dimensions or magnitude of such function differs under different set of ecological conditions even for same species. This role is called as the ecological niche of the species.

In view of the multiplicity of driving forces on each species, singly on individual and collectively on their populations, it becomes very difficult to measure the niche dimensions. When we fail to make measurements, we simply describe ecosystems in qualitative terms. But when we are able to translate the complexities of interactions between the living and non-living and non-living components of ecosystems in the language of mathematics i.e. in terms of formulae, equations, graphs, etc. then we enter into the realm of systems analysis. Representation of structure and function of ecosystems in different forms are known as models.

Despite application of powerful computers modelling is only a simplified and incomplete version of the actual picture of any ecosystem. Out of hundreds of interacting pathways, only a few physical and biological characteristics are possible to be translated into mathematical relationships or models.

In ecomodelling we must at first define the boundaries of the ecosystem to be studied. The components are then identified. Picture models are commonly made up of boxes and arrows. We then put the components and subcomponents, into boxes at appropriate places and connect them through arrows showing the direction and type of interaction between the boxes. Such a model is also called as a compartment model. Interactions may be through transfer of matter and energy or other kinds of influences.

In identifying the components and level of splitting into subsystems, we are guided by the nature of information we need. If we are to know the ecological efficiency, we have to measure input/output ratios of energy of the boxes, of primary producers, herbivores and carnivores. For nutrients we record the quantity and rate of transfers of the particular element in different boxes such as the boxes for soil subsystem, litter fall, litter disappearance, etc. For resource conservation such as of particular plant or animal species, or for water or soil we take into account that resource levels in relation to physical forces of environment on a time scale.

These are more often represented in mathematical models in terms of equations instead of picture models. If the information on transfers between compartments are set in tabular arrangement we call it a transfer matrix model. The mathematical application of correlating the rates and kinds of changes in plant or animal growth performance, or population behaviour or other functions in relations to physical forces, is through the use of multiple regressions.

With the use of appropriate regression equations an investigator can find out an approximate value of transfer functions between different compartments. But it must be borne in mind that even with the use of complex models, the values obtained are only near approximation of actual measured value of the real ecosystem.

Some ecomodels are prepared on lines of electrical circuit using different shapes of boxes and arrows representing energy sources, passive energy storage pools, heat sinks, etc. This is called electrical analog circuit model.

Model based on not real data is called simulation model. The input of such forces into an ecosystem that affect the component (like sunlight or rainfall affecting plants), but not being themselves affected by the components are called forcing functions, modelling becomes very complex and difficult and such a model is called stochastic model. But is chance variation in the forcing function is ignored, the model becomes mathematically less complicated and is called deterministic model.

Ecological Energetics

Energy is the basic force responsible for running the machine of life. In fact energy is the capacity to do work and all living things must work and all living things must work. A virus, a bacterium or a tall tree, or any other organism, needs energy for metabolic activities. Green plants capture solar energy through photosynthesis and convert it into a chemical form. In this stored from the energy is passed on from organism to organism as food. Thus the sun is the source of all energy for organisms on the earth. Even the coal and petroleum which we use to provide energy for many purpose is solar energy which was converted and stored by organisms in the geological past and which became fossilized.

In the sun exceedingly high temperature acts a nuclear furnace where nuclear rearrangement like continuous transmitation of hydrogen atoms into helium is taking place. This process releases fantastically high amounts of energy which radiates out in all directions in the form of electromagnetic waves called *solar radiation.* A very minute portion of this radiation i.e. about one fifty millionth of the suns energy output, reaches the earth's atmosphere. The green plants have evolved a mechanisms of trapping and converting it into the chemical form of energy.

The chief form of energy are:

(1) radiant energy (2) energy (3) chemical energy and (4) mechanical energy.

Solar energy travelling in form of radiation is called *radiant energy.* Part of this radiant energy in wavelengths between 40 mu to 760 mu is the visible ranger or *light energy.* Light energy is the readily converted into heat energy as can be seen in the open sun when light waves striking the ground is converted partly into heat which warms the soil or other objects. The light energy entering the chloroplast machinery of the plant body is taken up by electrons of the chlorophyll molecule and passed on through series of stages into chemical energy in the bonds of organic compounds.

Mechanical energy may be of two forms; *potential energy* or the stored energy which remains in reserve and which can be used only on conversion into *'free'* form or *kinetic energy of useful* energy.

Different forms of energy interchange their forms under certain set rules. These are defined in two laws of thermodynamics.

The first law may be stated; Energy cannot be created or destroyed but may be converted from one form to another.

The second law, for the purpose of ecology, may be stated "processes involving energy transformations will not occur spontaneously unless there is a degradation of energy from a non-random to a random form."

The unit of energy measurement is the erg. The work done in lifting 1 gram of weight to a height of 1 cm against the force of gravity is equal to 981 ergs. One crore ergs (10 ergs) is equal to one joule. All forms of energy can be completely converted into heat energy. For a better and uniform expression, in ecology, therefore, energy is measured not in terms of ergs but in joules or units of heat measurement. Heat is measured in calories. One calorie is equal to the heat energy required to raise to temperature of 1 gram of water from 14.50°C to 15.5°C, and one calories 10^{3}) make one Kilo calorie or kilogram calorie (Kcal or Call). Thee is now a trend of expressing energy in ecology literature in terms of Kilojoules.

In ecological energetic we are interested mainly in the (i) quantity of energy reaching an ecosystem per unit area (say a square metre) per unit time (say one hour, day or year) and (ii) quantity of energy trapped a by green plants and converted to a chemical form and (iii) the quantity and path of energy flow from green plants to organisms of different trophic levels over a period of time in a known area. There is a progressive loss of energy among the biotic components of the ecosystem and hence at successive stages from producers to consumers of different orders the quantity of energy available decreases.

In the earth's atmosphere about $15x10^8$ calories/metre^2/year of solar energy is received. A large fraction of its does not reach the ground surface and is scattered by suspended particles in the atmosphere. The constant rate at which solar energy reaches the earth's outer atmosphere is called solar constant. About 33% of the incoming solar radiation is reflected by clouds and another 9% by suspended dust particles. Nine per cent is further held by ozone, water vapour and other atmospheric gases, thus leaving about 47% to reach the surface of the earth. The quantity of solar radiation received at any place. Therefore, depends upon the clarity of the atmosphere. Latitude is another important factor and the equatorial region receives maximum solar radiation followed by other regions of the trophics. The quantity of the energy goes on decreasing with increase in latitude both in the Northern and Southern hemispheres. This figure is a approximately 2.5×10^8 cal/m^2/yr in Britain, 4.7×10^8 cal/m^2/yr in Michigan and 6.0×10^8 cal/m^2/yr in Georgia. U.N. Singh has estimated it to be 7.5×10^8 cal/m^2/yr at Varanasi. Only about one to five per cent of the energy reaching the ground is converted by the green plants into chemical energy which is used in building up to the plant body and indeed metabolic activities. The rest of the energy is lost in various forms especially as heat dissipated and heat used to evaporate water. Part of the chemical energy in plant tissues is passed on from organism to organism as they are successively eaten. Ultimately the entire energy trapped by a green at one time is lost from the ecosystem in several stages. One important concept of ecological energetic is that energy always flows in one direction in the ecosystem while materials like carbon, nitrogen, hydrogen, oxygen, etc. are repeatedly used in the ecosystem cycling from inorganic to organic and back to inorganic forms.

Energy does not more in a free condition, but rather through organic materials from one trophic level to another. In assimilation of energy the green plants build up organic material from inorganic substance and gain in their organic biomass. This gain in weight (dry matter weight) as a result of conversion of solar energy into chemical energy per unit area of space for unit time is called primary productivity. Out of this increase in organic production some energy is lost by way of respiration of the primary producers and the net gain in weight is only the difference between the gross production and the quantity lost due to respiration. Animals derive their food from the net primary production. Only a very small quantity of the organic matter eaten by herbivores is actually used in their weight and the rest is lost as heat or is lost unassimilated through excretion.

28

Systems for Renewable Energy

Natural resources are the raw materials obtained or derived from nature. They belong to two broad categories:

1. **Renewable resources :** These are mostly biological in nature and include forestry, agriculture, animals (biomass-based) etc. These can reproduce itself in nature and we may harvest them continuously through a sustained proper planning and management. Solar energy, wind energy, water energy (tides etc.), geothermal energy also belong to this category, since these are available in an inexhaustible form in nature.
2. **Non-renewable resources :** They are physical resources like coal, oil deposits, natural gas, minerals, soil, metals etc. These are available in nature only in limited amounts and can not be reproduced. Coal, petroleum and natural gas are the common sources of energy. They, being of organs origin, are also called fossil fuels. These account for 90% of the world's production of commercial energy, hydro-electric and nuclear power accounting for only 10%. The figures are :

Oil	-	39.5%
Natural gas	-	19.6%
Coal	-	30.3%
Hydro-electric	-	6.7%
Nuclear	-	3.9%

Energy is an important input for development. It aims at human welfare covering household, agriculture, transport and industrial complexes. Countries all over the world have begun to think about a policy on energy and look into a possibility of having energy systems with no or very limited environmental impacts. The coal and oil deposits are to exhaust one day. Moreover, most of the non-renewable sources cause more pollution problems than the renewable sources of energy. The energy crisis has shown that for sustainable development in energy sector, we must conserve the non-renewable sources and also replace/supplement them by non-polluting renewable sources. The renewable ones are more or less pollution-free, environmentally clean, and socially relevant. Moreover, no nation can afford to depend on only one form of energy. It shall have a mix of at least seven forms (biomass, solar, coal, petroleum, natural gas, hydro and nuclear).

The commercial energy consumption in the developed world has increased during last 3-4 decades. More than 80% of the total world consumption of energy is by them which accounts for only 30% of the world population. On the other hand, 20% of the energy is consumed by 70% of the world population in developing and social countries. There has been a shift from coal to oil.

The energy scene in India is very complex. Important sources of non-commercial energy are firewood, agricultural wastes, and animal dung besides draught-animal and human power, whereas those of **commercial energy** are coal, lignite, oil, hydro-electricity and to some extent atomic energy. According to Advisory Board on Energy (ABE) the various forms of energy are likely to make the following contribution in India by 2004/2005 : Coal 450-540 Mt; Oil 90.110 Mt; Electricity 501-592 billion kWh and Non-commercial energy 500 Mt. The energy consumption is agricultural, industrial and transport sectors in primarily based on electricity, coal/coke and hydrocarbons (Table 28.1). The coal does not contribute significantly to household energy, though we have large deposits of coal. According to ABE (1985), in rural sectors 84% lighting is by kerosene and 94.5% cooking is on non-commercial fuel. In urban sectors, lighting 53% from electricity and 45.2% from kerosene, cooking 58.1% through non-commercial and 26.5% through kerosene.

The **non-commercial sources** of energy belong to two categories :

(i) New, renewable sources, like solar, wind, geothermal, microhydel etc. and (ii) Biomass-based renewable systems, like agricultural residues, biogas etc.

The systems in category (ii) constitute more than 56% of the energy consumption for almost 80% of the energy consumption (ABE, 1985). Increased application of category (i) leads not only tor reduced consumption of fossil fuels (coal; petroleum and its products; natural gas) but also biomass-based firewood etc.) systems. In turn this helps in conservation of resources as also in reducing environmental degradation.

Table 28.1 : Estimated energy consumption in 2004/2005 base case (ABE-1985)

Fuel	Units	Sector of consumption				
		Household	Agriculture	Industry	Transport	Others
Fuelwood	10^6t	300-330	-	-	-	-
Dungcake	"	199-221	-	-	-	-
Agricultural residue	"	90-104	-	-	-	-
Kerosene	"	16-18	-	-	-	-
LPG	"	2-4	-	-	-	1-15
Biogas	10^6m^3	169-188	-	-	-	-
Softcoke	10^6t	20-22	-	-	-	-
Charcol	"	2.8-3.1	-	-	-	-
Coal	"	-	-	178-198	5.4-63	-
Electricity	10^6kWh	96-107	42-45	302-366	9-12	52-60
HSD/LDO	10^6t	-	7.4-8.1	(80)	21-40	-
MS	"	-	-	-	7-8	
ATF	"	-	-	-	2.7-3.5	
FO/LSHS/HHS	"	-	-	16-20	-	-

LPG-Liquified petroleum gas; HSD-high speed diesel; LDO-Light diesel oil; MS-Motor spirit; ATF-Aviation turbine fuel; FO-fumace oil; LSHS-low sulphur heavy stock; HHS-Hot heavy stock.

Biomass-based Energy

Biomass is a general term used for all materials originating from photosynthesis. Thus biomass includes all new plant growth, residues and wastesd (wood, short rotation trees); herbaceous plants; fresh water and marine algae; aquatic plants; agricultural and forest residues (straw, husks, bagasse, corn robs, bark, sawdust, wood shavings, roots, animal droppings); wastes (garbage, night soil, sewage, industrial refuse etc.) Biodegradable organic effluents from industries like cannaries, sugar mills, slaughter houses, meat packing plants, breweries, distillaries etc. are also included in this category. Biomass can also be produced from hydrocarbon plants, oils etc.

Biomass energy systems are renewable, and a sink for CO_2 helps to conserve soil and water and helps also in water run-off in halting desertification march.

(I) Firewood

One must combine environmental supply of firewood and other biomass energy sources. Besides we need technologies for total utilisation of biomass and/or conversion to solid (densification), liquid (liquification) and gaseous (gasification) fuel. According to ABE (1985) the demand for firewood is going to go up the order of 300-330 Mt in 2004/05 against the present level of 120-130 Mt.

Table 28.2 : Estimated fuelwood consumption (Mt) in 2004/05 (ABE, 1985)

Region/State/UT	Fuelwood consumption		
	Rural	Urban	Total
Northern Region			
1. U.P.	37	10	47
2. Rajasthan	17	4	21
3. Punjab	3.5	0.5	4.0
4. Haryana	2.0	0.5	2.5
5. Himachal Pradesh	3.0	0.5	3.5
6. J&K	2.5	0.5	3.0
Eastern & N.E. Region			
7. West Bengal	9.5	0.5	10
8. Bihar	19.5	1.5	21
9. Orissa	13	3.5	16.5
10. Assam	11.5	1.0	12.5
Southern Region			
11. Andhra Pradesh	26.5	8.5	35
12. Karnataka	17.5	7.5	25
13. Tamil Nadu	20	10.5	30.5
14. Kerala	11.5	4.5	16
Western Region			
15. Madhya Pradesh	20.5	6.5	27
16. Maharashtra	22.5	5.5	28
17. Gujarat	10	3.0	13
Total	247	69	316

It may be seen that nearly 70% of firewood demand pertins to the rural areas. Only 50 Mt of the fuelwood may become available from natural forests. According to NCA for next 15 years or so the av. fuelwood contribution from natural forests would be 0.75 t/ha/year and the rest is to be met from plantations. Whole of the required plantation is to be non-agricultural land, degraded forest forest land, culturable wasteland, barren/inculturable land, permanent pasture and grazing lands. Roughly 60 M ha of such land are available, and it may be difficult to bring more than 50% of this land under plantation.

Briquetting : The Indian Institute of Technology, New Delhi has developed the technology of briquetting saw dust into smokeless fuel.

Gasification : Gasification of biomass is an important means to harvest energy through thermo-chemical conversion. It yields biogas, producer gas and pyrogas.

Improved chullahs : Present day chullahs (about 112 million) have a very low, 2-10% efficiency. This results into waste of wood, forest and environment degradation and health problems. Department of Non-conventional Enery Source (DNES) could design the improved stoves with thermal efficiency to about 15-25%. These stores besides wood can also use coal cakes, cow dung, pellets etc.

II. Petroplants

There are attempts to identify potential plant species as sources of liquid hydrocarbons, a substitute for liquid fuels. The hydrocarbons present in such plants can be converted into petroleum hydrocarbons. The plants belong to Euphorbiacese, Asclepiadaceae, Apocynanceae, Urticaceae, Convolvulaceae and Sapotaceae and over 385 species have been screened for hydrocarbon content. Fifteen species are found promising. There is need to increase the biomass of these plants, and conversion of their hydrocarbons into petroleum fractions. The Indian Institute of Petroleum, Dehra Dun has done excellent work in this area, particularly on hydrocracking of the crude products. The products obtained from their latex processed biocrude were, gases, naphtha, kerosene, gas oil, heaviers, coke.

III Biogas

Biogas is an important solution to present energy crisis, especially in rural areas. This is an environmentally clean technology. There is plenty of animal dung every year, from which about 22,425 million m^3 gas can be produced. This can replace kerosene oil used in villages mainly for lighting and partly for cooking. Besides gas, slurry can produce 206 million tonnes of organic manure every year that can replace 1.4 million tonnes of N_2 1.3 million tonnes of P_2O_5 (phosphate) and 0.9 million tonnes of K_2O (potash). However, this source of bioenergy is not being fully tapped. During recent years DNES has been working in implementing biogas technology on large scale. As many as 1,50,000 biogas plants have been set up in the country during 1984-85. At present there are over 330,000 biogas plants. It is estimated that 150,000 plants of biogas will save 600,000 tonnes of wood equivalent every year. Nearly 16-25 million rural households can meet their cooking energy needs through biogas. About 75% of the dung would be available for biogas. All India average for dung production is 11.30 kg. for cattle and 11.60 for buffaloes with 67.10 m^3 of gas per t of wet dung. Biogas is composed of methane, CO_2, H_2 and N_2. At 40% methane content, calorific value is 3,214 k cal/m^3, at 50% is 4,429 k cal/m^3 and at 55% is 4,713 k cal/m^3. In biogas technology, the important factors are, dung excrement and availability, gas yield, calorific value and appliance efficiency.

In order to make best use of biogas technology, two things are relevant : restricted use of water and better strains of methane generating bacteria. For normal microbial activity 90% water content is needed, whereas in bovine manure and human wastes this value is 80%. Thus we need additional water, which may be critical to this technology in drier areas of the country. This is to be developed a dry process, that requires less of water. Also, there is need for methanogens, able to operate at temperatures lower than 20°C. Land area is also a factor. The Chinese use an underground design. The ideal temperature for methane generation is 35°C or more. In Northern areas, gas production stops at 10°C. In China, such problems have been solved, by the use of a feed stock of 10% human excreta along with the liquid fraction, 40% pig, cow and animal manure dung along with stalks and grass and 50% of water. Biogas can also be generated from sludge obtained from primary treatment of raw sewage and one such plant is in operation at Okhla, Delhi. Besides gas, the rest matter from sewage is good manure.

IV. Electric energy

The various sources of this energy are (i) atomic (nuclear) power reactors, (ii) lignite, (iii) lean gas, (iv) hydro-power plants, (v) coal plants, and (vi) small (micro-hydel) plants.

1. Atomic power : This is of course a principal source of energy, when the fossil fuel reserves are depleting very fast. A small quantity of radioactive material can produce an enormous amount of energy. For instance, one tonne of Uranium^{235} would provide as much energy as by three million tonnes of coal or 12 million barrels of oil. Besides electricity, atomic power is also used as fuel for marine vessels, heat generation for chemical and food processing plants and for space crafts.

For atomic energy, we need a **nuclear reactor.** The decay of fissionable matter produces enormous heat. This is used to make steam and channeled through a turbine connected to an electric generator. There are different types of nuclear reactors.

(a) Light water reactor (LWR) : Here we use ordinary water for cooling and moderation. These are of two basic types. (i) Boiling water reactor (BWR) and (ii) Pressurised water reactor (PWR).

There are also high temperature gas cooled reactors (HTGCR) which are basically of LWR-type.

(b) Heavy water reactor (HWR) : Here we use heavy water. The most popular one has been Canadian Deuterium - Uranium (CANDU) reactor. Here the design is different from that of LWR-type. The fuel is arranged horizontally rather than vertically as in LWR.

(c) Liquid metal fast breeder reactor (LMFBR). Here we use liquid sodium as the coolant.

There are over 300 atomic power plants, operating in the world. The maximum are in U.S.A. (83), followed by USSR (40), U.K. (35), France (34), Japan (25), F.R. Germany (15) and Canada (13).

In India, Nuclear Power Corporation (NPC) is engaged with establishment of nuclear power plants. We are the seventh nation in the world to have mastered the nuclear fuel cycle. At present there are six nuclear power plants in operation generating 1230 MWe (individual plant capacity ranging 210 to 235 MWe). These plants are located in states of Maharashtra (Tarapur), Rajasthan (Kota), and Tamil Nadu (Kalpakkam). Some more units are to come up in Gujarat (Kakarapar), Karnataka (near Karwar) and Uttar Pradesh (Narora) in near future increasing the present installed capacity of 1230 MWe to 10000 Mwe by 200 A.D.

2. Coal : About 6000 billion tonnes of coal lies under the earth and by now over 200 billion tonnes had been used. Though we too have much coal deposits in India, increased use of coal has threatened rapid depletion of this conventional energy source. If used at same rate, it may not last for more than 2-3 hundred years.

3. Hydro-power : This is the most important commercial renewable energy source. It is cheapest source of power generation as land is free and environmental costs are never calculated. It is regenerative, thus leading to conservation of fossil fuels. Substantial portion of this unexploited energy is located in the Himalayan region in the North and North-east India. This potential for hydro-energy is being developed through large-scale projects. However, these projects, during recent years have caused environmental concerns. The infrastructure developement leads to deforestation and accelerated soil erosion. Such issues of environmental impact assessment of river valley and multipurpose irrigation projects have been discussed by Maudgal (1985).

4. Micro-hydels : There is substantial potential for micro and mini hydel power schemes in India. Micro-hydels are appropriate means of generating electricity in remote areas. Such installations are usually developed on the run-off-the-river schemes and do not involve much infrastructure. They are more safe environmentally as they do not involve approach roads, no pondage, no loss of vegetative cover, and are free from pollution hazards. They are very useful for local people.

Non-conventional Renewable Sources

Due to rapidly depleting conventional energy sources, countries all over the world have been forced to concentrate over tapping the vast potential of non-conventional energy sources like dendrothermal, solar, wind, ocean (tidal) and geothermal energy.

While the responsibility for harnessing the potential for energy by conventional technology lies with the Department of Power/Central Electricity Authority and the State Govts, the Department of Non-conventional Energy Sources (DNES) in actively involved in R & D activity for developing non-conventional know-how technology. Non-conventional sources can augment power in specific areas, in a decentralized manner. Their potential is actually for local use.

I. Dendrothermal energy

This method of generating energy from energy plantation has been experimented particularly in Philippines. There is claim of 3.75 kW from wood growing on 7ha of land. In our country, there is so far no experience and technology to run dendrothermal plants. Even if we could design such plants, the availability of required amount of wood is limiting factor. Even if land and wood become available, this method may serve local people only. Dendrothermal plants may be set up in Rajasthan and M.P. on experimental basis. There is not much pressure on land in these states.

II. Solar energy

Domestic heating, water supply can be met by this source. In Israel, such systems of heating homes and water supply are already in operation. In U.S.A., commercial solar heaters are available in Florida, and California, In Asia, Africa and Australia where number of sunshiny days is high, this method has promising future. India receives abundance sun-shine with about 1648-2108 kW/m^2/year with nearly 250-300 days of useful sun-shine in a year. The daily solar energy incidence is between 5 to 7 kWh/m^2 at different parts of the country. This enormous solar energy resource may be converted into other form of energy through **thermal** or **photovoltaic conversion** routes. The solar thermal route uses radiation in the form of heat that in turn may be converted to mechanical, electrical or chemical energy. Solar thermal devices like solar cookers, solar water/an heaters, solar dryers, solar wood seasoning kilns and silicon systems have been developed. A solar cooker consisting of an aluminium reflector (10 sq. ft in area) has been tried and introduced by DNES in extremely cold and remote areas of Ladakh. IITs have been successful in developing the technology to trap solar and wind energy on the icy continent of Antractica and remote areas of

Leh and Ladakh. In a solar cooker, nearly one m^2 collector area at 17.3% efficiency would give a saving of 663 kg of wood at $4708x10^3$ kcal/kg.

The photovoltaic conversion systems convert solar radiation directly into electricity through silicon solar cells. These may be single crystal silicon cells polycrystalline cells, amorphic solar cells etc. Such systems are used for community lighting, radio and TV sets, light houses, offshore platforms and installation in remote areas. An important application of these systems is pumping of water for micro-irrigation and drinking purposes. Such systems are decentralized on the spot electricity generation systems and help to replace diesel-utilising systems. Thus photo voltaics are totally without chemical and noise pollution. They may be installed in remote areas as forests, deserts etc. Under the National Solar Photovoltaic Energy Demonstration Programme (NSPEDP) the DNES has since 1980 installed a number of photovoltaic devices with a target of producing 1 MW of electricity. Such systems are used for pumping water, village electrification use in TV sets etc.

The Rural Electrifications Corporation has identified 90,000 villages in remote areas where electric supply is very costly and physically tedious. In such cases photovoltaics can be used as also biomass-based systems. According to ABE (1985), nearly 30,000 villages would be benefited by photovoltaics a demand for 75 MW. More than 300 villages in Rajasthan, Gujarat, U.P. and Haryana are supplied with street lighting units. India is to set up its first solar power station near Delhi very soon. The working groups of the Asian Solar Energy Network sponsored by the DNES/ UNESCO had meetings for development of technology for solar heaters, drying, desalination etc. The Department of Metallurgy, College of Engg., Pune designed a solar furnace that can generate temperature of 2,000°C. This can be used for heat treatments like surface hardening of steel and cast iron etc. In Haryana, first solar water heater was installed in 1982 at Haryana Breweries Ltd., Murthal. The system is able to heat 15,000 litres of water per day at average temperature of 65°C. At Medical College, Rohtak, the system is capable of heating 25,000 litres of water at 65°C. At other places in Haryana, there are such systems in operation. The first solar desalination plant has been installed at Haryana Tourist Complex at Dubchick. It will treat 2,000 litres of potable water daily. The first multipurpose solar dryer-cum-warehouse is being constructed at Ganaur (Haryana) to be used for dry chillies, potato chips etc. More than 100 domestic solar heaters of 100.1. capacity are installed in different parts of Haryana. Photovoltain irrigation pump set has been installed at Haryana Agricultural University, Hissar.

III. Wind Energy

In the country there are areas which are quite windy. Av. annual wind density of 3 kW/M^2/ day are prevalent at a number of places in Peninsular India as also along coast line in Gujarat, Western Ghats and parts of Central India. The wind densities are even more than 10 kW/m^2/day during winter and wind densities exceeding 4kW/m/day are available for 5-7 months in a year.

Wind energy may be converted into mechanical and electrical energies : Todate wind energy has been utilised for pumping water in rural areas and may also be useful in remote areas. About 20,000mW electricity can be generated in Indian from wind. According DNES, the wind farms with a total capacity of 3.3 mW have already been set up in Mandvi; 1.1 mW in Kutch; 550 kW in Okha (Gujarat); 550 kW in Deogarh (Maharashtra): 550 kW in Tuticorin (Tamil Nadu) and 550 kW in Puri (Orissa). The types of the wind mills developed are, 12 PU-500, Sail Type, Vertical Axis type etc.

Wind energy is useful in remote areas, helps saving fossil fuels, would deliver on the spot small quantity of energy which is free from pollution and environmental degradation. DNES has installed 924 wind pumps (PU-500 type) with pumping capacity of about 20 m throughout the country. Gujarat is the first state in the country to start using wind power.

IV. Ocean (tidal) energy

Tidal power generation depends on the harnessing of rise and fall of sea level due to tidal action. Small tidal power plants have been constructed in China and USSR. The most important application of tidal power is in electricity generation. In 1966, France constructed the first major tidal electric plant. In India, prospective sites for exploitation of tidal energy are gulfs of Kutch and Cambay and Sunderbans. In Kutch, French assistance has been received. Other suitable sites are near Lakshadweep Islands and Andaman and Nicobar Islands. At both these places, the cold water layers (20°C or lower) at 1,000 m depth are available near the shore. In India, tidal power potential of the orde of 9,000 mW has been identified, of which 800-1000 mW in Gulf of Kutch; 7000-8000 mW in Gulf of Cambay and the rest in Sunderbans.

India could intensify work on **Ocean thermal energy conversion** (OTEC) and **wave energy,** that would go a long way in meeting the power requirements in remote oceanic islands and coastal towns. The country is already experiencing with exploiting tidal energy. The Central Electricity Authority and the Gujarat Electricity Board carried out site studies for establishment of tidal plants in the Gulf of Kutch. India has an excellent OTEC potential and some of the best sites in the world are known to be located off the Indian mainland and near the islands of Lakshadweep and Andaman and Nicobar. The total OTEC potential around India is nearly 50,000 mW, which is about 150 per cent of the present total installed power generating capacity in the country.

1. OTEC : We may utilise the temperature difference existing between warm surface sea water (28-30°C) and the cold deep sea water (5-7°C) which is available at a depth of about 800 m to 1000 m in tropical waters. The advantage is that power is continuous, renewable and pollution free. A floating OTEC plant can generate power even at mid-sea, and can be used to provide power for operations like off-shore mining and processing of manganese nodules. We have already designed a one mw plant for Lakshadweep islands.

2. Wind wave energy : The incessant motion of the sea surface in the form of wind waves constitutes a source of energy. About 1.5 percent of the incoming energy from sun is converted to wind energy. Part of this is transferred to the sea surface, resulting in the generation of waves. This then is carried to coastal lines where it is dissipated as the waves break. Extract of energy from waves is more efficient than direct collection of power from wind, since the wave energy is concentrated through the interaction of the wind and the free ocean surface. For India whose coastal line extends to about 6000 Km. in length wave energy potential is estimated to be around 60,000 mw. There are many other benefits in extraction of wave energy. A multipurpose wave regulator system (WRS) in the form of a long barrier results in the formation of a calm pool between the barrier and shore, and this can be used as; a harbour, space for aqua-culture, space for coastal transportation with lighter and faster crafts and shore protection against the erosion by sea. A WRS can be placed at a depth of about 10 m, a distance of about 500 m from the shore (available at most places along the Indian coasts).

Gujarat may become the first state of the country to make use of tidal power.

V. Geothermal energy

Here we utilise the heat in the interior of the earth for power generation. This is possible in volcanic regions or where hot springs and geysers occur. For developing countries the overall projected potential is estimated at 400×10^{18} joules for geothermal energy with a projected capacity at 2300 mW in 1990. The total potential works out to be nearly 1/5th of 1971 recoverable resources of natural gas. Efforts are being made to sue this energy for generating power and creating refrigeration etc. A cold storage unit and 5mW power plant have been set up at Manikaran (H.P.). Geological exploration for prospective location for geothermal energy for electricity generation are to be undertaken. Drilling into deep levels is necessary where there is a possibility of availability of geothermal fluids. At present nearly 350 geothermal springs have been located into country. Puga (Ladakh), geothermal area is a good site for power generation. Other areas to be screened, are, Tattapani (Sarguja Distt. In M.P.), Cambay Basin (Alaknanda valley, U.P.) and Parvati valley (H.P.).

29

Environmental Monitoring (Bioindicators)

In every sphere of human endeavour, the use of biological systems provides on elegant device as these do not disrupt the stability of natural ecosystems. A few examples would make the point clear. Bioenergy is proving gradually as good alternative to nonrenewable sources of energy causing serious pollution problems. Biofertilisers are becoming popular over chemicals in agriculture. For control of diseases, chemicals used as pesticides in agriculture are being replaced by biocontrols agents.

In order to assess the changes caused by human activities, effective and reliable monitoring systems are required to recognise and predict hazardous effects. Biological methods can be successfully applied in predicting the impact of human activities particularly of pollutants well in advance since they present effective and reliable method of evaluating the effect of anthropogenic substances on the living organisms. Thus microbes, plants animals, cell organelles, organs, individuals, populations, biotic communities and ecosystems show different levels of sensitivity and can be successfully employed as **ecological indicators (bioindicators)** to assess and predict environmental changes in a timely manner.

Thus organisms, chiefly plants, species communities or even systems serve as a **measure** or **inde** (indicator) of the environment. If plants serve as indicators, they are called **plant indicators.** Each response is the effect of some factor or factor complex (interacting factors) acting as a cause and is, therefore, the indication of this factor. It is thus evident that every plant is a product of the conditions under which it grows and is therefore, a measurement of environment. Dominant species in an area are most important indicators, as they receive the full impact of the habitat for over longer periods. Consequently, plant communities are reliable indicators than individual plants. Plants are indicators of conditions, processes and uses. Large species serve as better indicators than small species; usually '**steno**' species serve much better indicators than '**eury**' species. Numerical relationship between species, populations and whole communities are more reliable than single species. Some of the obvious cases where plants and to some extent animals also, serve as indicators of some characteristic types of environmental conditions are as follows:

1. Indicators of potential productivity of land : Forests serve as good indicator of land productivity. For example, vegetative growth of trees like species of *Qurcus (Q. marilandica, Q. stellata)* is comparatively poor on lowland or sterile sandy soil than the normal soil in which they grow under natural conditions.

2. Indicators of agriculture : Native vegetation of a particular region is the safe criterion of agricultural possibilities. Thus, plants growing under natural conditions provide information on capabilities of land for crop growth than those obtained through meteorological data or soil analysis.

3. Indicators of climate : Plant communities characteristic of particular region provide information on the climate of that area. For instance, **evergreen forests** indicate high rainfall in winter as well as summer; **sclerophyllous** vegetation indicate heavy rainfall in winter and low during summer; **grasslands** indicate heavy rains during summer and low during winter; **xerophytic** vegetation indicate a very low or no rainfall in the year.

4. Indicators of soil type and other soil characteristics : Luxuriant growth of some taller and deeply rooted grasses like *Psoralea* indicates a sandy loam type of soil, whereas the presence of grasses as *Andropogon* indicates sandy soil. *Rumex acetaella* indicates an acid grassland soil, whereas *Spermacoce stricta* the iron-rich so in the area. Plants like *Chrozophora rottleri, Heliotropium supinum* and *Polygonum plebejum* grow better in low-lying lands. *Shorea robusta, Cassia obtusifolia, Geranium* sp. and *Impatiens* sp. indicate proper aeration of soil. Grasses like *Saccharum spontaneum* prefer to grow in poorly-drained soils. Plants as *Artemisia ridentata, Kochia vesrita, Salicornia utahensis* and *S. rubra* indicate saline soils. *Capparis spinosa* and *Carissa spinarum* indicate intense soil erosion.

5. Indicators of fires : Some plants as *Agrostis hiemalis, Epilobium spicatum, Pinus contorta, Populus termuloides, Pteris aquilina* and *Pyronema confluens* (Fungus) dominate in area destructed by fires. *Pteridium* spp. in particular indicate burnt and highly disturbed coniferous forests.

6. Indicators of petroleum deposits : Some protozoans as *Fusilinds* indicate petroleum deposits in the area.

7. Indicators of adequate oxygen in water : Burrowing may fly (*Hexagenia* sp.) indicates proper oxygen regimes in the water.

8. Indicators of pollution : Plants like *Utricularia Chara, Wolffia* prefer to grow in polluted waters. Bacteria, like *Escherichia coli* also indicate water pollution, presence of diatoms in water indicates pollution by sewage. Movement of fish like *Catla catla, Labeo gonius, L. bata, L. rohita* and *Natopterus natopterus* away from the water indicates industrial pollution of water.

9. Indicators of overgrazing : Annual weeds and short-lived perennials like *Amaranthis, Chenopodium* and *Polygonum* etc. grow better in overgrazed areas.

Frequent visits of the areas by animals as cattle, horses, sheeps, goats etc. also indicate that the area is under intense grazing.

Biological methods of monitoring may provide information about the state of environment due to their following characteristic features at different levels:

(1) Microbes, plants and animals have the ability to accumulate a hazardous substance occurring in the environment. They may thus indicate the presence of such a substance.

(2) Life processes of different organisms can be used to evaluate the action of environmental pollution and that of a given pollutant.

(3) Changes in the population of species and in the structure of ecosystem can indicate the level of environmental deterioration.

Biological systems as indicators of the environment, therefore, have a remarkable potential in forecasting of disasters, prevention of pollution, exploration and conservation of natural resources all aiming at a sustainable development with minimal destruction of the biosphere.

Biological Monitoring Programme

Realising the urgent need of biological monitoring the International Union of Biological Sciences (IUBS), in its XXI General Assembly, held in Ottawa, Canada in 1982, decided to initiate a worldwide programme for identifying and applying biological indicators in environmental monitoring particularly to evaluate the effects of hazardous substances on ecosystems. The IBUS recognising the importance of Bioindicators Programme constituted an International Steering Committee headed by Professor J. Salanki of Hungary. The objectives of this programme are as follows:

(1) to encourage scientists, as well as scientific bodies to develop and improve methods indicative of hazardous substances occurring in the environment.

(2) to collect information on existing methodology of bioindicators.

(3) to promote exchange or experience between different laboratories to help dissemination of recent knowledge among different countries.

(4) to provide literature about bioindicators including different reference lists, general description of methods and practical manuals suitable for distribution to interested bodies.

(5) to promote interdisciplinary and international cooperation in standardizing and extending the use of agree methods.

(6) to stimulate scientific bodies to encourage the presentation of new results concerned with bioindicators at international meetings and

(7) to organise special regional or international symposia, workshops and seminars in cooperation with national bodies or other organisations on the methods, new results and there application.

Keeping in view of the increasing importance of bioindicators in environmental monitoring the Indian National Science Academy (INSA), New Delhi in association with the IUBS organised an International Symposium on Biological Monitoring of the State of the Environment (Bioindicatiors) as a part of its Golden Jubilee Year Celebrations, from 11-13 October, 1984 at New Delhi.

Bioindicators and Environmental Monitoring

Bioindicators of soil, water and air pollution provide information for developing suitable programme for biomonitoring. Bioindicators provide a practical way of assessing the health of environment. The term **bioindicators** covers a wide spectrum of organisms serving as indicators of environment. The most rational way of use of biological system as indicators of environment is to employ those organisms which in their presence or obsence and in all features of their phenotype and physiology, serve as an index of their environmental status. Undoubtedly, they indicate the level of pollution, but more important is the fact that they provide a clear insight into the composition of their substratum. Thus both the positive and negative facets of the environment can be monitored through biological systems. A bioindicator actually indicates the general toxicity of the environment without telling the exact physical or chemical factors responsible for this toxicity. The appropriate physical and chemical methods are to be supplemented to understand the exact chemical responsible for pollution.

There are a variety of biological systems which can be used as indicators of harmful anthropogenic substances. The IUBS programme divides these systems into following six groups : (i) Microbiology, (ii) Botany, (iii) Zoology, (iv) Cell Biology and Genetics, (v) Comparative Physiology and (vi) Hydrobiology.

We shall consider each of the systems briefly here.

I. Microbial systems

Microbes are rapid detectors of environmental pollution both in water, and soil. There are microbes, specifically sensitive to some substances while others take part in decomposition of pollutants. The elimination (of sensitive spp) or abundance (tolerant and involved in breakdown of pollutant) of species can indicate changing environments. Alteration of microbial communities and reduction of species diversity can be the result of the presence of specific toxic agents. Microbial muds from continental and intercontinental water bodies serve as ideal tools for detecting several compounds including sulphur. The detection of polar lipids of archeobacteria in petroleum belts and sediments is an elegant monitoring system. *Solmanella typhimurim,* other bacteria and fungi as *Neurospora* and *Aspergillus* provide excellent device for monitoring genetic effects of physical and chemical agents. Several microbes are used in assessment and prediction of changes in marine environment induced by human activities. These include *E. coli, Zilrio* spp, *Aeromonos* sp, *Pseudomonas, Clostridium, Streptococcus* etc. Cyanabacteria area used as bioindicators of soil pesticides. For instance *Nostoc microcopicum, Haplosiphon welwitschii* and *H. confervaceus* indicated the pollution of Dithane, Deltan, Aldrex, BHC-, Rogor, Phorate etc.

Some filamentous fungi, yeasts, actinomycetes and bacteria are used to monitor oil pollution (oil spillage). *Scolecobasidium, Mortierella* spp, *Humicola, Vertillium* sp. are able to utilise waste oil. Fedorak *et al* (1984) isolated about 74 yeasts and 224 filementous fungi from marine water and sediment samples. These included *Penicillium* spp, *Candida quilliermondii* and *Aureobasidium pullutans,* which utilise oil fractions.

II. Lower Plants

Both, from pollution as well as survey of resources viewpoints, different groups of plants indicate the nature of the environment. The susceptibility or resistance towards a substance in the environment varies with species. For instance, lichens due to their susceptibility and resistance to different environmental effects, are ideal monitoring agents. There are some lichens which can thrive only in the unpolluted air whereas others are resistant even to the most polluted systems. The wide variety of lichens including different species of *Lecanora* are good indicators of a broad spectrum of environment. The presence of SO_2 and fluorine in atmosphere is indicated by lichen thalli. Even the dead thalli are capable of absorbing fluorine and heavy metals including lead. Lichens are also utilised for survey of long life nucleids like cesium 137, strontium 30, released from nuclear explosions.

III. Higher Plants

Various groups of higher plants serve as bioindicators. Sensitive species are employed to detect and monitor specific air pollutants. Tolerant (indicator) species are used to determine the incidence of particular soil condition. Studies on plants have more often been directed specifically to monitoring heavy metals in waters. There is extensive literature on higher plants as indicators of air and water pollution. Different species have been used to detect and monitor gaseous (as SO_2, O_3, nitrogen oxides) as well as heavy metal pollutants. (A1, Ca, Co, Cu, Zn, Cd, Pb, Hg, Mo, Mn, Ag). the zinc tolerance of *Anthoxanthum* spp, copper tolerance of *Agrostis,* lead tolerance of *Festuca* and cadmium tolerance of *Impatiens* are well known example. These plants are not only indicators but also as pollutant scavangers. Studies are very useful in locating new areas of mineralisation.

A wide spectrum of phenotypic, metabolic and anatomical changes in the plant system reflect the nature of the compounds to which the plants are exposed. The weathered flakes of tobacco or chlorotic flakes of pine needle are good examples of zone damage. The collapse, glazing and bronzing of leaf cells are products of damage by peroxycetyl nitrate.

Several physiological and anatomical parameters are taken into account. For example, inhibition of photosynthesis and enolases is associated with fluorine damage. Bleaching of perianth and stamen injury are indicators of mercury poisoning. The nature of stomata, pigmentation, chlorosis and bleaching are generalised effects through bioindicators. Activities of several metabolities and enzymes are excellent indicators of environmental effects. Tiagi and Aery (1985) have presented a list of some of the well established plants as indication of heavy metals.

IV. Animal Systems

Individual species as well as whole community provide data on accumulation of chemicals in animals. Accumulation occurs to different extents in different organs. Accumulation of chemicals within food chains and consequently higher levels of toxicity in human food stuffs are indicated by selecting a suitable species for routine study.

Fish, *Daphnia* silver carp etc. are used to monitor heavy metal and pesticide pollution levels in water. Zooplanktons as rotifires and cladocerans are used as indicators of freshwaters. Earthworms are good bioindicators of soil radioactive pollution.

V. Aeroallergens

These include aerial flora and fauna. The air is medium of transport of flying animals germs of infectious diseases, plant and animal parts, fungal spores, pollen grain etc. The monitoring of airborne pollen and spares and related microphytoplankton such as trachieds, cuticles, algal filaments, insect scales and wings has received special attention of aerobiologists, mainly for its bearing on human allergy and plant pathogenicity. Pollen grains are pollutants causing biopollution. Pollen grains are considered as omnipresent (occurring everywehere around the globe)_ and thus good bioindicators in monitoring programmes. Nair (1985) presented an account of biomonitoring of airborne plant materials.

VI. Human System

Blood and urine along with others are widely used as exposure to toxic compounds. However, such an analysis bristles with the limitation of monitoring capacity being restricted to recent exposures. On the other hand human hair from head, can trap metallic vapour and dust over a long period of time due to their affinity with hair protein. Thus samples as old as 2000 years could be successfully analysed.

VII. Cell Biology, genetics and comparative physiology

Cellular and sub-cellular components, including chromosomes, adapted to specific environmental conditions, forms an excellent parameter for bioindicators. Both, short and long-term test systems have been developed **in vitro** as well as **in vivo** to monitor changes caused by different environmental agents. Grover *et al.* (1985) presented an account of monitoring environmental chemicals by chromosomal aberrations in plants.

Many animals show behavioural responses following the detection of environmental changes by their sense organs. A chemical may influence the functioning of endocrine, nervous, muscular, cardiovascular and excretory systems. Such changes may be investigated at morphological, biochemical or physiological levels and can indicate the presence of toxic substances.

30

Environmental Priorities in India

Environmental Management

Environment has been defined as "the sum total of all conditions and influences that affect the development and life of organisms". The interplay of material cycles and energy flows in natural ecosystems generates a self-connecting homeostasis with no outside control or set points. The ecosystem is capable of self-maintenance. However, the equilibrium is very sensitive to external stimus, such as human activities promoted by socio-economic goals.

The socio-economic system of man in contrast to natural ecosystem is founded on a material base, which of course is a partly finite. Increasing demands and aspirations of human society create a rising demand for goods and services. Thus human being imposes changes on natural ecosystems and increasing control of his environment often creates conflicts between his goals and natural processes. In humankind's present day economic system only money circulates within and between economic systems. A free exchange takes place between the natural ecosystem and the socio-ecosystem. The mis-allocation of resources, thereby leading to over-production of wastes eventually poses a threat to survival of mankind.

The **Central theme** of environmental management is "the reduction or minimisation of the impact of human activities on the environment, thus an endeavour to avoid the overuse, misuse and abuse of environmental resources." The cornerstones in any environmental strategy are (i) Environmental planning (ii) Environmental status evaluation (iii) Environmental impact assessment and (iv) Environmental legislation and administration.

In this chapter attempt is made to identify the priority areas and the strategies, in Indian context, to solve the current problems that should lead to sustainable development. Contrary to Western concept of **conflict** with nature, our basic philosophy has been of **harmony** with nature. However, during last 150 years or so we have been following the former one.

Today, the whole world, particularly the developing countries face a near crisis situation, both economic and environmental. The words **economics** and **ecology** have the same root, **oikos** (meaning house). While economics deals with financial house keeping, ecology deals with environmental housekeeping. Time has now come when sustainability in development has to enter in our planning process.

There are two major threats to the environments : (i) possible nuclear war, and (ii) unprecedented demographic pressure due to population explosion. This led to over-utilisation of natural resources and harnessing the forces of nature for the use of growing human population. The net result is spread of desertification, soil maladies, floods and droughts, urban congestion, extinction of species and pollution of land, water and air. Therefore, this is high time we work for environmental management i.e. proper resource use and resource management.

In Indian context the following are the environmental problems where priority action is needed :

(1) Population stabilisation
(2) Integrated land use planning
(3) Healthy cropland and grassland
(4) Woodland and revegetation of marginal lands
(5) Conservation of biological diversity
(6) Control of water and air pollution
(7) Development of non-polluting renewable energy systems
(8) Recycling of wastes and residues
(9) Human settlements
(10) Environmental education and awareness
(11) Updating environmental law, and
(12) New dimensions to national security.

These are the very issues that are posing threat to our environment as a whole, and to sustainability, in particular.

Population Stabilisation

Every sixth person born in the world in an Indian. Twenty-six Indian born each minute add to the world population increase of 154 individuals every 60 seconds. With 15.69 per cent of the world's population, India registers an annual increase of 13.7 million people (1981 census).

There has been abrupt rise in economic pressures due to demographic pressures all over the world. The cropland (food) woodland, (fuel) grassland (fodder), etc. that support human population are, thus largely affected. This could be disastrous to our economy. Therefore, population stabilisation as a major strategy. In many countries human population continued to grow rapidly and aggravated such environmental and natural resource problems as soil erosion, desertifiction and deforestation, which affect food and agricultural production.

In 1980, the birth rate was around 34/1000. It is to come down from over 30/1000 to about 21/1000 and nearly 11 in 2020, with a death rate 9/1000. There is need for safe effective, simple inexpensive and acceptable contraceptive methods of birth control. Today India accounts for 15% of the world population, while land area is only 2.4%. Thus, per capita availability of land is 0.48 ha as against 4.14 ha in USA, 8.43 ha in USSR, 1.91 ha in Burma and 0.98 ha in China and Pakistan. The population is concentrated in well-watered plains, thus heavily dependent on monsoon-oriented agriculture. Man-land ratio taken in relation to land is only 0.27 ha which for such predominantly agricultural country is indeed low.

There is thus an urgent need to take up critical studies on relationship between population growth and the life support system and the availability of resources including energy. Though ours is resource-rich country. The situation will change with enhanced demographic pressures. Population explosion is a time-bomb which has yet to be defused. Zero population growth (ZPG) holds the key to the well being and prosperity of the nation and the government must take effective measures to check population growth.

Integrated Land Use Planning

Land is one of the most important components of the life support system. Unfortunately land has been overused and even abused over the centuries. Mrs. Indira Gandhi in 1972 said "we can no longer afford to neglect our most important natural resource. This is not simply an environmental problem but one which is basic to the nature of our country" In a predominantly agricultural country like ours, land comes first. Due to exploding population, soil is used increasingly which poses threat to its productivity. Due to demographic pressures land is under stress due to sprawl in agriculture, industry and urbanisation. Cropland is losing fast fertile top soil.

Our land use statistics are confusing. For proper planning we need authentic figures agreed upon by the concerned departments (agriculture, forestry, revenue records etc.) using modern methods as remote sensing. For instance, good agricultural land in trans-yamuna area, from Ghaziabad onwards is going to industrial estates and for urban development. Thus valuable cropland is lost to agriculture for ever. This is not in national interest. The best land use planning has been done by Japan, being hard pressed for land. France has a most effective programme of land preservation. India has one of the lowest man-land ratio-hardly 0.48 ha/per capita. A strategy must be developed to cure the past damage and to save the country from future damage to land. This can be achieved as follows:

(i) to prepare accurate land use data through remote sensing etc., (ii) a time-bound nation-wide survey programme of micro-level land use planning giving short and long term scenarios, (iii) to prepare land use classes, (iv) to review all existing legislations and to update them, and (v) management plans for land amelioration are to be prepared. This would lead to a dynamic **Land-use Policy.** The Government has constituted an apex body headed by the Prime Minister, called the National Land Use and Wastelands Development Council (NLUWDC). At the second level there are two boards : National Land Use and Conservation Board (NWDB) (Ministry of Agriculture) and National Wasteland Development Board (Ministry of Environment and Forests). NWDB was set up in 1985.

I. Water resources

Water is an integral part of land/soil productivity base and its misuse can cause soil degradation and soil erosion. Water management is thus necessary for crop yields and other activities. Primary channel flow originates in upper catchments and these watersheds are very important for future. Therefore, health of watersheds is the most important.

II. Watershed management

Availability of water in a given soil environment is a critical factor and is related to erosion, siltation, loss of plant cover and productivity. In India, floods bring much havoc causing loss of life and property each year. Due to floods, the plains have become silted with mud and sand, thus affecting the cultivable land areas. Extinction of civilization in some coastal areas is mainly due to such natural calamities as flood. The National Commission on floods has calculated that the land area prone to floods has doubled from 20 million hectares in 1971 to 40 million hectares in 1980. Flood damage cost the country Rs. 21 crore in 1951 which increased to Rs. 1,130 crore in 1977. It was Rs. 128 crore per year during the decade 1960-70, increasing to Rs. 739 crore per year during 1970-80. The worst suffering states are Assam, Bihar, Orissa, U.P. and Wet Bengal. These aspects are to be given serious thought to save further destruction of mankind. Through modern technology and scientific knowledge, there is need of a proper understanding of the ecosystem so that such changes could be forecast well in time. Thus management of rainfall and resultant runoff is very important. Such management can be best based on a natural unit called watershed. A watershed is an area bounded by the divide line of water flow. Thus it may be a drainage basin or stream. The Himalayas are one of the most critical watersheds in the world. Our water regimes in the mountain ranges are threatened resulting in the depletion of water resources. The damage to reservoirs and irrigation systems and misused Himalayan slopes is mounting as are the costs for control measures during the "flood season" every year. The vast hydroelectric power potential can be harnessed from Himalayan watersheds only when proper control measures are taken. These include soil and land use survey, soil conservation in catchments of River Valley Projects and flood prone rivers, afforestation/social forestry programmes and Drought Prone Area Development Programmes and Desert Development and Control of Shifting Cultivation.

III. River Valley Projects

The environmental side effects of river valley and hydel power projects could be classified in three categories, viz.

(1) Impacts within and around the area covered by the dam and reservoir.

(2) Downstream effects caused by alteration in hydraulic regime, and

(3) Regional effects in terms of overall aspects including resource use and socio-economic aspects.

The impacts caused by construction of dams and reservoirs include changes in the microclimate, loss of vegetal cover, soil erosion, variation in water table and enhanced seismic activities due to pressure of water. The nature and magnitude of the impacts vary with the project locations and the conditions therein. For instance, in hilly tracts, blast in operations for road construction can cause considerable damage to the environment through loosening of hill sides and resultant landslides, sedimentation of reservoirs, drying up of springs and flashfloods. The creation of new settlements for the workmen and rehabilitation of project oustees in the watershed areas may aggravate the seriousness of advance impacts.

From the above it becomes clear that necessary mitigative and environmental safeguards are to be guaranteed before clearance of a project.

The safeguards for the following are to be guaranteed :

(1) Submergence of valuable agricultural and forest areas

(2) Siltation of reservoirs due to degraded catchment conditions

(3) Satisfactory rehabilitation of oustees

(4) Loss of flora and fauna

(5) Reservoir induced seismicity and

(6) Water borne and soil borne diseases.

A number of big, medium and minor dams have been envisaged under different river valley projects. These dams have been undertaken for irrigation, power generation and water supply. These dams hailed as the **Temples of Modern India** by the country's first Prime Minister Jawaharlal Nehru, have increased agricultural production, power generation and reduced dependence on imports.

But according to some experts, the social, environmental and even economic costs of these dams, however, far outweighs their benefits. According to them, most important social consequence of big dams has been the displacement of millions of tribals from their homeland and their eventual influx into urban areas, almost as refugees. Scientists, environmentalists, journalists, social activists, lawyers and bureaucrats have now taken up the cry against big dams.

Mounting opposition from scientists and environmentalists has forced the Government to review a number of proposed dams in the light of their impact on local tribals, flora and fauna. A glaring example that had effect on Government policy was the scrapping of the silent valley project in Kerala. Another example is Koel and Karo project in Bihar. This was also abondoned due to opposition from local people as it would have displaced several thousands of santhal tribals in the area.

The four major projects which have generated much controversy are, (i) Sardar Sarovar Project, Gujarat, (ii) Narmada Sagar Project, Madhya Pradesh, (iii) Bodhghat Project, M.P. and (iv) Tehri Dam Project in U.P. Though these have been given environmental clearance, struggle is still on to force the Government to drop these projects. In the dailies, headlines as "Loses exceed Theri dam benefits", "Government forced to review dam projects", Big dams spell doom...", "How green was my valley". "The displacement factor" etc. have been much common.

Sardar Sarovar (SS) project, near Navagam in Bharuch district of Gujarat is one of the costliest projects affecting villages in three states - M.P., Maharashtra and Gujarat. About 245 villages will be submerged, of which about 193 in M.P. alone. Over 75,000 (nearly 50,000 in M.P. alone) people will be evicted. Additional displacement is likely to be caused during social and environmental rehabilitation work undertaken to repair the dislocation and damage caused by the project. Thus compensatory afforestation and setting of wildlife sanctuary will displace or affect other villagers in the area. It has been officially admitted that nearly 43,000 ha of land will be needed for rehabilitation of SS oustees.

Narmada Valley Project (NVP), the world's largest river valley project has attracted the greatest attention. The 30 big dams and over 3,000 medium and minor dams are envisaged at cost of Rs. 25,000 crore. It would displace over one million people, mostly tribals, submerge 56,000 ha. of fertile agricultural land. Total forest area of nearly 60,000 ha. will be destroyed. About 25 species of birds will be deprived of their habitats.

Bodhghat project on Indravati river in M.P. is in heavily forested Bastar district. The project will destroy teak and sal forests, and spell doom for the last surviving wild buffaloes. The criticism of the project forced the Government and the World Bank to reconsider it.

Tehri Dam on the Bhagirathi river in U.P. at the foothills of Himalayas, is Soviet-financed and challenged in the Supreme Court. This Dam will displace over 85,000 people, and will totally immerse the Tehri town and completely or, partly submerge nearly 100 villages. The site is prone to intense seismic activity. The 3,200 million tonnes of water that the Dam would impound could cause a major earth tremor. In th event of a disaster, the entire religious townships of Deoprayag, Haridwar and Rishikesh would be devastated. Thousands of hectares of rich, agricultural land will be drowned. The project costs about Rs. 3,000 crores. A major drawback is the site, situated barely 100 km. from the Chinese border.

IV. Irrigation (Waterlogging and Salinity)

Rain water is conserved through wells, reservoirs, bunding of streams and canal systems. Since independence, there has been increase in net irrigated area of the country. Since independence, there has been increase in net irrigated area of the country. Since the start of the Five Year Plans the area increased to 38.0 million ha. Of this 60% is by wells and tube wells.

Though there has been phenomenal growth in irrigation potential, it has not given returns in terms of yield or money commensurate with the investments. Moreover, large areas became submerged under reservoirs and have been destroyed by construction of roads and colonies. Aded to these are the problems of salinity and water logging. Net result is loss of land productivity. Canal irrigation is harmful to soils which degrade due to water logging and salinity. In India, more than 20% of irrigated land is damaged due to water logging and salinity. It is estimated that fertility between, 30 to 80% of the world's irrigated land has been affected due to salinisation. In high rainfall areas of Assam, Bihar, M.P., Orissa, W. Bengal and Kerala, canal water is difficult to manage during rainy seaons. Water logging and increased salinity are mostly the outcome of poor drainage in areas with massive irrigation projects.

Based on above, following conclusions, could be drawn.

(1) Benefits from canal irrigation can be sustained in low rainfall areas for the first 15-20 years and thereafter occur water logging and salinity.

(2) Canal distributaries and field channel are to be lined.

(3) Water needs proper management at the farm level for regular and timely supplies.

(4) Proper land levelling and land shaping to increase water use efficiency.

(5) A critical review of the extent and nature of irrigation in high rainfall areas.

V. Wastelands development

Degraded, mined and other type of wasteland should be reclaimed and put to some productive use. Wastelands are those which for one reason or the other do not fulfill their life sustaining potential. Increasing misuse of land resources through shortsighted development policies has resulted into waste lands. About half of the land area of the country is lying as wasteland of varying intensity of degradation.

1. Degraded land : More land is needed for agriculture and forestry due to increasing population. Morover, our good land is shrinking both in quantity and quality. The former is due to unexpected demands and the latter due to erosion, desertification, water logging, salinity, alkali soils and toxic effects of agrochemicals and industrial effluents.

The only way to raise land resources is by reclamation and developing degraded land.

2. Mined areas : Most mining work has been unscientific with no environmental portection. There have developed large tracts that lost productivity. There have been water and air pollution, despoilation of land and deforestation, noise and ground vibration problems etc. Mined areas are to be reclaimed for agriculture, forestry, fisheries and recreation.There exist standard methods of reclamation. During last 20-30 years, a number of mining operations have been started in the country. These affected forest and cultivated land areas. Such operations have been taken mainly in U.P., Bihar, M.P., Orissa and Andhra Pradesh. The use of land on large scale for townships, communication, excavation and transport affected the socioeconomy and ecology of these areas. In Ranchi several hundred sq. km. of land has been converted to **bad lands.** In Singrauli complex forests and hillocks havebeen erased due to construction of high power transmission lines, roads and rail tracks. Establishment of other factories as cement and super-therml power stations around coal mines have resulted into environmental degradation.

The abovesaid problems were realised and there was set up **The National Wastelands Development Board** (NWDB) in 1985 with main objective of preventing good land from becoming wasteland. Besides this NWDB also looks for regeneration of degraded forest areas and reclamation of ravines, usar lands, arid tracts, mine spoils etc. Similar activities are also being carried out by **Eco-Task Forces** of ex-servicemen, a joint venture of the Ministry of Environment and Forests, Ministry of Defence and the concerned State Govts. Three such forces viz. Eco-Task Force (TA-127) U.P., Eco-Task Force (TA-128), Rajasthan and Eco-Task Force (TA-129), J&K are responsible for regeneration activities.

The Ministry of Envirornment and Forests along with NWDB have selected five districts in the country for going to the ground and drawing up action plans for reforestation and amendment of degraded land there. Detailed maps of 146 districts in the country representing every state had been prepared for identifying the wastelands and plantations. The five districts chosen are Almora (U.P.), Purulia (W. Bengal), Bellary (Karnataka), Durgapur (Rajasthan) and Sundargarh (Orissa).

During last three years, a total of five million hectares of land is said to have been planted by NWDB. Degraded forest land will be turned to green cover by planting tree species for fuel and fodder.

To augment fuelwood production in rural area, such programmes have been undertaken. Social-forestry programmes are currently under implementation in 14 states with financial assistance from the World Bank, USAID, SIDA and CIDA.

There is a project called the "IFFCO Farm Forestry Project" taken up by IFFCO in ten states to cover a total of 50,000 hectare of wasteland and similar land in seven years commencing from 1987. Rajasthan has a share of 5,000 hectare of such land. IFFCO has set up a nursery for ten lakh plants.Some saplings are to be supplied by State Forest Department.

The plant species selected fro plantation are those which could withstand moisture stress. The species to be planted are babul (*Acacia nilotica*), Israeli babul (*Acacia tortilis*), Subabul (*Leucaena leucophela*), Khejri (*Prosopis* spp), Khair (*Acacia catechu*), Karanj (*Pongamia pinnata*), neem (*Azadirachta indica*) and bamboo (*Bambusa arundinacea*).

Healthy Cropland and Grassland

India made good progress in agriculture, production increasing from 50.8 Mt in 1950-51 to 155.5 Mt in 1984-85. Since population is increasing at alarming rate, it has to go up to 275 Mt by 2000. There is need to boost productivity.

I. Cropland

The cropland is under stress due to urbanisation, village expansion, high ways, industrialisation etc. It is already indicated for trans yamuna region from Ghaziabad onwards. The same is also happening in NOIDA area in U.P. With such threat to cropland, the only alternative is to increase productivity per unit area/time. At present agriculture in India, involves about 64% of people and 47% of the total land area. The increase in food grains production to 155.5 Mt (1984-85) has been due to advances in agricultural research, increase use of fertilisers and pesticides, extensive use of irrigation, increase in agricultural credits, expansion of Cooperatives and better agricultural extension and integrated rural development. In U.S.A., there are only 2.4% people and about 20% cropland in agriculture production.

The following are the possible solutions to our agricultural problems:

(1) Closing the gap between present and potential yield (i.e. to go vertical rather than horizontal in agriculture, the latter not possible as no more land is available). This is possible by integration between already available and time-tested plant breeding tools and biotechnology. This can be done by enhancing photosynthetic efficiency conversion of C_3 to C_4 plants, biological nitrogen fixation by upgrading rhizobia or transfer of genes. There can also be developed varieties suitable for stress conditions.

(2) The Western model of high-input (fertilizers, pesticides, machinery etc.) high-output agriculture has increased production but brought problems of loss of top soil and its degradation and other environmental problems. However, India at this stage can not afford to switch over to low-input, low-output agriculture due to its population factor. The only alternative is to go for **sustained** high-output agriculture through organic farming, multiple cropping, good water harvesting programmes etc.

Soil erosion (loss of top soil). Today soil erosion is a global concern and in our country it is eating into our resource base very rapidly. Soil loss is maximum in regions with high population densities due to pressure on food production.

II. Grassland and over-grazing

In poor countries like ours increase in human population is accompanied by increase in livestock population. Due to this there follows over-grazing with consequent denudation of the forest and vegetable cover. People move to new areas and the land for grazing is reduced, and the number of grazing animals increase, which causes over-grazing, sol erosion and desertification. North Africa, once a granary for Roman Empire, turned today into a grassless and treeless landscape. This is general picture in African continent.

Though over-grazing was natural, but the scale and rate grow beyond measure. Human and livestock populations have outgrown the biosphere. The major concern is, therefore to protect our vegetal cover from grazing animals.

In India, due to religious and cultural taboos, there are mixed the issues of grasslands, grazing and problems of stray cattle. Prohibition against cow slaughter, beef eating etc. are the chief reasons, and animals in our culture are for draught power, milk, dung etc.

The number of animals increased, especially in hilly regions. These are not productivity and are maintained free of cost, more for dung (uses as fuel and fertiliser). The chief consequence is over-grazing in forests leading to their degradation. About 88% of our forests are open to grazing.

Grassland in India is not a climax vegetation. Moderate grazing is good in some areas as it keeps down inflammable grass and clears ground for particular forest trees like pine, teak and sandalwood to colonise. But excessive grazing can reverse the succession, and unpalatable shrubs may overtake the forest cover. Following measures are suggested to check over grazing.

(1) Updating the data base for grassland and grazing land areas.

(2) Livestock development programme be launched, so that villagers have few but productive cattle.

(3) Elimination of unproductive cattle.

(4) Number of animals to be kept a appropriate carrying capacity of the area.

(5) Rotational grazing to or have feed animals.

(6) Fodder to be taken from wild grass cover.

(7) Improved fodder grasses/nerbs/shrubs/trees ar to be developed.

(8) To demarcate migration routes of animals in plains of different states, developing grazing areas on such routes.

(9) Planting of fodder or multiple purpose trees in large areas close to villages.

(10) Culturable or unculturable land is to be used for Agro-Silvi-Pastoral System for food, fuel and fodder.

(11) Fodder development programmes around forests.

(12) Proper management with legal steps to regulate use of grass cover.

Forest Cover (Woodland) and Revegetation

The effective forest cover at present is very low, hardly 14%. It should have been 33% in the plains and 60% in the hills. The reasons of forest destruction are many and varied. Moreover, there is wide gap between demand and supply of wood for firewood (87.5%), timber and industrial raw material including paper, board, newsprint etc. The strategy to meet this shortage has two major objectives of forestry : (i) affording long term ecological security, and (ii) supply of goods and services to people and industry. To achieve these, three main types of forestry, viz. Conservation, Production and Social Forestry is needed to be practiced.

There is need for considerable inputs in forestry with regards to science and technology, education and training and demonstration and extension. Agriculture, forestry and animal husbandry (three land-based vocations) are to be integrated and village level to afford protection and ensure production and socio-economic benefits to wood based industry and rural areas. That this can be done and deforestation brought down has been very well shown by China and South Korea.

Conservation of Biological Diversity

We are rich in biological wealth with nearly 50,000 plant and 80,000 animal species. There are about 20,000 flowering plants and 67,000 insects, 1,000 molluscs, 6,500 other invertebrates, 1,400 fishes, 140 amphibians, 420 reptiles, 1,200 birds and 340 mammals. We have so far given much attention to preserve large mammals, crocodiles, birds etc., ignoring plants, particularly forest trees. We have to take steps to keep the species away from decline. The conservation policy is to be made broader being on ecosystem basis rather than species basis. This could lead to development of **biosphere reserves.** In addition there have also been developed **sanctuaries** and **national parks** for preservation of some species and a particular animal species respectively. We will present details of these later under Wildlife Management.

There has been done much work on threatened plants and animals by Botanical and Zoological Surveys. Another way to conserve important species is to declare them as National or State mammal, bird, tree and flower by Acts.

Control of Pollution in Water and Air

Pollution is the outcome of development. The major sources of pollution in our country are : domestic waste (dispersed as well as point source), thermal power, industry (particularly distillaries), irrigation, auto-exhaust emissions, abuse of agrochemicals etc. Dept. of Environment and Central and State Pollution Control Boards have been able to control pollution from about 50% of the polluting industries during last 6-7 years. Minimum National Standards (MNAS) with respect to most industry have been established. Implementation of MNAS would control pollution at source. This should be applicable to thermal power plants, distillaries and industries dealing with toxic chemicals and vehicular exhaust.

Water pollution needs control River Basin-wise like the Ganga Action Plan. In the coming decades the important problems needing attention and solution are :

(1) maintenance of water and air quality through control of domestic waste from point and dispersed sources.

(2) ensuring drinking water quality and control of contamination of waterbodies.

(3) control of air pollution due to SO_2, NO_x, SPM, toxic chemicals, hydrocarbons, and

(4) control of noise, vibration etc. from industry, agriculture, transport etc.

Development of Non-polluting Renewable Energy Systems

For development, energy is an important input. Non-renewable forms of energy are regarded as energy capital. The energy crisis has shown that to sustain energy sector, we must conserve the non-renewable sources and also replace/supplement them by non-polluting renewable ones which are more or less pollution-free, environmentally clean, renewable and socially relevant. Most of the non-renewable sources are highly polluting. No nation can now afford to depend only on one form of energy. It should have a mix of at least seven forms biomass, solar, coal, petroleum, natural gas, hydropower and nuclear power, each supplying not more than 20% of total energy needs. In a country, like ours renewable forms of energy, particularly biomass is most suitable where there are over 5,76,000 villages. Firewood is mostly used and the crisis can be solved by suitable programmes.

Biogas is another major sources of energy in rural areas, if enough feed stocks (dung, human excreta, domestic and agro-industrial wastes) are available. Other sources are micro-hydel, solar, wind ocean and geothermal. There is an Advisory Board on Energy (ABE).

Recycling of Wastes and Residues

For stable economic growth, resources must be used carefully and technologies for recycling of wastes and residues are to be evolved.

The choice is between one-time use of materials i.e. "Throw-Away Society" and use of recycled materials i.e. "Sustainable Society". The latter will make available materials to future generations and will also save energy. Single use will lead to scarcity. Several developed countries like Japan and to some extent Holland, U.K., West Germany, Denmark and U.S.A. are evolving ways of recycling and even reuse. The amount of waste materials recycled in Japan increased from 16% in 1974 to 48% in 1978. During 1981, out of 43 Mt of garbage (at 0.828 Kg/person/day) 65% was incinerated, 32% was used as landfill and 3% for compositing and feeding animals. There are over 50 treatment plants. This work had most marked effect on automobile industry. Sweden also recycles newspapers. Clean Japan Centre aims at recycling the accumulated waste. This conserves resources and also keeps environment clean. The centre, supported by Government and private sectors has done excellent work in developing technologies for recycling a number of organic and inorganic materials including discarded tyres, plastics, chemicals, home appliances etc.

Recyling techniques followed in Japan are -

(i) crushing of materials as discarded furniture, appliances etc. (ii) selection for magnetic, non-magnetic and specific gravity based materials (iii) thermal decomposition for having organics in form of gas and oil, (iv) food sources from organic wastes for livestock, (v) compositing (vi) methane fermentation plus residual sludge, (vii) conversion into solid fuel (wood chips and saw dust, (viii) melting blast furnace slag for making artificial jewellery etc., (ix) alcohol fermentation, (x) use of incineration heat, (xi) land fill, (xii) prevention of secondary pollution.

In U.K., there is major campaign by Warmer Bulletin - "Warmth and Energy from Refuse" The benefits obtained are popularly called *Dustbin Energy*. The major use is for community heating systems. Refuse derived fuel (RDF) is becoming common. Other aspect is to "Keep Britain Tidy". At Edmonton, about 400,000 tons of garbage is disposed off every year. In U.S.A. New York generates 27,000 tons of garbage/day, enough to fill 15 city blocks, 20 storeys high. There are several plants to utilise solid wastes in Chicago, Iowa, New York.

Waste in India

National Committee on Science and Technology (NCST) prepared a report on utilisation and recycling of wastes in 1975. We can not afford to waste any wastes from agriculture, forestry, animal husbandry, industry, minerals, urban and rural community and building materials. By proper utilisation of these materials we would conserve the natural resources, reduce the production cost and solve the pollution problem. "What is waste to one industry maybe raw material for another".

Depending on their origin, the waters are grouped under **seven** categories. These are given below :

1. **Agricultural wastes :** (i) 80 species of non-edible oil seeds, sal seeds, groundnut cake after oil extraction, rice bran, rice husk, rice straw, tea waste, bagasse, molasses, press mud, tobacco waste, mahua flowers, wood wastes saw dust bark. (ii) coconut waste by products, coconut shell, coconut husk, coconut water (iii) lac waste, (iv) arecanut waste/by products, arecanut husk, lead sheath, chogaru liquid. (v) cashew wastes/by products, cashew apple, cashew sesta, cashew shell liquid (vi) coffee waste, (vii) rubber waste/by products, rubber seeds, rubber wood (viii) cotton wastes/by product, cotton stalks cotton leaves, cotton linters, (ix) jute wastes and by products, jute sticks, jute mill waste.
2. **Fruit and vegetable processing industries :** (i) mango peels and kennels (ii) citrus peels, pomace etc. (iii) apple pomace. (iv) pineapple peels, ears and eyes (v) banana pseudostems (vi) waste from vegetable canning industry.
3. **Animal waste :** (i) cattle shed waste (ii) fisheries waste, (iii) prawn waste (iv) squilla, (v) frog wastes (vi) carcass (vii) slaughter house waste (viii) wool wax
4. **Aquatic weeds :** These may be converted into food, (for fish, duck, donkey,, pig, sheep etc.), fertiliser, pulp and paper, fibre, energy (biogas and waste water treatment.
5. **Community waste :** (i) night soil, sewage, sludge, cattle dung and urine (ii) city refuse.
6. **Industrial waste :** (i) paper waste (ii) mica waste, (iii) flyash, (iv) salt industry byproducts, (v) steel plant waste (slags)
7. **Construction material from wastes :** (i) waste from sugar factories (ii) agricultural wastes, (iii) coconut, arecanut and cashewnut after oil extraction (iv) redmund from aluminium industry (v) water works silt.

Utilisation of fly ash

Fly ash from coal fired power stations is a major industrial waste. India is surplus in coal and there are several coal fired power stations in operation. About 22.016 Mt of fly ash at 80% collection was available during 1985-86.

Fly ash may be used for cement and concrete, land and land foundations. Fill type dam materials, asphalt filters, blocks for civil engineering, agriculture and fisheries (soil improvement, potash fertiliser, gathering ground for fish), building materials (rock wool, glass wool) and sewer treatments.

Fly ash is very good for bricks, since less heat is needed to fire bricks. The advantages of bricks made from fly ash are : these are fired at lower temperature; bricks take less time to dry and fire; fly ash is cheaper than clay; landscape takes longer to get spoiled; can be used in pottery industry.

In developing countries garbage is sorted out by hand for reuse as tin, glass, paper, plastics, rags, bones etc. which generates new materials and also employment. The reasons for waste utilisations are : economic, environmental, resource conservation, employment generation and provision of basic necessities of life.

In India, several materials are recycled. Waste utilisation contributes to economic development. For example in paddy only 10% of biomass is grain and 90% is straw. In the grain 22-75% is husk and 5-7% is bran. Unused waste may be a potential environmental hazard in spreading diseases and leaching of toxic chemicals into life support system. Many useful products would generate from waste utilisation which are of value (as food, fuel, fertiliser, fibre, paper pulp, industrial chemicals, metals etc.) from agricultural and agro-industrial, wood and forest residues, animal residues, mineral wastes, textile and textile processing wastes and metallurgical, metal utilities, paper and plastic wastes.

At present in India, we make three major uses of waste; (i) compositing for agricultural use is being practiced at several places like Jaipur, Baroda, (ii) energy generation (biogas from sewage at Okhla, Keshavpurand Padrona); heat generation at Delhi; incineration at many places (iii) sanitary land fill; most common use.

Human Settlements and Slum Improvement

Human settlements can not be treated in isolation. It is a total concept applicable to a city, town, village etc.

The man-land ratio is lowest in India. Nearly 76% of the population lives in rural areas. Many of them go for open-air defecating. Insanity conditions are created due to waste water from bathing and cooking. Open drains do not function in rural areas.

There is an urgent need for shelter particularly for houseless rural workers, villagers, and pavement and slum dwellers and squatters. The country faces acute shortage of at least 25 M houses, of which 75% are needed in rural while 25% in urban areas. Maximum shortage is in Bihar (4 M) followed by U.P. (1.7 M), M.P. (1.9 M), W. Bengal (1.6 M), Karnataka (1.2 M), Tamil Nadu (1.04 M). about 91% housing is needed in rural sector of Bihar and Assam respectively.

Slum settlements arises as part of city, particularly in metros, due to shortage of houses. Due to influx of rural poor into urban areas in search of means of subsistence livelihood they tend to occupy any vacant land outside the city, where they build a dwelling with any available material. In slums there are no facilities of water supply, drainage, road, transport etc. that lead to several social evils. China could totally eradicate slums within 25 years.

The slums are in a chaotic state. They become so due to environment of insanitation which breeds social implications and health and economic problems.

There are nearly 25 M people in slums of which 40% live in metro cities of Ahmedabad, Pune, Mumbai, Nagpur, Bangalore, Chennai, Calcutta, Lucknow, Delhi and Jaipur. About 67% of population Calcutta and over 45% in Mumbai are slum dwellers. Slum dwellers are important component of urban/metropolitan economy as small traders, contractors, workers, hawkers, drivers, carpenters etc. They have skill, creativity and energy. Most valuable section in slums is the children due to poverty, malnutrition, poor drinking water and insanitary environment together with sickness and disease and lack of education. We talk of mrain drain, but do not bother about genetic drain in slum areas. Thus they need basic amenities which may be looked as minimum needs like drinking water, sewerage, sewage, storm water drainage, paved lanes, street-lighting, bathrooms and latrines and above all education.

Due to poverty slum dweller or a rural poor can not afford to buy even cheap dwellings, due to expensive building material. Use of traditional materials as stone, mud, timber and thatch, which are less expensive should be followed. Mud buildings are suited to hot climate, with proper insulations against heavy rains and dampness. There are many other advantage also of stone, mud, timber and thatch. These are intensive family labour short period of construction, less capital, greater variety and choice etc. Bricks, cement and steel cannot satisfy our stupendous housing needs. The Central Building Research Institute (CBRI) at Roorkee has designed Sarvatogrhia Technology which avoids costly materials. CBRI has also devised designs for self-owned economical rural tenements for Haryana.

With the situation in human settlements being unsatisfactory, we must plan for ideal or near ideal human settlements in rural/slum areas which among other things should have following characteristics :

(1) Architectural designs should be compatible with local ecosystems.

(2) Domesticated animals co-exist with humans in rural areas.

(3) Building material of local origin and of low cost.

(4) Access roads and transportation facilities.

(5) Recycling to domestic solid and liquid waste.

(6) Use of pollution - free renewable energy systems (biogas, solar, wind, firewood).

(7) Water management and conservation.

(8) Food, fodder, fertiliser and fuel need to be raised and residues recycled.

(9) Creation of adequate green belt around settlements.

(10) Landscaping to be done in keeping with availability of irrigation water and other inputs.

(11) Proper technologies to keep air, water and land clean, and all systems are well integrated.

(12) Proper land use plan to be evolved for future expansion and development.

Environmental Education and Awareness

With the acceptance of the Tiwari Committee Report (1980), the country has accepted the need for environmental education. Since people themselves are are newable resource base, they are to be trained so that they could help to restore the environment. Many people recognise the urgent need for environmental education. Besides introducing the subject of Environmental Science at all levels of education, we must give much emphasis on a new approach to environmental education which cuts across various subjects, thus the idea should be to bring in environmental concerns in all subject areas. It will bring environmental concerns to all facets of life.

The area of environmental education includes both, formal and non-formal education. In the formal education four distinct but inter related components are recognised : awareness, real-life situation, conservation and sustainable development. These are to be suitably matched with Primary, Lower Secondary, Higher Secondary, College and Higher University levels. Under the University system, there should be four major areas : Environmental Engineering, Conservation and Management, Environmental Health and Social Ecology. The non-formal education must cater to adult education, rural youth and non-student youth, tribals and forest dwellers, children, public representatives, senior executives and administrators and foundation courses for probationary officers from different services including armed forces and trained manpower (teachers, professionals, technicians, legal experts etc.).

There are several governmental and non-governmental organisations to educate people and to generate awareness towards environment. Department of Environment has set up in 1982 an Environmental Information System (ENVIS) for this purpose. There is a Centre for Environmental Education (CEE) at Ahmedabad. There are over 200 non-governmental organisations involved in environmental education, nature conservation, pollution control, afforestation and social forestry, rural development, wildlife conservation, waste utilisation, eco-development and floristic and faunal studies.

We will elaborate this issue later in this book as a separate chapter.

Updating Environmental Law

Though, there are over 400 Central and State legislations relating to environmental protection, none except Water (Prevention and Control of Pollution) Act, 1974 and the Air (Prevention and Control of Pollution) Act, 1981 take care of short and long term effects on environment. It has been sugggested that environmental activities with national implications be included in the Union List. These include to lay down environmental policies and standards, and to oversee and monitor their implementation. Other related aspects including implementation in most cases should be left to states. The laws must be amended from time to time, watching the progress made and priority areas for future. Some of these have already been amended. For example, Forest (Conservation) Act, 1980 was amended in 1988. The Air (Prevention and Control of Pollution) Act, 1981 was also amended in 1988. Some of the environmental concerns that need proper legal support and/or updating of existing legislation are, land use planning, environmental impact assessment, catchment area protection, over-grazing, desertification, biosphere reserves, mines and menerals, Act, 1957, recycling and rouse, marine conservation, hazardous substances management, Convention on International Trade in Endangered Species and Wild Fauna and Flora (CITES).

New Dimensions to National Security

We give maximum emphasis to defence to meet threats to our national security. This concept of national security (defence biased) could be widened to include food security which ensures physical and economic access to food to all people at all times. Food security has been further refined to include nutritional security.

However, we have always ignored a bigger security - the long range **environmental security**, which is infact a non-military threat. We are waging a war against our life support system-the biosphere. In hills, real enemy are the land slides and not a foreign country. We need trees to give us healthy environment so that we may survive. In the whole Himalayan Belt, people have now become aware of the dangers to their very existence and demand proper action. The future of our 400 M people in lowlands of India, Pakistan and Bangladesh depends on the well being of Himalayas wherein live over 46 M people in Pakistan, India, Nepal, Bhutan and Burma.

History is replete with instances where civilisations have fallen just because concerned countries did not take proper care of environmental base. We, need, therefore to mount a full scale assault through science-based programmes to ensure our national security.

31

Environmental Organizations and Agencies

There are a number of international organisations, agencies and programmes involved in different areas of environment, forestry, wildlife and other relevant aspects. Some of the important bodies of this type are given below.

International Bodies

1. Earthscan : An agency, founded by UNEP in 1976, that commissions original articles on environmental matters and sells them as features to newspapers and magazines, especially in developing countries.

2. Convention on International Trade in Endangered Species (CITES) : An international forum, whose membership for agreement is open to all countries. For India, the Ministry of Environment and Forests functions as nodal agency for participation in international agreements.

3. Environmental Protection Agency (EPA) : This is an independent Federal Agency of the U.S. government established in 1970. It deals with protection of environment by air, water, solid wastes, radiation, pesticides, noise etc.

4. European Economic Communtiy (EEC) : It is a community of 12 European nations, with sound political, economic and legal base. The community has joint agricultural and scientific programmes between its members. It also assists other countries in environmental programmes. It has programmes of framing and implementation of coordinated policy for environmental improvement and conservation of natural resources. CPCB, India has recently taken up a project on air quality monitoring with assistance of EEC.

5. Human Exposure Assessment Location (HEAL) : The project is a part of the Health Related Monitoring Programme by WHO in co-operation with UNEP. This project has three components, viz. (i) air monitoring, (ii) water quality monitoring, and (iii) food contamination monitoring on a global basis.

6. International Council of Scientific Unions (ICSU) : A non-governmental organisation, based in Paris, that encourages the exchange of scientific informations, initiates programmes requiring international scientific co-operation and studies and reports on matters related to social and political responsibilities in treatment of scientific community.

7. International Union for Conservation of Nature and Natural Resources (IUCN) : An autonomous body, founded in 1948, with it Headquarters at Morges, Switzerland, that initiates and promotes scientifically based conservation measures. It also cooperates with United Nations and other intergovernmental agencies and with sister bodies of World Wildlife Fund (WWF).

8. International Marine Consultative Organisation (IMCO) : It regulates the operation of ships in high seas, from marine water pollution viewpoint.

9. South Asia Co-operative Environment Programme (SACET) : This has been recently set up for exchange of professional knowledge and expense on environmental issues among member countries - Afghanistan, Bangalesh, Bhutan, India, Iran, Pakistan and Sri Lanka.

10. United Nations Educational, Scientific and Cultural Organisation (UNESCO) : An United Nations agency, found in 1945 to support and implement the efforts of members states to promote education scientific research and information, and the arts to develop the cultural aspects of world relations. It also holds conferences and seminars, promotes research and exchange of information and provides technical support. Its Headquarters are in Paris. Independently as well as in collaboration with other agencies like UNEP, it supports activities related to environmental quality, human settlements, training to environmental engineers and other socio-cultural programmes related to environment.

11. Untied Nations Environment Programme (UNEP) : A UN agency, responsibile for co-ordination of inter-governmental measures for environmental monitoring and protection. It was set up in 1972. There is a voluntary United Nations Environment Fund to finance environmental projects. There is an Environmetnal Coordination Board, to coordinate the UNEP programmes. Its Headquarters are in Nirobi, Kenya. UNEP was founded to study and formulate international guidelines for management of the environment. UNEP is a assisting many such programmes in India.

12. World Commission on Environment and Development (WCED) : This is a 23 member commission, set up in 1984 in pursuance to a UN General Assembly resoulution in 1983 to re-examine the critical environmental and development issues and to formulate proposals for them. The is a call for political action to manage better environmental resources to ensure human progress and survival. The commission makes an assessment of the level of understanding and commitment of individuals, voluntary oganisations and govermental bodies on environmental issues.

13. Earthwatch Programme : A worldwide programme, established in 1972 under the terms of the Declaration on the Human Environment. It monitors trends in the environment, based on a series of monitoring stations. Its activities are coordinated by UNEP.

14. Man and the Biosphere Programme (MAB) : The programe is the outcome of International Biological Programme (IBP) that has already concluded its activities. MAB was formally launched by UNESCO in 1971. There are 14 project areas under this programme. We shall provide here the details of MAB with special reference to major activities in our own country so far done under the same, and the priority areas for furure.

Man and the Biosphere Programme (MAB)

MAB in fact is the outcome of the experience of those involved in the International Biological Programme (IBP). It was realised that several problems require collaboration of natural and social scientists, planners and managers and the local people. MAB was conceived at the International Biosphere Conference of UNESCO in 1968 adn was officially given shape by the General Conference of UNESCO at its 16th Session in 1970. The programme was formally launched by UNESCO in

November 1971, when the MAB International Coordinating Council held its first session and identified 13 project areas of cooperative research. One more project area was added in 1974.

I. General Objective

MAB is an interdisciplinary programme of research and training which emphasizes and ecological approach to the study of interrelationships between man and his environment. The general objective of the programme is to develop the basis within the natural and social sciences for the rational use and conservation of the resources of the biosphere and for the improvement of the global relationships between man and the environment; to predict the cosnequences of today's actions on tomorrow's world and thereby to increase man's ability to manage efficiently the naural resources of the biosphere. MAB is not a programme of management, but seeks to provide the scientific knowledge and trained personnel needed to manage the natural resources in a rational and sustained manner.

II. Approach

MAB is a problem-oriented programme rather than discipline-oriented. It is interdisciplinary in approach and emphasises the collective efforts by natural and social scientists. The programme aims to provide scientific basis to solve real practical problems of resource management through understanding the environmental problems in the ecosystem context. Finally, it intends to follow a system approach to find the best possible solution under a given set of conditions. Thus MAB helps bringing together the planners, policy makers, managers and scientists for arriving at a rational decision on the issues of resource management.

III. Organisation

At international level MAB is guided by International Coordinating Council (ICC) which is inter-governmental and representative in character. ICC is made up of 30 countries elected by the General Conference of UNESCO. It meets after every two years to review and evaluate the progress being made in implementing the programme. MAB National Committees, so far have been set up in 101 countries, including India. A national committee has scientists from universities and research institutions and public or private bodies' representatives concerned with environmental management. This committere directs and supervises the programme.

MAB cooperates closely with several other organisations as United Nations Environment Programme (UNEP), Food and Agricultural Organisation (FAO), World Health Organisation (WHO), International Council of Scientific Unions (ICSU) and International Union for Conservation of Nature and Natural Resources (IUCN).

In India, the MAB programme is serviced and funded by the Department of Environment, Forests and Wildlife. The national committee, Indian National Man and the Biosphere Committee, was first constituted in 1972, which supervises and directs the programme. The usual term of the members is two years. The members are taken from universities and other scientific organisations. This committee functions as an advisory/recommendatory body to the Department of Environment, Forests and Wildlife for the MAB programme and catalyses actions on the issues related to this programme.

IV. Project Areas

Research under the MAB Programme is divided into 14 broad themes which are given below. major activities in our country-Indian scenario, under each project are also summarised. Project areas 1 to 7 focus on man's interaction with particular type of ecosystems (e.g. tropical forests, arid zones, fresh water and marine water and mountain ecosystems). Project areas 9 to 14 focus on particular impacts of major processes (engineering works, demographic changes, urban systems, pesticides use, pollution etc). Project area 8 is related to establishment of Biosphere Reserves. For actual research activities among these 14 areas each country has its own priorities.

Project Area 1. : Ecological Effects of Increasing Human Activities on Tropical and Subtropical Forest Ecosystems :

The broad themes include :

- baseline studies in natural tropical forest ecosystems;
- problems involved in the management and natural regeneration of exploited tropical forest ecosystems;
- investigations into the methods of regeneration of exploited tropical forest ecosystems; including studies on agrosilviculture;
- Effects of various methods of cultivation and land use on soil structure and fertility of agro-ecosystems derived from tropical forest ecosystems;
- investigations on other uses of tropical forest ecosystems including non-conventional uses.

The work is in progress on forest management in M.P., jhum cultivation in north-east India, man-forest interaction in Orissa and M.P., and deforestation in Kerala. In Himalayas and Western Ghats also there is work going on in eco-development areas.

Project Area 2. : Ecological Effects of Different Land Uses and Management Practices on Temperate and Mediterranean Forest Landscapes.

The broad objectives are :

- to identifiy and assess changes due to man's activities on such landscapes, and the effects of these changes on man;
- a comparative study of structure, function and dynamics of natural, modified, and managed forest ecosystems and their replacement systems;
- a comparative study of interrelationship between natural and man-made forest ecosystems and socio-economic processes and the impact of changes in human population and technology on the future viability of these systems;
- to develop ways and means of measuring quantitative and qualitative changes in the forest environment for setting up the cirteria of rational use and management of forest resources.

Among other research themes, the following areas were considered of special interest:

- ecological and socio-economic consequences of the use of introduced tree species in plantation forestry;
- impact of tourism and other recreation activities on forest landscapes;

- ecological impact of forest fires, including their use in management;
- effects of air pollution on forest ecosystems and forest influences on air quality;
- ecological impact of free-range grazing on forest ecosystems.

There has been completed on project on ecological effects of land use and management practices of forest landscapes of Srinagar mountains.

Project Area 3. : Impact of Human Activities and Land Use Practices on Grazing Lands : Savanna, Grassland (from Temperate to Arid Areas).

The general objectives are to secure, quantify, synthesise, distribute and apply information from natural and social sciences research on grazing lands, in order to provide guidelines for opitmal managemet of these lands under different climatic and socio-economic conditions and to provide more effective means to achieve optimal management of this resource.

In India, a project on marginal lands at Rajkot in Western India was undertaken to study typical arid, semi-arid and sub-humid eco-climates. Possible means for improving grazing, land productivity and human welfare are to be developed. Research work on natural and modified temperate grassland ecosystems of Western Garhwal Himalayas has also been undertaken to assess hte actual and potential productivity of these lands.

Project Area 4. : Impact of Human Activities on the Dynamics of Arid and Semi-arid Zone Ecosystems with Particular Attention to the Effects of Irrigation.

The aims are general ecological improvement of arid and semi-arid regions with special emphasis on the effects of irrigation i.e. impacts on water recame, waterlogging, salinity and alkalinity, lands use, soil and atmospheric conditions, flora and fauna and human populations with reference to the problems related to health, migration and socio-economic changes.

Research work on impact of human activities on sand-dune ecosystem of the Indian desert has been undertaken at Jodhpur. A project on wind erosion in dry crop lands in plain district of Tamil Nadu is also initiated.

Project Area 5 : Ecolocial Effects of Human Activities on the Value and Resources of Lakes, Marshes, Rives, Deltas, Estuaries and Coastal Zones.

The main human activities having ecological impact on aquatic systems are : land use (agriculture and forestry), large engineering works, industrial and domestic pollution, tourism and recreation, conservation and introduction of species and resource utilisation particularly in the Coastal Zones. The wetlands and mangroves are important elements of this project due to their important ecological function, their generally high productivity and serious threat of their own exploitation.

Lake Ecology : A number of projects have been completed. These include studies on Nainital lake (U.P.), Chilka lake (Orissa), Bharatpur and Ramgarh takes (Rajasthan), some lakes in Hyderabad (A.P.) and Loktak lake (Imphal). Work is in progress on Dal Lake and some rural lakes (Kashmir), lower lakes of Bhopal (M.P.) Rawalsar and Renuka lakes (H.P.) and Udaipur lake (Rajasthan). The work includes analysis of demographic and socio-economic problems (tourism, industry and human settlements) related to structure and function of these lakes.

Rivers : There is coordinated research programme for the integrated study of the river Ganga and its immediate basin. The main areas are :

- origin, nature and extent of pollution and its effect on aquatic life, other animals and man (indicator organisms for pollution monitoring); ecological and pollution modelling;
- sewage problems;
- harnessing the river for agriculture, animals husbandry, fisheries, drinking and industrial water supply;
- tidal movement effects on ecosystem;
- problems on human settlements on bank;
- creating environmental awareness among people.

The river has been divided into three sectors : upper (Gangotri to Narora), middle (Narora to Batha) and lower (Batha to Hooghly).

Upper — Universities of Garhwal, Roorkee and Gurukul Kangri.

Middle — Universities of Kanpur, Banaras, Aligarh, Ranchi, Patna, Agri. Uni, Kanpur, Allahabad, Rajendra Ag. Uni., Patna.

Lower — Universities of Burdwan, Kalyani, Jadavpur and B.C. K. Vishwa Vidyalya, Calcutta.

Mangroves : These include deltaic mangroves along the eastern coast, estuarine mangroves along the western coast and insular mangroves on the Andman adn Nicobar Islands. There is a National Mangroves Committee in the Dept. of Environment. There is developed an All India Coordinated Project on Mangrove Research. Studies on mangrove ecosystems of Sunderbans, Andaman Islands, Cochin back-waters are in progress.

Project Area 6 : Impact of Human Activities on Mountain Ecosystems.

The chief objectives are :

- effects of land use alternatives from conservation and restoration viewpoint (effect of shifting agriculture, over-grazing, deforestation);
- impact of large-scale technology such as hydroelectric power and water storage dams, extractive processes such as strip mining, extraction of sand and gravel, oil and gas exploitation; roads and other communication links; buildings and other structures related to tourism etc.;
- effects of tourism and recreation;
- human settlements (emphasising population migration and impact of inroads of modernisation on the culture and life style of people).

Following programmes have been undertaken:

(a) Coordinated eco-development research-cum-field action programme in the Himalayan region. (i) Watershed management (Dachigam in Kashmir; Gamohar and Kotgarh areas of Sutlej catchment in H.P.; gola catchment in Kumaon Himalyas and Rangoon and Singtum in Sikkim). (ii) Ecosystem studies. (iii) Landslide hazad zonation and mitigation (Kumaon-Garhwal region). (iv) Regeneration of the Himalayan foothills (Parandal region in J & K; other aras in H.P., Punjab, Haryana and W. Bengal).

(b) Establishment of the Himalayan Institute of Environment and Development. The Gobind Ballabh Pant Institute of Himalayan Environment and Development has already been established by the Ministry of Environment Forests. The institute has started operating from its temporary office at Kosi, near Almora from August 1988. It consists of a chain of centres of advanced studies and field stations along the entire Himalayan belt.

(c) Coordinated eco-development research-cum-fied action programme in the Western Ghats. (area of 134,000 Km^2 in states of Maharashtra, Karnataka, Tamil Nadu, Kerala and U.T. of Goa). There are 49 research projects in 21 organisations under this major project;

(d) Project on Impact of technological and economic development programmes on bio-cultural heritage of Gaddis and Pangawala population groups of Himachal Pradesh has been completed and another one on Impact of human activities on ecosystems and vice-versa with reference to Sikkim-Himalays in in Progress.

Project Area 7. : Ecology and Rational Use of Island Ecosystems.

The objective is to preserve the traditional life style of the natives of such areas. Emphasis is given to impact of tourism, resources exploitation, and alien plants and animals on such systems.

In India, there are two such groups os islands; one on the eastern flank (the Andaman and Nicobare Archipelago in the Bay of Bengal), and the second on the Western flank (the Lakshadweep group of islands in the Arabian sea). A study has already been done on Value of floral resources of the Great Nicobar More studies are being undertaken.

Project Area. 8 : Conservation of Natural Areas and the Genetic Material They Contain.

The main areas of interest are :

- preparation of inventory of the biomas and their genetic diversity;
- preparation of an inventory of the rare and theratened plant and animal species;
- protection of representative ecosystems in the form of Biosphere Reserves and National Parks etc.;
- studies on population dynamics of endangered/rare plants and animals species in the important natural ecosystems;
- development of techniques for rapid multiplication of threatened and rare plant and animal species. Domesticated plants and animals and microorganisms of direct application to science, technology, medicine, agriculture are excluded for genetic diversity studies.

(a) Biosphere reserves : There are about 45,000 plant and 65,000 animal species in Indian region. A major step towards conservation of its flora and fauna was taken by enactment of Indian Wildlife (Protection) Act in 1972. Today India has 67 national parks and 394 sanctuaries. to supplement efforts in this direction, Indian MAB committee identified a network of 13 representative ecosystems to be protected as Biosphere Reserves. Of these four have already been declared as Biosphere Reserves.

(b) conservation of endangered plant species programme on seed biology and tissue culture: This programme si initiated with special emphasis on medicinal plants, with the objective of conservation in *situ* and mass multiplication of plant species. Endangered species are identified on regional basis and studies are made on their gene poll conservation, artifical propagation in nature, development of tissue culture techniques etc. An all India coordinated project is being carried with the help of two expert committees, one in tissue culture and the other on seed biology. The centres are, Garhwal University Srinagar, Regional Research Lab., Jammu, Botany Dept., Calcutta University, Botany Dept., Univ. of Delhi, National, National Chemical Lab., Pune, Botanical Survey of India, Calcutta nad N.E.H.U., Shillong.

(c) Study of plant and animal resources : Studies are being made on biological resources in different ecological zones in the country. Notable among them are, cytogenetic survey of the flora of Silent Valley in Kerala; endangered and threatened species of animal and birds and Nanda Devi Sanctuary (U.P.), floral and faunal resources of Bisophere Reserve areas of Namdapha Wildlife Sanctuary (Arunachal Pradesh) and of Sacred groves of Dakshin Kannada and Kodagu Distt. in Karnataka and Bioecological Studies on the grey pelican and horseshoecrab.

Project Area 9. : Ecological Assessment of Pest Management and Fertiliser use on Terrestrial and Aquatic Ecosystems.

The programme lays special emphasis on development of ecological safe methods for control of pests of agriculture and public health importance. Other areas of research are, bio-magnifiction of pesticides through specific food chains; effects on non-target organisms; movement, cycling and degradation of pesticides within selected ecosystems; vector control and integrated pest management. Most work in such areas is being done by the Indian Council of Agricultural Research and Agricultural Universities. MAB keeps in touch with them.

Project Area 10. : Effects of Major Engineering Works on Man and his Environment.

The project aims to study the impact of major engineering works such as irrigation and hydro-electric project. A project on long-term impacts of Beas-Sutlej hydro-electric project has been done, where the following are involved: H.P. Ag. Univ. Palampur; B.S.I.; Z.S.I.; Central Inland Fisheries Res. Inst., Barrackpore. another project is on long-term environmental and ecological impacts of Idukki Hydel Project in Kerala by D.O. En. A project on assessment of the soil deterioration due to irrigation by Saryu Canal project in U.P. is carried out to study salinity and alkalinity problems, waterlogging, seepage, depletion of micro-elements etc.

Project Area 11. : Ecological Aspects of Urban Systems with Particular Emphasis an Energy Utilisation.

The areas of research are, energy and material flows into the urban areas and pattern of their utilisation; interrelations betwen urban and rural areas; ecology of human settlements with emphasis on conceptual modelling and systems analysis.

Project Area 12. : Interactions Between Environmental Transformation and the Adaptive Demographic and Genetic Structure of Human Populations.

Here man's interaction with environment is considered. In India, an intersectroal study on tribals of the country is in progress. There is an All India initiated by the D.O. En. There are about 38 million tribals of diverse ethnic groups with distinct cultural heritage in India. The above said project has following objectives:

- folklore survey and collection and identification of plants and animals used by tribals;
- conservation of plants used by them;
- impact of development and technology on the life style, skills and resources.
- impact of tribal culture on vegetation and wildlife;
- phamacological study on useful chemical components of plants/mammals materials used by tribals;
- technology of the development of tribal communities.

The following are participating in the programme : Regional REs. Lab. (CSIR), Jammu Tawi; Central Drug Res. Inst., Lucknow; N.B.R.I. Lucknow; B.S.I., Howrah; Z.S.I., Calcutta; Birbal Sahni Insti. of Paleobotany, Lucknow; University of Garhwal, Srinagar.

Project area 13 : Perception of Environmental Quality.

The following research themes might form the basis of cooperative research between countries on different aspects of the project area:

perception of environmental hazards;

perception of the environment in isolated or periphereal ecological areas;

perception of National Parks and other relatively unmodified natural areas;

perception of typical man-made landscapes of ecological, historical and aesthetic importance;

perception of quality in urban environments.

Project Area 14 : Research on Environmental Pollution and its Effects on the Biosphere.

The major areas in the project are :

- analysis of pathways and transformations involved in the movement of pollutants through the ecosystems (a locally relevant pollutant will be emphasised);
- to identify the environmental indicators;
- acute and long-term effects of pollutants from domestic, industrial, agricultural or other sources on bioata;
- to monitor the levels of different pollutants in the environment.

A number of research project on these aspects have been sponsored. A Unit of Genetic Toxicology is established at Calcutta University to study (i) the chronic effects of sub-toxic doses of pollutants on human systems in relation with different modifying factors as genetic polymorphism, nutrition etc. and (ii) identify the species of vegetation genetically tolerant to various pollutants.

A project is initiated to study the effects of SO_2 and particulate matter on vegetation in the urban and industrial areas. The project is designed to focus on specific (industrial emissions) and unspecific air pollution. the organisations involved are, B.H.U., Varanasi; J.N.U., New Delhi; N.B.R.I., Lucknow; N.E.E.R.I., Nagpur; University of Calcutta; Vikram Univ., Ujjain and B.S.I., Calcutta.

National Organisations

There are a large number of governmental as well as non-governmental organisations, agencies, and programmes engaged in environmetnal studies. A number of non-govenmental, voluntary organisations have been doing good job in this area.

Most of the governmental bodies involved in environmental studies are either put under the administrative control of, or assisted by the Department of Environment, Forests & Wildlife (D.O.En) in the Ministry of Environment and Forests, Government of India.

The role, allocation of business and organisation of the D.O.En. is given below:

Department of Environment, Forests & Wildlife (Govt. of India)

The Department of Environment was set up in 1980 to serve as the focal point in the administrative structure of the Government for planning, promotion and coordination of environmental programmes.

The present integrated Department of Environment, Forests and Wildlife (E.O.En) in the Ministry of Environment and Forests was created in September 1985. The Ministry serves as the focal point in the administrative structure of the Central Government for the planning, promotion and coordination of environmental and forestry programmes. The Ministry's main activities are, the survey and conservation of flora, fauna, forests and wildlife, prevention and control of pollution, afforestation and regeneration of the degraded areas of the environment. These are to be achieved through impact assessment, support to organisations implementing the programmes, environmental and forestry research, extension, education and training to augment the requisite manpower, collection, collation and dissemination of environmental awareness.

The following items of work are allocated to the Ministry :

(1) Environment and ecology
(2) Botanical Survey of India and Botanical Gardens
(3) Zoological Survey of India
(4) National Museum of Natural History
(5) The Water Pollution Acts, 1974, 1977
(6) The Air Pollution Act, 1981
(7) The Environment (Protection) Act, 1986
(8) Biosphere Reserve Programme
(9) National Forest Policy and Forestry Development
(10) Forest Policy
(11) Indian Forest Service
(12) Wildlife Preservation
(13) Fundamental Research in Coordination with Higher Education in Forestry
(14) Padmaja Naidu Himalayan Zoological Park
(15) National Land Use and Wasteland Development Council

(16) National Wastelands Development Board

(17) Central Ganga Authority.

Ganga Action Plan

The Central Ganga Authority was constituted in 1985 to guide and oversee the implementation of a programme for restoring the quality of the river, Ganga. The Ganga Project Directorate (GPD) of the Ministry coordinates the implementation. The State Governments of U.P., Bihar and W. Bengal are to coordinate the Action Plan work.

National Wastelands Development Board (NWDB)

NWDB was set up in 1985 with the following main objectives:

- to increase tree and other green cover on wastelands;
- to prevent good land from becoming wastelands;
- to formulate within the overall Nodal Policy, perspective plans and programmes for the management and development of the wastelands in the country.

Divisions

There are 18 Divisions. There are three wings for Impact Assessment Divisions. These are, Impact Assessment I (A-I), II (A-II) and III (A-III).

There are 7 Associated Units under the administrative control of the Ministry. Of these, two (Botanical Survey of India; Zoological Survey of India) are under Conservation and Survey Division; two (Indian Council of Forestry Research and Education; Indira Gandhi National Forest Academy) under Forest Research Education and Training Division; one (Forest Survey of India) under Forest Survey and Utilisation Division; one (National Zoological Park) under Wildlife Conservation Division; and one (National Museum of Natural History) is under the Education and Information Division of the Ministry. There are five Institutes under te Indian Council of Forestry Research and Education.

There are 11 Autonomous Agencies assisted by the Ministry. These are attached with different Divisions. The Divisions to which they are attached are shown in parentheses below:

(1) Salim Ali Centre for Ornithology and Natural History (Conservation & Survey)

(2) Andaman & Nicobar Islands Forests & Plantation Dev. Corp. Ltd. (Forest Survey & Utilisation)

(3) Wildlife Institute of India (Wildlife Conservation)

(4) Padmaja Naidu Himalayan Zoological Park (Wildlife Conservation)

(5) Central Pollution Control Broad (Control of Pollution)

(6) Centre for Ecological Research & Training (Research and Eco-Regeneration)

(7) Centre for Mining Environment (Research and Eco-Regeneration)

(8) Gobind Ballabh Pant Himalayan Paryavaran Evam Vikas Sansthan (Research and Eco-Regeneration)

(9) Indian Institute of Forest Management (Forest Research, Education & Training)

(10) Centre for Environment Education (Education & Information)

(11) C.P. Rama Swami Aiyar Environment Education Centre (Education & Information)

Other National Organisations

There are many other governmental and non-governmental organisations/agencies involved in environmental issues. Some of the important ones are as follows:

(1) Advisory Board on Energy (ABE)

(2) Bombay Natural History Society (BNHS)

(3) Central Forestry Commission (CFC)

(4) Department of Non-Conventional Energy Sources (DNES)

(5) Industrial Taxicology Research Centre (ITRC)

(6) National Environmental Engineering Research Institute (NEERI)

(7) National Dairy Development Board

(8) National Natural Resources Management System

(9) National Wetland Management Committee

(10) State Pollution Control Boards (SPCB)

(11) Tata Energy Research Institute (TERI)

(12) Several Research Institutes under I.C.A.R. including I.G.F.R.I., Jhansi, Central Soil Salinity Research Institute, Karnal.

BIBLIOGRAPHY

Abraham, C. (1989) Report on the conservation and management of sea tutles on the Madras coast, 1988 report, Students sea turtle conservation Network, Chennai.

Abraham, C. (1990) Conservation and management of sea turtles on the Madras coast 1989-90, report Students sea turtle conservation Network.

Aditya, A.K. and P. Haldar (2000) Biodiversity and Environment, Daya Publishing House, New Delhi.

Anon, (1995) Draft management plan for Bhadra Wildlife Sanctuary 1996-2000. DCF Office Chikmagalore, Karnataka State Forest Department, Karnataka.

Anon, (1991) Animal resources of India Protezoa to mammals, Zoological Survey of India, Calcutta, India.

Anon, (1987) Mangroves in India, Status report Government of India Ministry of Environment and Forests, New Delhi.

Agarawal, K.C. (1999) Biodiversity, Agrobotanical publishers, Bikaner, India.

Agarwal, K.C. (1993) Environmental biology, Agrobotanacial publishers, Bikaner, India.

Ali, S and Ripley S.D. (1969) Hand book of birds of India and Pakistan, Oxford University Press Delhi.

Ali, S. (1996) The book of Indian birds, BNHS, Oxford University Press, Mumbai.

Asad R Rahamani. (1993) Project Bustard last chance to save the great Indian Bustard. 73.75 Cited in Bird Conservation Strategies for the nineties and Bryon Editor Verghese A, Sridhar S and Chakravathi A.K., Bangalore, November 12.14 1993.

Asad R Rahamani. (1996) Strategies for long term conservation of the Great Indian Bustard, Ardiotis nigriceps in India, Journal BNHS, Vol. 440-458.

Asad R Rahamani. (1991) Birds of the Karela Birds Sanctuary, Madhy Pradesh journal's of BNHS, Vol. 88 170-194.

Asad R Rahamani. (1989) The Great Indian Busterd, Final report pp 234 BNHS, Mumbai.

Asad R Rahamani. (1987) Protection of Great Indian Bustard, Oryx 21(3) 174-179.

Asad Rahamani and Manakadan R. (1990) Past and present distribution of the Great Indian Bustard. BNHS 87(2) 175-194.

Arvind Kumar, (1997) A review of legislations and agrements on the conservation of threatened river dolphins, Journal nature Conservation 9(2) 223-234.

Arvind Kumar, (1996) Impact of Industrial pollution on the population status of endangered ganetic dolphin (Platinistagangerica) in the River Ganga in Bihar, Pol Arch Hyderabad, 43(9) 464-476.

Arvind, N.A. and Dinesh Rao, (2001) Biodiversity An Introduction, Cited in Trends in wildlife biodiversity conservation and management, Daya Publishing House, Delhi-35 India.

Arvind Kumar and A.K. Singh, (2001) Ethological studies in Trends in wildlife biodiversity conservation and management, Daya Publishing House Delhi-35 India Eds Hosetti B.B. and M. Venkateswaralu.

Bhatt, G.K. (1998) A field guide to the Caecilians of the Western Ghats India, Bioscience 23(1) 73-85.

Bhatt, G.K. (1997) Caecilian diversity of the Western Ghats In search of the rare animals, Current Science, 73(2) 183-187.

Blasco, F. (1977) Outlines of ecology botany and forestry of the mangals of the Indian subcontinents Cited in Ecosystems of the world. I West Coastal Ecosytems 241-260.

Blasco, F. (1975) Lets mangrooves de I Inde (the managroves of India) Institute Francaise De Pondichery, Travaus de is Scientifique et Technique. All India Press Pondichery, 14(1) 1-175.

Blastter, E.J. (1905) The mangrove of the Bombay presidency and its biology, J., BNHS, 16: 644-656.

Bhupathy S and Kannan P, (1997) Status of agmid lizard in the Western Ghats of Tamil Nadu, India, Salim Ali Centre for Ornithology and Natural History, Coimbatore.

Bora, P.J. and Yogendra Kumar (2003) Floristic Diversity of Assam, Daya Publishing House, New Delhi.

Cocke, T. (1908) The flora of the presidency of Bombay, Taylor and Francis, London.

Collar, N.J. and Andrew P. (1986) Birds in watch, ICBP Technical publication No. 8.

Colchester M. (1989) Pirates, syatters and poachers: The political ecology of dispossession of the native people of the Sarwak survival international London.

Daniels, R.J.R. (1997) Fields guide to frogs and toads of Western Ghats, India, Cobra 27: 1-25.

Daniels, R.J.R. (1991) The problem of conservation amphibians in the Western Ghats of India, Current Science, India 60(1) 630-632.

Daniels, M.S. and Toor, H.S. (1981) Some observations on nest of common crow pheasant, Centropus sinesis (Stephense) J. BNHS 78: 600-602.

Easa, P.S. (1998) Survey of reptiels and Amphibians In Derala part of Nilagari Biophere Reserve. KFRI Research report No. 148.

Ensminger, (1992) Poultry Science, International and Book Dist. Co., Lucknow.

Frank, F. (1990) Garden Birds of India, Royal Publications Delhi.

Gamble, J.S. (1936) Flora of the presidency of Madras. Botanical Survey of India, Calcutta (Printed edition, 1976) 409-1387 vol. I,II, and III.

Ganashiah, K.N. Umasankar R and Bawa K.S. (1997) Diversity of species assemblages in Islands: Predictions and their test using the species composition of shola fragments, Curr Science 73: 188-194.

Gina Caplen. (2000) The effectiveness of *In situ* conservation in India-with Bhadra wildlife sanctuary as a case study, M.Sc. Dissertation, University of salford Salford, England.

Gole, P. (1983) Observing the arus crane, Proc. International crane workshop, Bharathpur 107-114. Cited in International Crane foundation publication, Baraboo Wisconsin, Eds Archibald G.W. and R.F. Pasquier USA.

Gorirup, P.D. and Verdhan H. (1980) Bustards in decline, Tourism and wildlife society of India, Jaipur, India.

Green, A.J. (1992) Status and conservation the WWWD, IWRD Special publication No.17.

Gundappa, K.R, Balakrishna, T.A. and Katre S. (1981) Ecology of *Ichthyophis glutinosa* L (Apoda:Amphibia), Current Science 50: 480-483.

Hosetti, B.B. and Venkateshwarlu M. (2001) Trends in wildlife biodiversity, conservation and management. Daya Publishing House, Delhi-35, India.

Hosetti, B.B. (2002) Glimpses of Biodiversity, Daya Publishing House, New Delhi.

Hong, E. (1987) The natives of Sarwak. The third world network, Penang, Malayasia.

Inger, R.F. and Dutta, K.S. (1987) An overview of the amphibian fauna of India J. BNHS 83: 135-146.

Johnsingh A.J.T. (1983) Large mammalian prey predators in Bandipuir, BNHS 80: 1-57.

Joseph, G.K. and Ramachandran, K.K. (1998) Recent population trends and management of lion tailed macaque (*Macaca siletus*) in Silent valley National Park, Kerala, Indian Forester 10: 833-839.

Joslin, P. (1972) The environmental limitation and future of the asiatic lion. Chicago, Zoology Society Braclfield, Illinois USA.

Joslin, S.K. (1996) Medicinal plants, National Book Trust of India.

Juliet, V. (2000) Some behavioural observations on Rousettus leschenaulti at Cheranmahadevi, Tamil Nadu, CCINA News letter 1(2) 4pp.

Kameshwar Rao, C. (2000) Database of medicinal plants, Karnataka State Council of Science and Technology and Department of Forests Environment and Ecology, Bangalore.

Kanamadi, R.D. Nandihakl, H.N. Saidapur, S.K. and Patil, N. S. (1996) Parental care in frog Phylautus varabilis Gunther, J. Adv. Zool, 17(2) 68-70.

Keshavamurthy, K.R. and Yoganarshiman, S.N. (1990) The Flora of Corg. Kodagu 56. Kumar A. 1997: The liontailed meaque Cited in In Danger, Ed. Paola Manfredi, New Delhi Ranathambore Foundation.

Lakshmi, H.S. Halamma, D.R. and Mallappa, G. (1997) Study on insect fauna of B.R. Project, M.Sc. Dissertation, Kuvempu University, B.R. Project Shimoga Karnataka India.

Macnae, W. (1968) A general account of the fauna and flora of Mangrove swams and forests in the Indo-west Pacific region, Adv. Mar. Biol, 673-270.

Mac Arther, R.H. and Wilson, E.O. (1967) The theory of island biogeography, Princeton University Press. Princeton.

Manjappa, H. (1999) Distribution of macrobenthos in the brackish water pond along Netrravathi estuary, M.F.Sc. Dissertation, Fisheries College, Mangalore.

Mukti, N. Shrestha. (1989) Musk deer Moscehus leaucogaster. Musk extraction from live deer, J. BNHS 86: 438-440.

Murugan, A.E. (1988) Ecological diversity and its management, Croom Kelm.

Murthy, T.S.N. (1985) A field guide to the lizards of the Western Ghats, Rec. Zool. Surv India Occasional Paper No. 72.

Mysers, N. Mittermeier, R.A. Mittermeier, G.A.G.A.B. Fonseca and Kent J. (2000) Biodiversity Challenges: Expanded hot spot for conservation priorities, Nature, 403. 853-858.

Myers, N. (1990) (1988) Threatened biota: Hotspots in tropical forests, Environmentalists, 8. 1-20.

Natarajan, V. (1997) Breeding biology of southern crow pheasant Calimere, Tamil nadu, J. BNHS 94: 56-63.

Navalkar, B.S. (1951) Succession of mangrove vegetation of Bombay and Salsette islands. J. BNHS 50: 157-160.

Nectoux, F. and Kuroda, Y. (1989) Timber of south seas: An analysis of Japanese tropical timber trade and its environmental impact WWF. International Gland, Switerzeland.

Prater, S.H. (1971) The Book of Indian Mammals. BNHS Mumbai.

Pritchrd P.C.H. (1980) The conservation of sea turtles: The Practices and Problems. American Zoologist, 20:609-617.

Raghavendra, L (1997) Biodiversity in some beats of Kudnur section of Lakkavalli, State forest region of Bhadra Wildlife sanctuary. M.Sc. Dissertation, Kuvempu University, Shimoga.

Rao, T.A. (1998). Conservation of wild orchids of lidagu in the Western Ghatt Karnataka. Association for the advancement of Science, Bangalore-1 India.

Ramchandran, N.K. and Vijayan, V.S. (1994) Distribution and general ecology of the Saras crane Grus antigone in Keoladeo National Park, Bharthpur, Rajasthan, J. BNHS 91: 211-223.

Ravi Challam and John Singh, (1993) Management of Asiatic lions in the Gir forests. Wildlife Institute of India Publications. Dehradun.

Riki, K. (2000) Fruit bats and their management orchards, CCINA News letter, 1(2) 4p.

Rodgers, W.A. and Panwar, H.S. (1988) Planning a wildlife protected area network in India. Volumes. Project PO:Ind/ 82/003 FAO. Dehradun, 267-339.

Salim Ali, (1996) The book of Indian Birds BNHS Mumbai publications.

Saldhana C.J. and Nicolson D.H. 1976: Flora of Hassan District Karnataka, India. Amerind Publishing Co-New Delhi.

Santhapau, H. (1999) Common Trees, National Book Trust of India.

Sharma, Ravindra (2003) Medicinal Plants of India – An Encyclopaedia, Daya Publishing House, New Delhi.

Silas, E.G. (1984) Proceedings of the workshop on sea turtle conservation, Central Marine Fisheries Research Institute, Spl. Publication No. 18. Cochi.

Silas, E.G. (1984) Sea turtle research and conservation. CMFRI Bulletian No. 35 Cochin, Kerala.

Singh, H.S. (1997) Population dynamics group structure and natural dispersal of the asiatic lion Panthors leo perica, BNHS 94(2) 65-70.

Singh, M.P. and Vinita Vishwakarma (1997) Forest Environment and Biodiversity, Daya Publishing House, New Delhi.

Taylor, E.H. (1968) The caclians of the world: A taxonomic review, Lawrence Kansas University Press. USA.

Townsend, C.R. Harper, J.L. and Begon, M. (2000) Essentials of ecology, Blackwell Science Inc, USA.

Venkteswarlu, M. Personal Communication (2001).

Whitekar, R. (1978) Common Indian Snakes: A field guide. The Mac millan Company of India Ltd New Delhi.

Yadav, C. (2000) Conservational scenario of Indian marine turtle, Cited in : Conservation of Biodiversity, Published by Society of Bioscieces Muzaffarnagar, India.

Yadav, M.R. (1999) Conservational needs supports and priorities of Indian marine turtles. National conference on conservation of biological diversity. 27-28 November 1999 Muzaffarnagar, India.

Yadav, H.S.R. (1993) Conservation priorities for the white winged wood duck Carina stutulata in India, Cited in : Bird conservation strategies for the nineties and beyond Ornithological Society India.

INDEX

S

W

www.ingramcontent.com/pod-product-compliance
Ingram Content Group UK Ltd.
Pitfield, Milton Keynes, MK11 3LW, UK
UKHW050951290926
15470UKWH00005B/1768

9 788170 359883